Mercury Cadmium Telluride Imagers

A Patent-oriented Survey

HANDBOOK OF SENSORS AND ACTUATORS

Series Editor: S. Middelhoek, Delft University of Technology,
The Netherlands

Volume 1   Thick Film Sensors (edited by M. Prudenziati)
Volume 2   Solid State Magnetic Sensors (by C.S. Roumenin)
Volume 3   Intelligent Sensors (edited by H. Yamasaki)
Volume 4   Semiconductor Sensors in Physico-Chemical Studies (edited by L. Yu. Kupriyanov)
Volume 5   Mercury Cadmium Telluride Imagers (A. Onshage)

HANDBOOK OF SENSORS AND ACTUATORS 5

Mercury Cadmium Telluride Imagers

A Patent-oriented Survey

Anders C. Onshage
European Patent Office
The Hague, The Netherlands

1997
ELSEVIER
Amsterdam - Lausanne - New York - Oxford - Shannon - Tokyo

ELSEVIER SCIENCE B.V.
Sara Burgerhartstraat 25
P.O. Box 211, 1000 AE Amsterdam,
The Netherlands

ISBN: 0 444 82790 0

© 1997 Elsevier Science B.V. All rights reserved.

No part of this publication may be reproduced, stored in a retrieval system or transmitted in any form or by any means, electronic, mechanical, photocopying, recording or otherwise, without the prior written permission of the publisher, Elsevier Science B.V., Copyright & Permissions Department, P.O. Box 521, 1000 AM Amsterdam, The Netherlands.

Special regulations for readers in the U.S.A. - This publication has been registered with the Copyright Clearance Center Inc. (CCC), 222 Rosewood Drive, Danvers, MA 01923. Information can be obtained from the CCC about conditions under which photocopies of parts of this publication may be made in the U.S.A. All other copyright questions, including photocopying outside of the U.S.A., should be referred to the publisher, Elsevier Science B.V.

No responsibility is assumed by the publisher for any injury and/or damage to persons or property as a matter of products liability, negligence or otherwise, or from any use or operation of any methods, products, instructions or ideas contained in the material herein.

This book is printed on acid-free paper.

Printed in The Netherlands

Introduction to the Series

The arrival of integrated circuits with very good performance/price ratios and relatively low-cost microprocessors and memories has had a profound influence on many areas of technical endeavour. Also in the measurement and control field, modern electronic circuits were introduced on a large scale leading to very sophisticated systems and novel solutions. However, in these measurement and control systems, quite often sensors and actuators were applied that were conceived many decades ago. Consequently, it became necessary to improve these devices in such a way that their performance/price ratios would approach that of modern electronic circuits.

This demand for new devices initiated worldwide research and development programs in the field of "sensors and actuators". Many generic sensor technologies were examined, from which the thin- and thick-film, glass fiber, metal oxides, polymers, quartz and silicon technologies are the most prominent.

A growing number of publications on this topic started to appear in a wide variety of scientific journals until, in 1981, the scientific journal Sensors and Actuators was initiated. Since then, it has become the main journal in this field.

When the development of a scientific field expands, the need for handbooks arises, wherein the information that appeared earlier in journals and conference proceedings is systematically and selectively presented. The sensor and actuator field is now in this position. For this reason, Elsevier Science took the initiative to develop a series of handbooks with the name "Handbook of Sensors and Actuators" which will contain the most meaningful background material that is important for the sensor and actuator field. Titles like Fundamentals of Transducers, Thick Film Sensors, Magnetic Sensors, Micromachining, Piezoelectric Crystal Sensors, Robot Sensors and Intelligent Sensors will be part of this series.

The series will contain handbooks compiled by only one author, and handbooks written by many authors, where one or more editors bear the responsibility for bringing together topics and authors. Great care was given to the selection of these authors and editors. They are all well known scientists in the field of sensors and actuators and all have impressive international reputations.

Elsevier Science and I, as Editor of the series, hope that these handbooks will receive a positive response from the sensor and actuator community and we expect that the series will be of great use to the many scientists and engineers working in this exciting field.

Simon Middelhoek

Mercury Cadmium Telluride Imagers

Summary

The evolution of imager structures using Mercury Cadmium Telluride (HgCdTe) is described based on published patents and patent applications. The book is divided into two parts. The first describes monolithic arrays, with the detector elements and read-out means integrated in the same semiconductor body. The second part describes hybrid arrays, with the detector elements and the read-out means formed in separate semiconductor bodies.

The types of monolithic arrays discussed are charge coupled device imagers, ambipolar drift field imagers, static induction transistor imagers, charge injection device imagers and charge imaging matrices. The part on hybrid arrays has specific chapters on flip-chip arrangements, Z-technology arrangements, and arrays with the detector array directly contacting the read-out chip. The remaining hybrid arrays are separated according to whether the detector elements are provided with individual read-out leads or not.

The cited documents are presented with an introductory part followed by a descriptive part which in most cases includes one or more figures. An overview of the subject of this book may be aquired by reading only the introductory parts in combination with the figures.

A patent number index with information about the patent applicants, inventors, priorities and patent-families, an inventor index, a company index, and a subject index can be found at the end of the book.

Preface

This book is an attempt to describe the evolution of mercury cadmium telluride (HgCdTe) imager structures based on published patents and patent applications. Most of the patent documents cited in this book originate from the part of the systematic documentation at the European Patent Office (EPO) which corresponds to the International Patent Classification (IPC) H01L27/14 [1].

Focus has been directed to the structures of the imagers. The steps of manufacturing of the structures have been included when steps of specific interest have been disclosed. It has been necessary to exclude documents not explicitly mentioning mercury cadmium telluride to reduce the material to a manageable size. It should be noted that no verification of the feasibility of a device or its operation is needed to render it patentable. The analysis of certain Japanese patent applications have been limited to the published Patent Abstract of Japan.

The patent number, the name of the applicant and the date of publication are indicated for each document cited in the book. A patent number index, an inventors index, a company index and a subject index is found at the end of the book.

It is not unusual that a patent application has been filed in several countries referring to a common first filing *i.e.* priority. These applications are said to belong to the same patent-family. In this book English-speaking documents of the patent-families have been selected provided it has been possible. The patent numbers or patent application numbers of other documents belonging to the same patent-family as a cited document may be found in the patent number index at the end of the book.

---

[1] *H01L27/14: Devices consisting of a plurality of semiconductor or other solid-state components formed in or on a common substrate, including semiconductor components sensitive to infra-red radiation, light, electromagnetic radiation of a shorter wavelength, or corpuscular radiation of a shorter wavelength, or corpuscular radiation and adapted for the conversion of the energy of such radiation into electrical energy or for the control of electrical energy by such radiation.*

The book is separated into a first part with monolithic arrays and a second part with hybrid arrays. Each part comprises five chapters. The documents of each chapter are placed in a chronological order with the documents with earliest priority placed first. The documents are separated by a star '*'. Sometimes a document has been placed at a position which is earlier than that which corresponds to its priority date. This has been done when the document shares basic features with a document already presented in the chapter. The document then follows directly after the earlier document and the documents are separated by two stars '* *'.

I would like to express my appreciation to all the people who have assisted me in this work and especially to S. Behmo and A. Cardon, Directors at the EPO, L.J.L. Fransen, S. Greene, P. Gori and A. Visentin, Examiners at the EPO.

It is my hope that this monography will serve as a useful summary of the patents and patent applications in the field of mercury cadmium telluride imagers.

Rijswijk, 1995

A. C. Onshage

Contents

Introduction	xi
Part One: Monolithic Arrays	**1**

Detector elements and read-out means are provided in the same semiconductor body.

1.1	Charge Coupled Device Imagers	3
1.2	Ambipolar Drift Field Imagers	21
1.3	Static Induction Transistor Imagers	39
1.4	Charge Injection Device Imagers	49
1.5	Charge Imaging Matrices	71

Part Two: Hybrid Arrays   **83**

Detector elements and read-out means are provided in separate semiconductor bodies.

2.1	Detector Arrays with Individual Detector Element Read-Out Leads	85
	Detector elements are provided with individual read-out leads formed in or on a non-active supporting substrate.	
2.2	Detector Arrays without Individual Detector Element Read-Out Leads	123
	Individual detector elements are provided with bump connectors or with no connectors at all.	
	- Cross-Talk Preventing Measures	202
	- Passivation and Leakage Current Preventing Measures	253
2.3	Flip-Chip Arrangements	269
	- Thermal Stress Preventing Measures	289
2.4	Z-Technology Arrangements	309
2.5	Detector Arrays Directly Contacting the Read-Out Chip	329
	- Connections made by Through Hole Technologies	374
Patent Number Index		**401**
Inventor Index		**421**
Company Index		**433**
Subject Index		**437**

Note on Cited Patent Documents

An international two-letter country code is used for published patents and patent applications *i.e.*:

AT	⇔	Austria
AU	⇔	Australia
CA	⇔	Canada
CN	⇔	China
DE	⇔	Germany (Federal Republic)
EP	⇔	European Patent Office
ES	⇔	Spain
FR	⇔	France
GB	⇔	United Kingdom
IL	⇔	Israel
IT	⇔	Italy
JP	⇔	Japan
NL	⇔	The Netherlands
SE	⇔	Sweden
US	⇔	United States of America
WO	⇔	International Bureau of W.I.P.O. (PCT-application)

The country code is followed by a one-letter publication code, *I.e.*:

A	⇔	First Publication Level
B	⇔	Second Publication Level
E	⇔	Reissue Patent

Introduction

In 1958 it was discovered that mercury cadmium telluride alloys $Hg_{1-x}Cd_xTe$ are semiconductors, with a bandgap which can be varied from approximately -0.3 to 1.6 eV as the variable composition x goes correspondingly from 0 to 1 [1]. This has opened up the possibility of designing infrared detectors in mercury cadmium telluride tuned for the 3-5 μm or the 8-14 μm wavelength atmospheric window. The material has been thoroughly studied since its appearance and overviews of its properties have been published [2-4]. In figure 1, the energy bandgap/cut-off wavelength is shown versus temperature for various values of x.

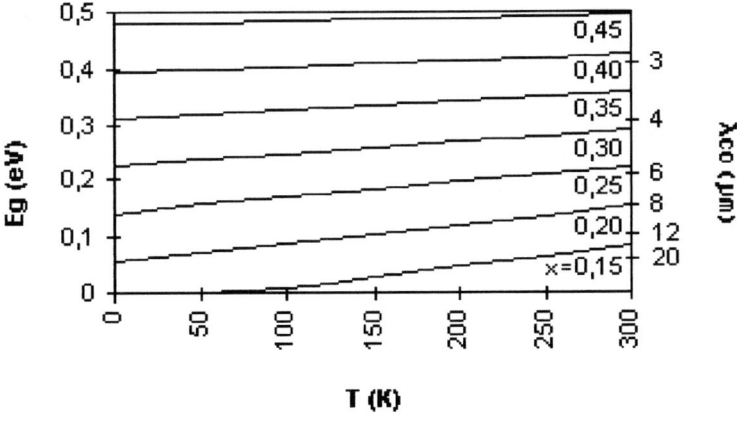

Fig. 1

Methods of manufacturing mercury cadmium telluride material have evolved from bulk melt growth to liquid phase epitaxy (LPE) technology, vapor phase epitaxy (VPE) and metal-organic chemical vapor deposition (MOCVD) [5-7]. These new methods have made it possible to manufacture large two-dimensional focal plane arrays [8-11].

Designing infrared detectors and infrared focal plane arrays is however a demanding task. The task is not made easier by the fact that thermal background radiation at room temperature is many orders of magnitude greater in the infrared range than in the visible range.

Monolithic arrays, in which read-out means are integrated in the same mercury cadmium telluride chip as the infrared detectors are presented in part one of this book. The chapters in this part, chapter 1.1 to 1.5, correspond to the operation of the arrays. The arrays discussed are charge coupled device imagers [12-14], ambipolar drift field imagers, static induction transistor imagers, charge injection device imagers [12,15,16] and charge imaging matrices [17,18]. An example of a charge imaging matrice is shown in figure 2 [19].

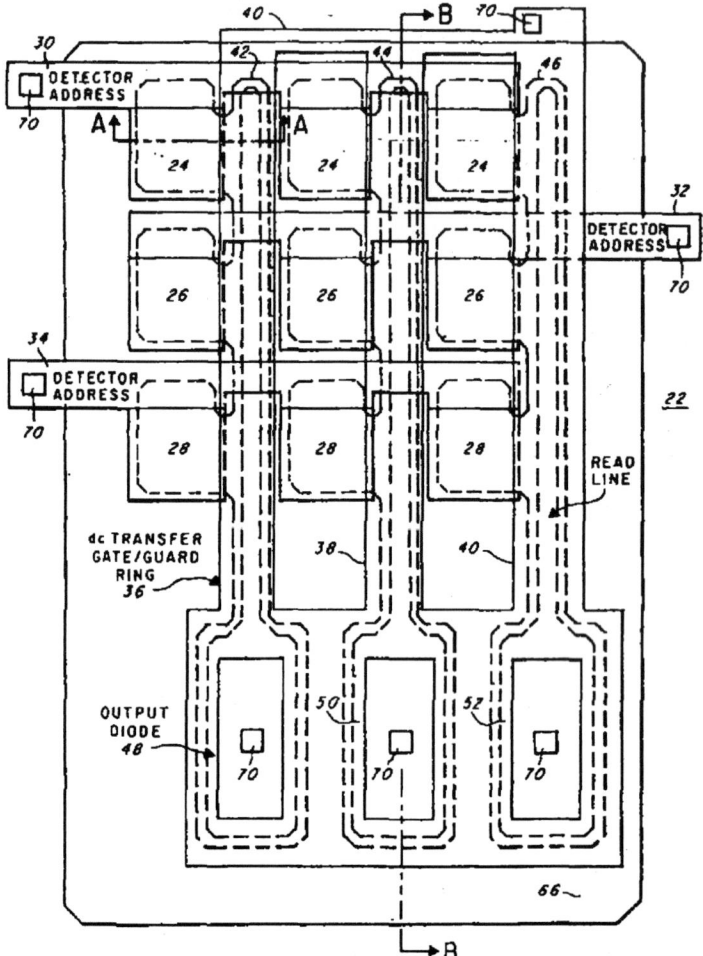

Fig. 2

Detector elements 24, 26 and 28, arranged in rows and columns, are addressed through detector address buses 30, 32 and 34. The columns of detector elements are connected by transfer gates 36, 38 and 40 to pn-junction read lines 42, 44 and 46 which are connected to output collectors 48, 50 and 52.

A problem with the monolithic arrays is that the techniques for building metal-oxide-semiconductor (MOS) devices in silicon cannot be transferred intact to narrow bandgap materials such as mercury cadmium telluride, mainly due to tunneling and avalanche breakdown occuring at very low voltages. A monolithic array, in which read-out electronics is integrated in the same mercury cadmium telluride chip as the infrared detectors, is therefore difficult to achieve.

An alternative approach is to form an hybrid array [20]. The detector array is then formed in a mercury cadmium telluride substrate and the readout circuits are formed in a silicon substrate. The hybrid arrays are presented in part two. Detector arrays having detector elements which are provided with individual read-out leads formed in or on a non-active supporting substrate are presented in chapter 2.1. The detector elements may however be directly connected to a read-out chip which is bonded to the detector chip. A flip-chip bonding technique using indium bumps may be used as shown in figure 3 [21].

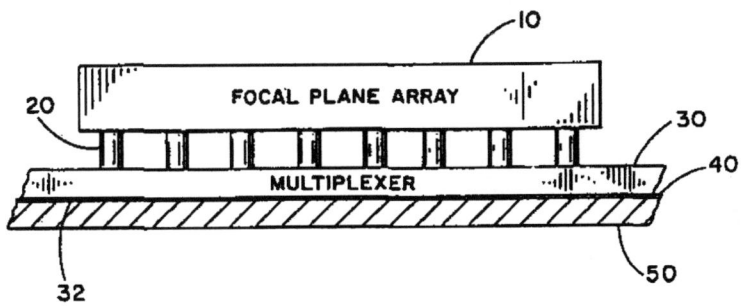

Fig. 3

A focal plane array is formed in a mercury cadmium telluride substrate 10. Individual detector elements are connected to a multiplexer circuit formed in a silicon substrate 30 by means of indium bumps 20.

These detector arrays, as well as detector arrays for which no connectors at all are disclosed, are presented in a chapter 2.2. Sections for cross-talk preventing measures and passivation and leakage current preventing measures are comprised in this chapter.

When a silicon read-out chip and a mercury cadmium telluride detector chip are bonded together by the flip-chip bonding process, the chips are exposed to a mechanical stress which may lead to damage, especially of the fragile detector chip. Damage may also occur due to mechanical stress generated when the temperature of the detector array is changed. Due to the small energy bandgap of mercury cadmium telluride, the detector array is normally cooled to cryogenic temperatures during operation to reduce thermally generated noise. The temperature cycles which are generated when the array is cooled from room-temperture to an operational temperature of for example 77 K create mechanical stress in the array due to the difference in thermal expansion coefficients between silicon and mercury cadmium telluride. The coefficients of thermal expansion for silicon and mercury cadmium telluride are 1.2×10^{-6} m/mK and 3.8-4.5×10^{-6} m/mK, respectively. Flip-chip arrangements for mercury cadmium telluride detector arrays are presented in chapter 2.3. The chapter comprises a section for thermal stress preventing measures.

A detector array having a more compact structure is formed by using the Z-technology [22-24]. These detector arrays are presented in chapter 2.4. An example of such an array is shown in figure 4 [25].

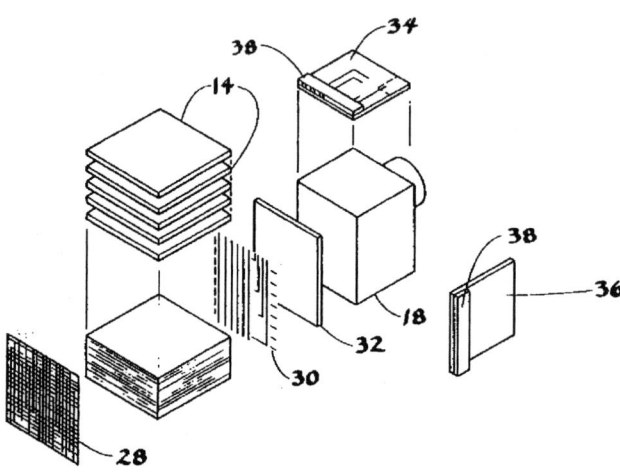

Fig. 4

A stack of silicon chips 14 are glued together. Each silicon chip comprises an integrated circuit and electrical leads which terminate at a focal plane. The leads contact individual photo-detectors of a detector "mosaic" 28. The array also comprises a back plane wiring 30, insulating boards 32, 34 and 36 and a cooling structure 18.

In another approach to form a detector array a mercury cadmium telluride detector chip is directly contacting or directly glued to a read-out chip. These detector arrays are presented in chapter 2.5. An advantage is that the assembly of mercury cadmium and read-out chip may be processed in a silicon-like fashion. A separate section is provided for detector arrays having through holes formed in the mercury cadmium telluride through which the detector elements are connected to the read-out chip [26]. Such a detector array is shown in figure 5 [27].

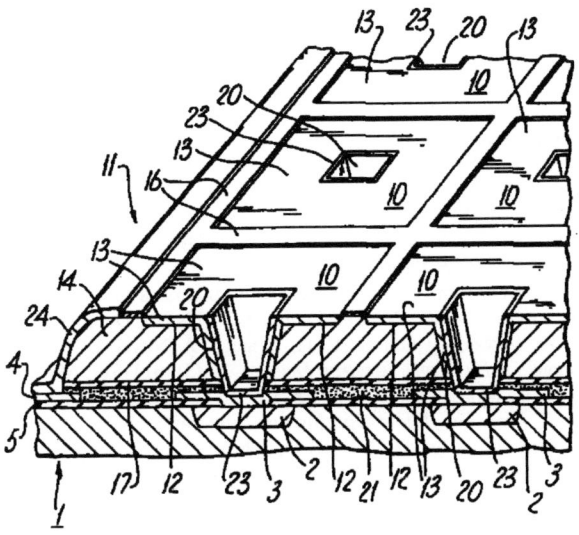

Fig. 5

A silicon substrate 1 comprises a read-out circuit with input regions 2 and metal electodes 3. Detector elements 10 are formed in a p-type mercury cadmium telluride body 11 which is glued to the silicon substrate. Each detector element comprises an n-type region 13. Apertures 20 extend through the mercury cadmium telluride body. The n-type regions 13 are electrically connected to the metal electrodes 3 by a metallization layer 23 formed in the apertures.

References:

[1] W. D. Lawson, S. Nielsen, E. H. Putley, A. S. Young, "Preparation and Properties of HgTe and Mixed Crystals of HgTe-CdTe", *J. Phys. Chem. Solids*, 9, 325-329, 1959.
[2] R. K. Willardson and Albert C. Beer, "Semiconductors and Semimetals, Volume 18, Mercury Cadmium Telluride", *Academic Press*, 1981.
[3] John Brice and Peter Capper, "Properties of Mercury Cadmium Telluride", *INSPEC (The Institution of Electrical Engineers)*, 1987.
[4] G. L. Hansen, J. L. Schmit and T. N. Casselman, "Energy gap versus alloy composition and temperature in $Hg_{1-x}Cd_xTe$", *J. Appl. Phys.* 53 (10), 7099-7101, October 1982.
[5] R. Balcerak, J. F. Gibson, W. A. Gutierrez and J.H. Pollard, "Evolution of a new semiconductor product: mercury cadmium telluride focal plane arrays", *Opt. Eng.*, 26 (3), 191-200, March 1987.
[6] D. D. Edwall, J. Bajaj and E. R. Gertner, "Material characteristics of metalorganic chemical vapor deposition $Hg_{1-x}Cd_xTe/GaAs/Si$, *J. Vac. Sci. Technol. A*, 8 (2), 1045-1048, March/April 1990.
[7] *European Patent Application: EP-A-0135344*, The Secretary of State for Defence, GB, Inventors: S. J. C. Irvine, J. B. Mullin and J. Giess, Published: 27.03.85.
[8] M. R. Kruer, D. A. Scribner and J. M. Killiany, "Infrared focal plane array technology development for Navy applications", *Opt. Eng.*, 26 (3), 182-190, March 1987.

[9] K. Vural, "Mercury cadmium telluride short- and medium-wavelength infrared staring focal plane arrays", *Opt. Eng.*, 26 (3), 201-208, March 1987.
[10] D. A. Scribner, M. R. Kruer and J. M. Killiany, "Infrared Focal Plane Array Technology", *Proc. IEEE*, 79 (1), 66-85, January 1991.
[11] P. R. Norton, "Infrared image sensors", *Opt. Eng.*, 30 (11), 1649-1663, November 1991.
[12] R. A. Chapman, S. R. Borrello, A. Simmons, J. D. Beck, A. J. Lewis, M. A. Kinch, J. Hynecek and C. G. Roberts, "Monolithic HgCdTe Charge Transfer Device Infrared Imaging Arrays", *IEEE Trans. Electron Devices*, ED-27 (1), 134-145, January 1980.
[13] A. A. Cederberg and S. R. Borrello, "HgCdTe CCD Area Arrays for Infrared Imaging in the 3-5 µm Spectral Region", *IEDM Tech. Dig.*, 149-152, December 1982.
[14] T. L. Koch, J. H. De Loo, M. H. Kalisher and J. D. Phillips, "Monolithic n-Channel HgCdTe Linear Imaging Arrays", *IEEE Trans. Electron Devices*, ED-32 (8), 1592-1598, August 1985.
[15] A. J. Lewis, R. A. Chapman, E. Schallenberg, A. Simmons and C. G. Roberts, "Monolithic HgCdTe Charge Injection Devices", *IEDM Tech. Dig.*, 571-573, December 1979.
[16] C.-Y. Wei, H. H. Woodbury and S. C.-H. Wang, "A Novel CID Structure for Improved Breakdown Voltage", *IEEE Trans. Electron Devices*, 37 (3), 611-617, March 1990.
[17] S. R. Borrello, H. B. Morris, R. A. Schiebel and C. G. Roberts, "Charge Imaging Matrix for Infrared Scanning", *Proc. Soc. Photo-Opt. Instrum. Eng.*, 409, 69-75, April 1983.
[18] J. P. Omaggio, "Monolithic HgCdTe MIS IR Detectors with a Floating-Diode Sense Mechanism", *IEEE Trans. Electron Devices*, ED-33 (10), 1494-1501, October 1986.
[19] *US Patent: US-A-4429330*, Texas Instruments Incorporated, USA, Inventor: A. R. Chapman, Published: 31.01.84.
[20] R. B. Bailey, L. J. Kozlowski, J. chen, D. Q. Bui, K. Vural, D. D. Edwall, R. V. Gil, A. B. Vanderwyck, E. R. Gertner and M. B. Gubala, "256 x 256 Hybrid HgCdTe Infrared Focal Plane Arrays", *IEEE Trans. Electron Devices*, 38 (5), 1104-1109, May 1991.
[21] *US Patent: US-A-4943491*, Honeywell Inc., USA, Inventor: P. W. Norton, J. A. Stobie and P. H. Zimmermann, Published: 24.07.90.
[22] J. Y. Wong and J. P. Rode, "The Advent of Three-Dimensional Imaging Array Architectures: A Perspective", *Proc. Soc. Photo-Opt. Instrum. Eng.*, 501, 128-135, 1984.
[23] J. C. Carson, "Applications of Advanced "Z" Technology Focal Plane Architectures", *Proc. Soc. Photo-Opt. Instrum. Eng.*, 930, 164-182, 1988.
[24] P. I. Zappella, W. L. Robinson, J. W. Slemmons, P. J. Redmond and F. J. Woolston, "Implementation of a Hybridization Station for Mating Grumman Z-Plane Modules with HgCdTe Arrays", *Proc. Soc. Photo-Opt. Instrum. Eng.*, 1097, 117-125, 1989.
[25] *US Patent: US-A-4551629*, Irvine Sensors Corporation, USA, Inventor: J. C. Carson and S. A. Clark, Published: 05.11.85.
[26] I. M. Baker and R. A. Ballingall, "Photovoltaic CdHgTe - silicon hybrid focal planes", *Proc. Soc. Photo-Opt. Instrum. Eng.*, 510, 121-129, 1984.
[27] *European Patent Application: EP-A-0061803*, Philips Electronic and Associated Industries Limited, GB, Inventor: I. M. Baker, Published: 06.10.82.

PART ONE

MONOLITHIC ARRAYS

Imagers having detector elements and read-out means provided in the same semiconductor body are presented in this part.

Chapter 1.1

Charge Coupled Device Imagers

Summary

The techniques for building MOS devices in silicon cannot be transferred intact to narrow bandgap materials, mainly due to tunneling and avalanche breakdown occuring at very low voltages. In a charge coupled device (CCD) infrared imager [1 -3] a perplexing problem is the fact that thermal background radiation at room temperature is many orders of magnitude greater in the infrared than in the visible. A second limitation is the requirement for very large storage capacitance. The integration time has a lower limit set by the maximum clock rate, and the amount of charge generated by the background in this integration time is large. Thirdly, the CCD itself must have an extremely large dynamic range such that when the background charge is subtracted from the total, the remaining charge is a true representation of the signal.

- A charge coupled device infrared imager having an HgCdTe substrate is disclosed in US-A-3806729.

- A uniphase, buried-channel charge transfer device is disclosed in US-A-4229752 wherein a portion of each cell includes an inversion layer, or "virtual electrode" at the semiconductor surface, shielding that region from any gate-induced change in potential.

---

[1] R. A. Chapman, S. R. Borrello, A. Simmons, J. D. Beck, A. J. Lewis, M. A. Kinch, J. Hynecek and C. G. Roberts, "Monolithic HgCdTe Charge Transfer Device Infrared Imaging Arrays", *IEEE Trans. Electron Devices*, ED-27 (1), 134-145, January 1980.
[2] A. A. Cederberg and S. R. Borrello, "HgCdTe CCD Area Arrays for Infrared Imaging in the 3-5 μm Spectral Region", *IEDM Tech. Dig.*, 149-152, December 1982.
[3] T. L. Koch, J. H. De Loo, M. H. Kalisher and J. D. Phillips, "Monolithic n-Channel HgCdTe Linear Imaging Arrays", *IEEE Trans. Electron Devices*, ED-32 (8), 1592-1598, August 1985.

- In US-A-4228365 an infrared focal plane array is disclosed, which has a heterostructure injection scheme that prevents CCD well filling by using a heterojunction barrier between an absorber and a transfer layer.

- In WO-A-8707082 it is pointed out that the device of US-A-4228365, discussed above, is relatively complex to fabricate and that the high voltage required to lower the electropotential barrier at the heterojunction between the wide and narrow bandgap materials could produce relatively large tunnel currents. Therefore, a different structure is proposed which also includes a buried channel and the capability of two colour operation.

- A CCD imager in which a native oxide layer is combined with ZnS insulation layers is disclosed in US-A-4231149.

- A heterojunction CCD imager is shown in JP-A-55102280.

- A charge transfer device formed in HgCdTe where the distances between transfer electrodes and a substrate are equal is disclosed in JP-A-57062563.

- To reduce the problem of interface states which may trap signal charge in a CCD, it is proposed in JP-A-57169278 to use an epitaxially grown cadmium telluride layer as an insulator between the transfer electrodes and an HgCdTe semiconductor layer.

- The HgCdTe CCD of US-A-4885619 comprises a passivation layer of CdTe which forms a heterojunction with HgCdTe. A gate insulator of SiO_2 or Si_3N_4 is provided between the CdTe layer and the CCD transfer gates.

To improve the signal-to-noise ratio of a CCD imager system an image is moved in parallel to the direction of charge transfer in a parallel register by the means of a mechanical scanner. The velocity of the image is made equal to the average velocity of the charge in the register. Thus, as a picture element is scanned in synchronization with the phase movement, each bit of data exiting the parallel register has seen the same infrared pixel during its movement through the parallel register. This time-delay and integration (TDI) method improves the signal-to-noise ratio since the radiation-background-limited noise and background-limited signal-to-noise ratio are directly proportional to the square-root of the integration time.

- A problem with imagers adopting the TDI method is that the charge packets in the parallel shift register potential wells increase in size from the first to the last stages and overflow after saturation. This problem is addressed in EP-A-0066020 where a larger capacity is obtained by using successively larger voltages and well depths down the parallel shift registers.

- An imager in which CCD registers are separated by regions having high impurity concentration (channel stops) and a method of manufacturing the imager is shown in JP-A-58171848.

- An alternative way of creating channel stops is disclosed in JP-A-60140869. The isolation regions are formed by exposing the substrate to Hg vapour and heating the regions using a laser beam.

- The dark currents of photovoltaic detectors, which are exponentially dependent on the bandgap of the semiconductor materials, are reduced in US-A-4791467 through the use of two semiconductor layers having different bandgaps.

- An imager may be formed by connecting a first substrate of HgCdTe comprising a detector array and a second substrate of silicon comprising a read-out circuit. An obvious problem is to electrically and mechanically connect the two substrates. In JP-A-63005560 an imager is shown in which detectors and read-out charge coupled devices are formed in two HgCdTe layers grown on top of a substrate which allows a rather simple connection between the detectors and the charge coupled devices.

- An imager having a similar design to JP-A-63005560, discussed above, is disclosed in JP-A-63046765.

- A detector element in which amplification of photon generated charge takes place by avalanche multiplication is disclosed in US-A-4912536.

Charge Coupled Device Imagers

The techniques for building MOS devices in silicon cannot be transferred intact to narrow bandgap materials, mainly due to tunneling and avalanche breakdown occuring at very low voltages. In a charge coupled device (CCD) infrared imager a perplexing problem is the fact that thermal background radiation at room temperature is many orders of magnitude greater in the infrared than in the visible. A second limitation is the requirement for very large storage capacitance. The integration time has a lower limit set by the maximum clock rate, and the amount of charge generated by the background in this integration time is large. Thirdly, the CCD itself must have an extremely large dynamic range such that when the background charge is subtracted from the total, the remaining charge is a true representation of the signal.

A charge coupled device infrared imager having an HgCdTe substrate is disclosed in US-A-3806729 (Texas Instruments Incorporated, USA, 23.04.74).

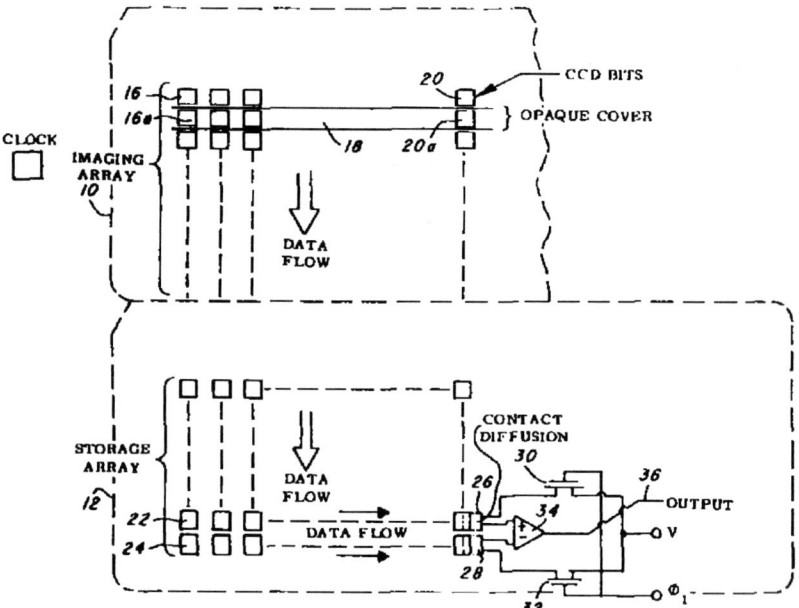

Fig. 1.1.1 (US-A-3806729 fig. 1)

The imaging array 10 defines a number of rows 16 of CCD bits. Each column of bits in the array is in essence a vertical CCD shift register, where data can be shifted from one bit to the adjacent bit by application of clock pulses from a clock generator. Alternate rows 16 are characterized as optically active, the remaining interlaced rows (such as row 16a) are characterized as optically inactive. The CCD bits in the optically inactive rows are covered by opaque strips. Radiation generated minority carriers are collected alternately from a scene and from a uniform background reference source. The signal charge and the reference charge are coupled to a differential detector. Since both the signal and reference charge are collected in the same detector element and are processed identically, material homogeneity requirements are greatly reduced.

*

A uniphase, buried channel charge transfer device is disclosed in US-A-4229752 (Texas Instruments Incorporated, USA, 21.10.80) wherein a portion of each cell includes an inversion layer, or "virtual electrode" at the semiconductor surface, shielding that region from any gate-induced change in potential.

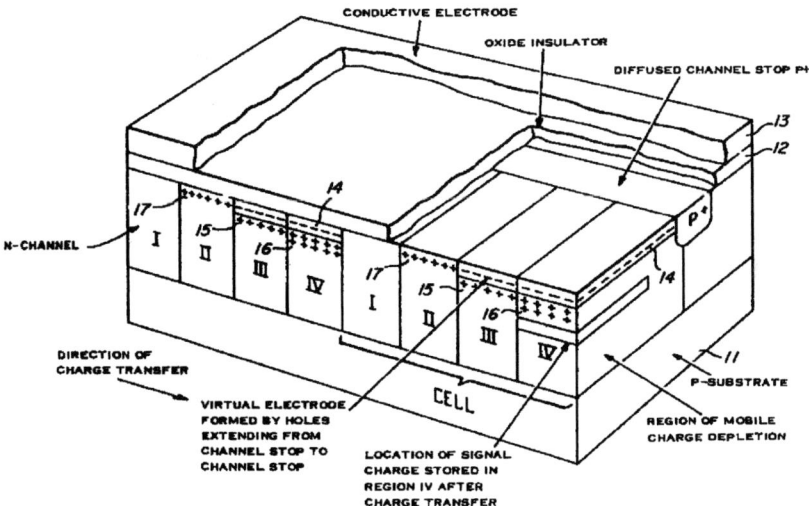

Fig. 1.1.2 (US-A-4229752 fig. 1)

A surface of a substrate 11 is covered by an insulating layer 12, which extends along the length of an n-type channel region. A continuous gate electrode 13 extends along the length of the channel and is connected to a clock pulse source. Spaced along the length of the channel are a plurality of cells, each of which includes a p-type inversion layer 14 at the surface of regions III and IV of each cell, which acts as a virtual electrode, shielding that portion of each cell from any gate-induced change in potential. Just below the inversion layer the buried channel potential maxima of regions III and IV are determined by selective donor implants 15 and 16. Each cell also includes regions I and II wherein the potential maxima are determined by the gate potentials, and the impurity profiles, including donor implant 17.

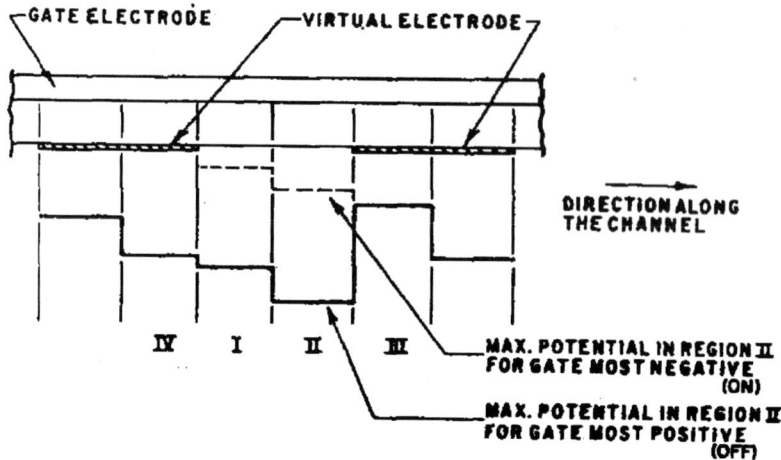

Fig. 1.1.3 (US-A-4229752 fig. 4)

Clocking the gate causes the potential maxima in regions I and II to cycle above and below a fixed potential maximum in regions III and IV beneath the virtual electrode. It is claimed that the substrate may be formed of mercury cadmium telluride.

*

In US-A-4228365 (The United States of America as represented by the Secretary of the Army, USA, 14.10.80) an infrared focal plane array is disclosed, which has a heterostructure injection scheme that prevents CCD well filling by using a heterojunction barrier between an absorber and a transfer layer.

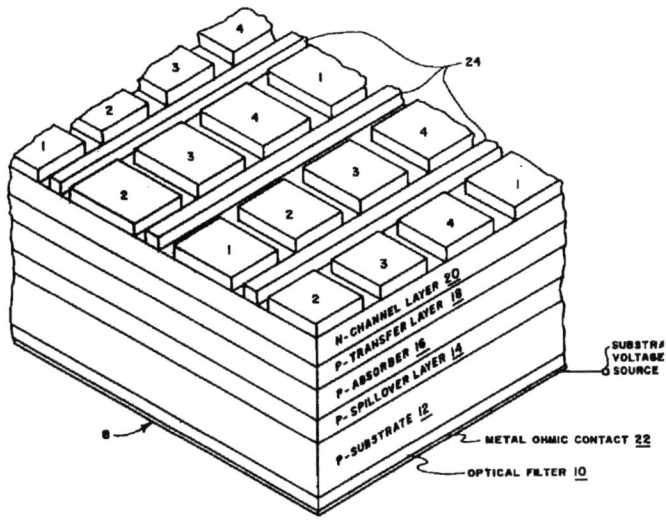

Fig. 1.1.4 (US-A-4228365 fig. 1)

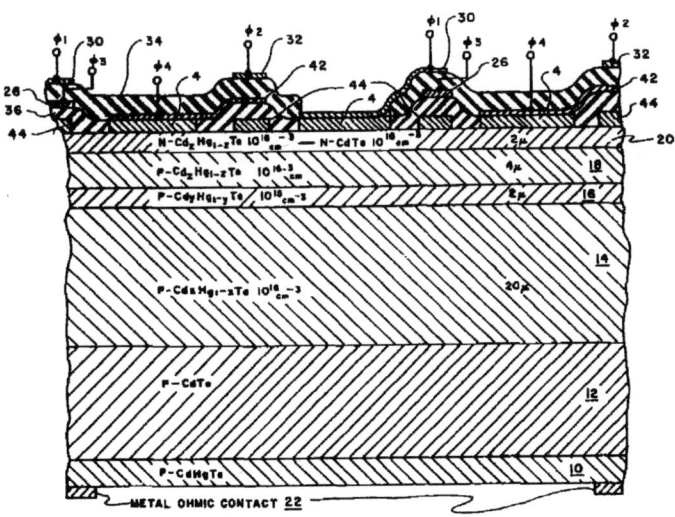

Fig. 1.1.5 (US-A-4228365 fig. 4)

A p-type CdTe single crystal wafer 12 is used to seed the epitaxial growth of a p-doped $Hg_{1-x}Cd_xTe$ spillover layer 14, a p-doped $Hg_{1-y}Cd_yTe$ absorber layer 16, a p-doped $Hg_{1-z}Cd_zTe$ transfer layer, and a p-doped $Hg_{1-z}Cd_zTe$ channel layer 20. A CCD gate structure is formed on the channel layer. Deposited on the side of the substrate 12 opposite to the side with the four epitaxially grown layers is an optical filter 10 made of p-doped HgCdTe, which defines the short wavelength cutoff of the image device 8.

The device is four phase operated with the various phase connections labeled as 1, 2, 3 and 4 on metal Schottky barrier gates. The incoming radiation is collected in an internal absorber layer 16 whereupon the photon wave energy is converted to electron charge. The imager does not have distinct detectors, but has a plurality of areas defined by the low bandgap absorber layer that exists under the activated Schottky barrier gates. If there is overflow of electrons in the absorber layer, the electrons will overflow into the spillover layer 14. In the operational condition, the electrons momentarily reside in and recombine in the absorber layer and possibly in the spillover layer if the absorber layer overfills. Therefore, the "well" that is formed between the substrate 12 and the transfer layer 18 will not overfill and spill into the CCD channel 20 because the thickness and bandgap of the spillover layer provides a large enough charge handling capability to handle any extra charge from the absorber layer. Hence, the generated electron charges are retained in the well that is formed by both the absorber and spillover layers until the electron charges are "punched down" into the CCD channel layer 20. Lattice matching at the p-p heterojunction between the absorber layer 16 and the transfer layer 18 is important for efficient injection across the p-p heterojunction.

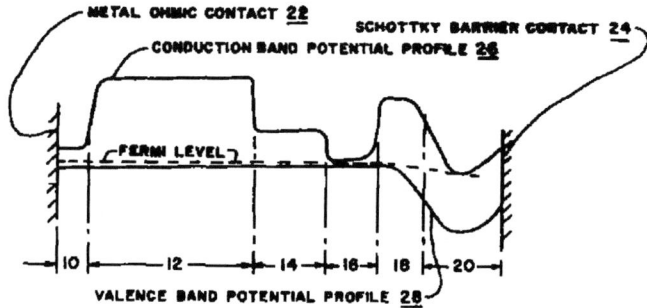

Fig. 1.1.6 (US-A-4228365 fig. 2a)

The equilibrium energy band diagram of the structure is shown above. The method of manufacturing the imager is claimed in US-A-4273596 (The United States of America as represented by the Secretary of the Army, USA, 16.06.81).

* *

In WO-A-8707082 (Santa Barbara Research Center, USA, 19.11.87) it is pointed out that the device of US-A-4228365, shown above, is relatively complex to fabricate and that the high voltage required to lower the electropotential barrier at the heterojunction between the wide and narrow bandgap materials could produce relatively large tunnel currents. Therefore, a different structure is proposed which also includes a buried channel and the capability of two colour operation.

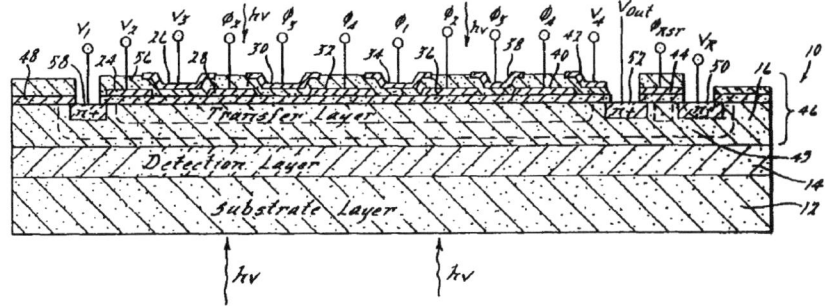

Fig. 1.1.7 (WO-A-8707082 fig. 1)

A substrate 12 of CdZnTe is formed. CdZnTe has a low dislocation density and a sufficiently close lattice match to allow the growth of a high quality epitaxial layer of HgCdTe. A detection layer 14 of $Hg_{0.8}Cd_{0.2}Te$ is grown on the substrate. Since the energy bandgap of the substrate is larger than the energy bandgap of the detection layer, a first heterojunction is created between them. On top of the detection layer, a transfer layer 16 of $Hg_{0.5}Cd_{0.5}Te$ is grown. The detection layer and the transfer layer are grown by using liquid phase epitaxial growth techniques. A second heterojunction is formed between the detection layer and the transfer layer due to the difference in the composition values.

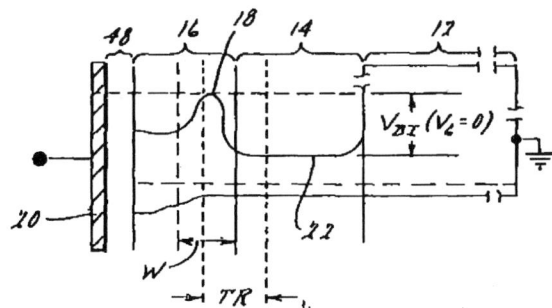

Fig. 1.1.8 (WO-A-8707082 fig. 2)

By suitably doping the detection layer 14 and the transfer layer 16 an electropotential barrier 18 is formed, which inhibits photoexcited electrons in the detection layer from migrating to the transfer layer. The electrons in the transfer layer are able to be displaced through the transfer layer without causing the electrons in the electropotential well 22 to migrate to the transfer layer. When the potential at gate 20 is increased, the electropotential barrier 18 is eliminated which allows the photoexcited electrons to migrate to the transfer layer. Because the transfer layer and the detection layer are sensitive to infrared radiation of different wavelengths, the imager can be used for two colour detection.

Prior to receipt of photoexcited electrons by the transfer layer, packets of electrons in the transfer layer are removed by clock signals applied to the electrodes 24-42 and drained away via region 50. After this process has been repeated a predetermined number of times, the transfer layer becomes depleted of electrons which causes the potential maxima of the transfer layer to move away from the interface between the transfer layer and the insulator layers 48 and 56 thereby creating a buried channel 49.

Most of the electric fields that are generated within the imager are located in the transfer layer because the depletion layer W is located substantially within the transfer layer. Since the transfer layer has a larger energy bandgap than that of the detection layer, it is able to sustain the relatively large electric fields associated with the buried channel without tunnel currents being generated.

*

A CCD imager in which a native oxide layer is combined with ZnS insulation layers is disclosed in US-A-4231149 (Texas Instruments Incorporated, USA, 04.11.80).

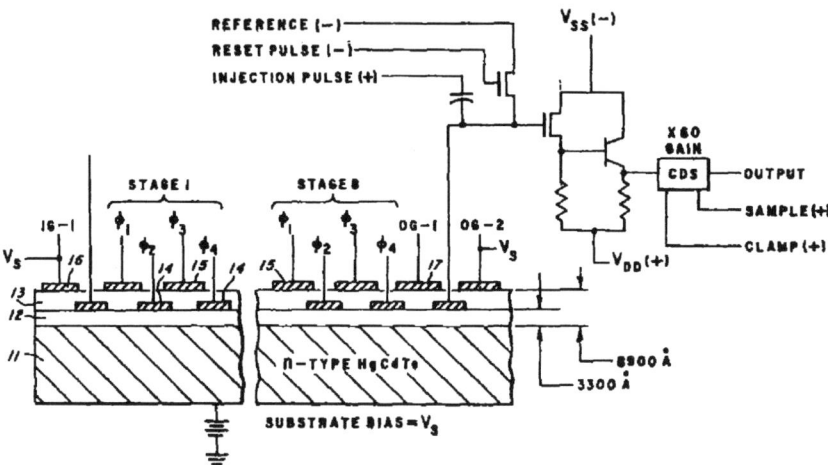

Fig. 1.1.9 (US-A-4231149 fig. 1)

A native oxide layer on an n-type mercury cadmium telluride substrate 11 is used in combination with ZnS to provide first level insulation 12. The ZnS is deposited by thermal evaporation and an opaque field plate is provided for signal channel definition. Second level insulation of ZnS is deposited. This layer is thicker than the first level, and is provided with a stepped or sloped geometry under the first level gates. Input and output diodes are provided with MIS guard rings to increase breakdown voltages.

*

A heterojunction CCD imager is shown in JP-A-55102280 (Fujitsu Ltd, Japan, 05.08.80).

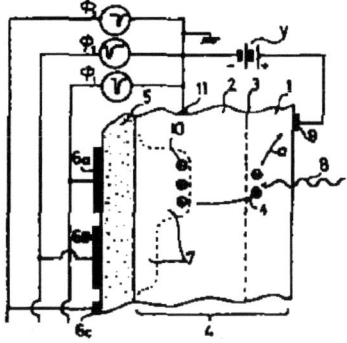

Fig. 1.1.10 (JP-A-55102280 fig. 1)

An n-type HgCdTe region 1 having a bandgap of 0.1 eV and an n-type HgCdTe region 2 having a bandgap of 0.3 eV are formed contacting each other. Thus, the bandgap of region 1 is narrower than the bandgap of region 2, and a heterojunction 3 is formed at the boundary between the two regions. A ZnS layer 5 is coated on region 2, on which transfer electrodes 6 are formed. Radiation is impinging onto region 1 and radiation generated charge is collected in a potential well 7 in region 2.

*

A charge transfer device formed in HgCdTe where the distances between transfer electrodes and a substrate are equal is disclosed in JP-A-57062563 (Fujitsu Ltd, Japan, 15.04.82).

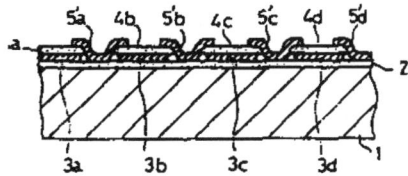

Fig. 1.1.11 (JP-A-57062563 fig. 7)

A first ZnS film 2, an aluminium film and a second ZnS film 4 are formed on top of each other on an HgCdTe substrate. The second insulating film 4 is selectively removed by a photoetching method and first transfer electrodes 3 are formed by removing the exposed aluminium film. Subsequently, a film of aluminium is deposited and patterned into second transfer electrodes 5 by a lift-off method.

*

To reduce the problem of interface states which may trap signal charge in a CCD, it is proposed in JP-A-57169278 (Mitsubishi Denki KK, Japan, 18.10.82) to use an epitaxially grown cadmium telluride layer as an insulator between the transfer electrodes and an HgCdTe semiconductor layer.

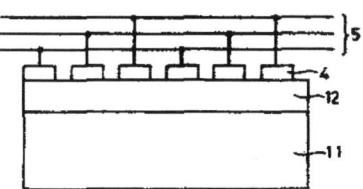

Fig. 1.1.12 (JP-A-57169278 fig. 2)

A cadmium telluride layer 12 is epitaxially grown on a mercury cadmium telluride layer 11. Transfer electrodes are formed on the cadmium telluride layer, which functions as an insulating film. The interface between layer 11 and 12 can be made with a low concentration of crystal defects.

* *

The HgCdTe CCD of US-A-4885619 (Santa Barbara Research Center, USA, 05.12.89) comprises a passivation layer of CdTe which forms a heterojunction with HgCdTe. A gate insulator of SiO_2 or Si_3N_4 is provided between the CdTe layer and the CCD transfer gates.

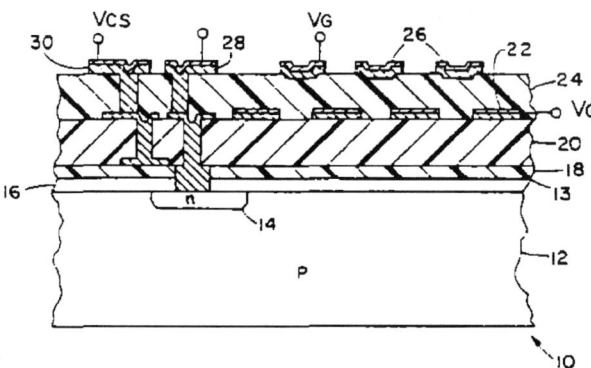

Fig. 1.1.13 (US-A-4885619 fig. 1)

An n-type region 14, which may act as a photodiode, is formed in a p-type HgCdTe substrate 12. A passivation layer 16 of epitaxially grown CdTe overlies the upper surface region 13 of the substrate. A gate insulator layer 18 of SiO_2 or Si_3N_4 is formed on the passiavation layer. A channel stop insulator layer 20 overlies the gate insulator layer on which metal gates 22 and an overlying insulating layer 24 are formed. The insulating layers 20 and 24 are formed of SiO_2. Metal gates 26 are formed on the insulating layer 24. The device also comprises a channel stop terminal 30.

*

To improve the signal-to-noise ratio of a CCD imager system an image is moved in parallel to the direction of charge transfer in a parallel register by the means of a mechanical scanner. The velocity of the image is made equal to the average velocity of the charge in the register. Thus, as a picture element is scanned in synchronization with the phase movement, each bit of data exiting the parallel register has seen the same infrared pixel during its movement through the parallel register. This time-delay and integration (TDI) method improves the signal-to-noise ratio since the radiation-background-limited noise and background-limited signal-to-noise ratio are directly proportional to the square-root of the integration time.

A problem with imagers adopting the TDI method is that the charge packets in the parallel shift register potential wells increase in size from the first to the last stages and overflow after saturation. This problem is addressed in EP-A-0066020 (Texas Instruments Incorporated, USA, 08.12.82).

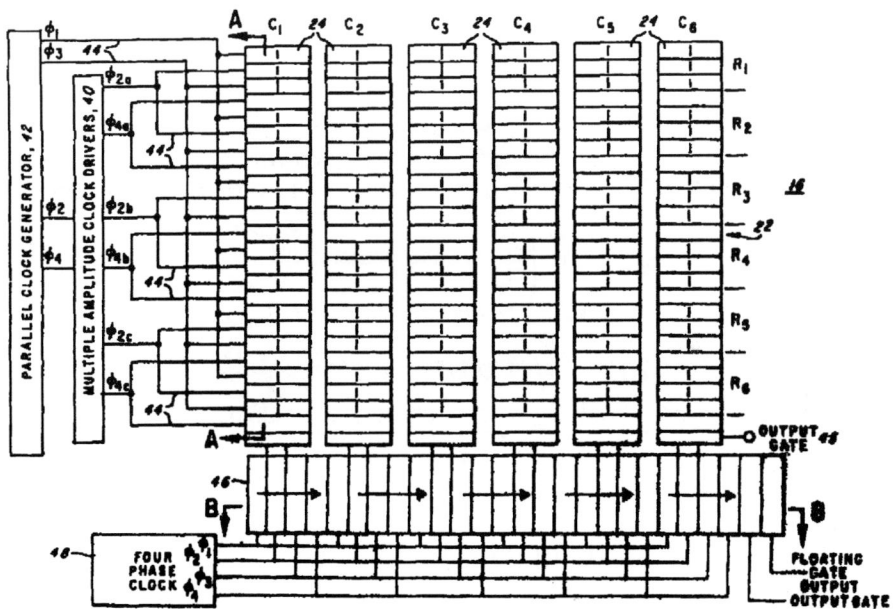

Fig. 1.1.14 (EP-A-0066020 fig. 2a)

A charge transfer device infrared sensor matrix 22 comprises CCD elements 24 arranged in rows and columns. Each element 24 is formed on a mercury cadmium telluride substrate. The electrodes of the CCD elements are connected by metallized bus lines 44 to a four phase parallel clock generator 42 or indirectly to the clock generator through a clock driver 40. The phase 3 and phase 1 electrodes are for transfer gates and the phase 2 and phase 4 electrodes are for charge storage wells. To provide charge storage wells of sufficient depth, a larger capacity is obtained by using successively larger voltages and well depths down the parallel shift registers. The infrared background provides an increasing background fat zero charge which can be set to keep the available charge capacity and electric field constant along the length of the shift register.

*

An imager in which CCD registers are separated by regions having high impurity concentration (channel stops) and a method of manufacturing the imager is shown in JP-A-58171848 (Fujitsu KK, Japan, 08.10.83).

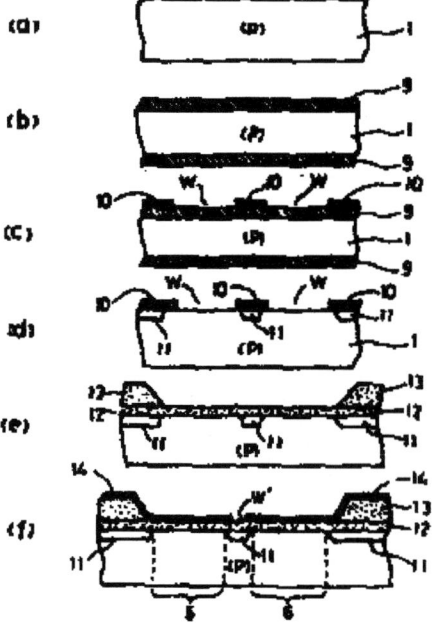

Fig. 1.1.15 (JP-A-58171848 fig. 2)

A p-type HgCdTe substrate 1 is exposed to Hg vapour at a high temperature. This treatment removes Hg atoms from the substrate and high carrier concentration layers 9 are formed. Then, windows W are formed in a mask 10. The device is subjected to Hg vapour at a low temperature to diffuse Hg atoms into the surface areas of the windows. These areas return to the same carrier concentration as the one of the substrate while the areas covered by the mask will keep the high carrier concentration and consequently form the channel stops of the device.

An alternative way of creating channel stops is disclosed in JP-A-60140869 (Fujitsu KK, Japan, 25.07.85).

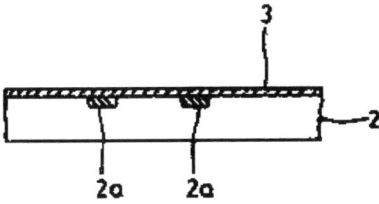

Fig. 1.1.16 (JP-A-60140869 fig. 3b)

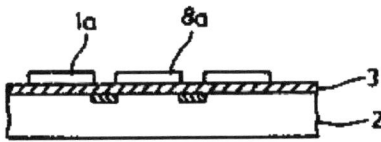

Fig. 1.1.17 (JP-A-60140869 fig. 3c)

An HgCdTe substrate has a high amount of Hg vacancies immediately after the crystal has been grown. Hg vacancies, which act as acceptors, are filled by heating the substrate and exposing it to Hg vapour, and the donor concentration is set to 10^{15}. Then the isolation regions 2a are formed by exposing the substrate to Hg vapour and heating the regions to approx. 400°C using a laser beam. The impurity concentration of the isolation regions 2a is set to 10^{17}. The CCD is completed by forming an SiO_2 film 3 and gate electrodes such as 1a and 8a.

*

The dark currents of photovoltaic detectors, which are exponentially dependent on the bandgap of the semiconductor materials, are reduced in US-A-4791467 (Commissariat A L'Energie Atomique, France, 13.12.88) through the use of two semiconductor layers having different bandgaps.

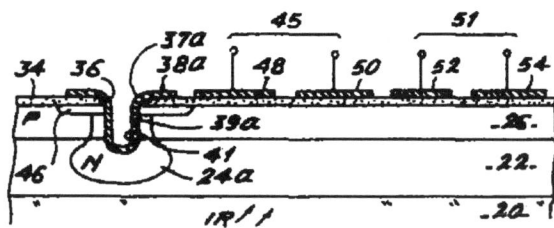

Fig. 1.1.18 (US-A-4791467 fig. 7)

The detector comprises a p-type layer 22 of $Hg_{0.8}Cd_{0.2}Te$ containing a diffused or implanted n-type region 24a and a layer 26 of $Hg_{1-x}Cd_xTe$ with $x > 0.2$. The layer 26 is reserved for the processing of the signal produced in the active zone of the detector. The composition x is chosen solely as a function of electrical requirements in order to make the processing circuits as uncritical as possible. The detector also comprises an insulator 34 provided with an opening 36 for contacting the sensitive zone of the detector by an electric contact element 38a. A NMOS transistor 45 is coupled to a charge transfer device 51. The NMOS transistor comprises an n-type source 46, a gate electrode 48, and a drain induced in layer 26 having a drain electrode 50. The charge transfer device comprises a gate electrode 52 and a multiplexing electrode 54.

*

An imager may be formed by connecting a first substrate of HgCdTe comprising a detector array and a second substrate of silicon comprising a read-out circuit. An obvious problem is to electrically and mechanically connect the two substrates. In JP-A-63005560 (NEC Corp., Japan, 11.01.88) an imager is shown in which detectors and read-out charge coupled devices are formed in two HgCdTe layers grown on top of a substrate which allows a rather simple connection between the detectors and the charge coupled devices.

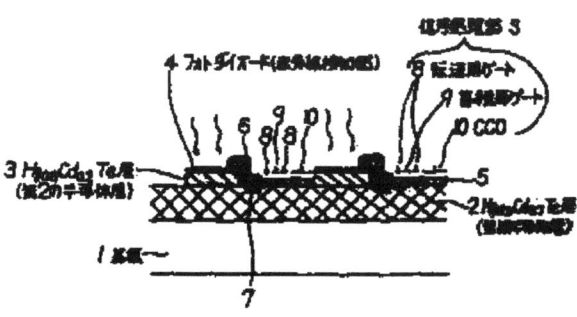

Fig. 1.1.19 (JP-A-63005560 fig. 1a)

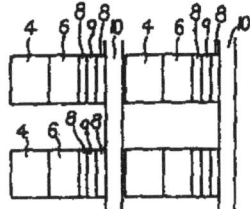

Fig. 1.1.20 (JP-A-63005560 fig. 1b)

A first HgCdTe layer 2 is epitaxially grown on a substrate 1, and a second HgCdTe layer 3, which is formed on the first layer 2, is formed into islands. An infrared detecting part 4 is formed on the islands and connected to a charge coupled device. The charge coupled device is formed in between the islands in the first layer 2. The energy bandgap of the first HgCdTe layer 2 is larger than the energy bandgap of the second HgCdTe layer 3.

*

An imager having a similar design to JP-A-63005560, shown above, is disclosed in JP-A-63046765 (NEC Corp., Japan, 27.02.88).

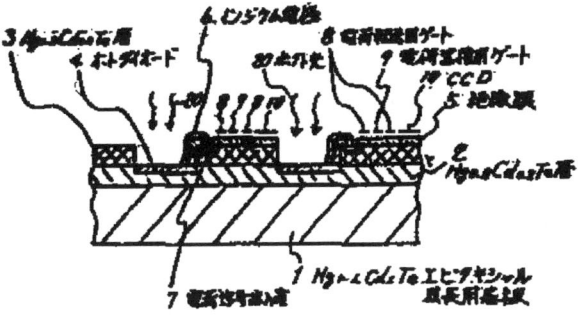

Fig. 1.1.21 (JP-A-63046765 fig. 1b)

A first layer 2 of $Hg_{0.8}Cd_{0.2}Te$ is grown on a substrate 1, and a second layer 3 of $Hg_{0.3}Cd_{0.7}Te$ is formed on the first layer 2. The layer 2 is exposed by partly removing the layer 3, and a photodiode 4 is formed. A read-out charge coupled device 10 is formed in the layer 3 and is connected to the photodiode by an indium electrode 6. The energy bandgap of the second layer 3 is larger than the energy bandgap of the first layer 2.

*

A detector element in which amplification of photon generated charge takes place by avalanche multiplication is disclosed in US-A-4912536 (Northrop Corporation, USA, 27.03.90).

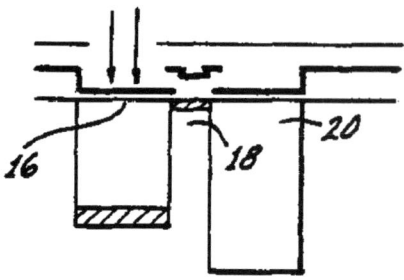

Fig. 1.1.22 (US-A-4912536 fig. 3b)

A first gate 16 is utilized as a photon detection and charge accumulator. In response to a photon flux, charges are generated in the gate 16. A second gate 18 is utilized to control charge transfer of accumulated charge from gate 16 to a third gate 20, the avalanche multiplication gate. The avalanche multiplication gate is utilized to provide one component of the internal gain of the detector whereas the time of transfer between the gates 16 and 20 under the control of the transfer gate 18 provides a further component of the gain of the detector.

*

Chapter 1.2

Ambipolar Drift Field Imagers

Ambipolar drift field imagers are presented in this chapter.

Summary

The principle of operation of ambipolar drift field imagers is presented in GB-A-1488258.

- A thermal imaging device having an elongate strip of HgCdTe is disclosed in GB-A-1488258.

- Apparatus for parallel-in to serial-out conversion is shown in GB-A-2007909. According to this invention a stationary image is focussed on the detector and a standing distribution of ambipolar carriers is allowed to accumulate. After an accumulation period a pulsed bias current is applied to sweep the carriers to the read-out region.

- In GB-A-1605321 devices using a plurality of read-out means are disclosed.

- The resolution is improved in GB-A-2019649 where generated photocarriers are caused to drift in a flow path in which their diffusive spread in a longitudinal direction is limited to a spread substantially less than the diffusion length characteristic of the material.

- A modified read-out structure is shown in GB-A-2201834 where each semiconductor strip branches into two parts separated from each other by a slot. This narrowing of the drift path results in a constriction of the bias current and so introduces a higher electric field which increases both the drift velocity and the sensitivity of the device.

- The read-out structure is further improved in EP-A-0061801 where each strip comprises a recess extending across a part of the width of the strip from the opposite side of the strip and towards a protruding connection. The protruding connection at one side of one strip extends into the recess at the opposite side of an adjacent strip. In this way a compact device geometry is achieved.

- In EP-A-0061802 the sensitivity is increased by forming a meandering drift path in the read-out area.

- A read-out structure which improves the frequency response and the spatial resolution is suggested in GB-A-2095906. The output bias contact is shaped to concentrate an electric field bias in the immediate vicinity of the contact. This concentrated field sweeps away minority carriers which otherwise accumulate near the output contact.

- A device providing both serial and parallel scan coverage of a scene is disclosed in GB-A-2207801.

- Sensitivity is improved in GB-A-2094548. The length of the detector element or the magnitudes of the bias and scan velocity are selected so that the time taken to scan the detector element from one end to a read-out region is greater than the lifetime of the photocarriers generated in the element.

- A detector involving an ambipolar drift of radiation-generated free minority carriers using the loophole technique is disclosed in GB-A-2207802.

- The detector of GB-A-2260218 comprises a photo-responsive material and an array of planar antennae formed thereon which concentrate radiation in fringe fields at antenna edges and extremities interacting with the photo-responsive material. This structure allows the photo-responsive material to be formed thinly, which reduces volume-dependent generation-recombination noise with consequent increase in responsivity and detectivity. Furthermore, reduced element thickness allows higher element resistance.

- In EP-A-0151311 a device having decreased time for read-out is disclosed. A first bias current is applied through the active region of the strip thereby producing an ambipolar drift of photocarriers. A second bias current is applied to the read-out region so that the drift of the photocarriers in the read-out region has a velocity which is greater than the velocity of the drift of the photocarriers in the active region of the strip.

- A four terminal device with increased spatial frequency response is proposed in EP-A-0188241. A problem with the three terminal devices is that the use of the ground bias contact as one of the two read-out electrodes can degrade the performance of the device. This degradation manifests itself as an additional smearing out of the image being scanned along the detector. It arises because the photogenerated excess carriers remain too long in the read-out zone as a result of either blocking of minority carriers at the ground contact or an increase in the effective length of the read-out zone due to contact resistance effects.

Ambipolar Drift Field Imagers

A thermal imaging device having an elongate strip of HgCdTe is disclosed in GB-A-1488258 (Secretary of State for Defence, GB, 12.10.77).

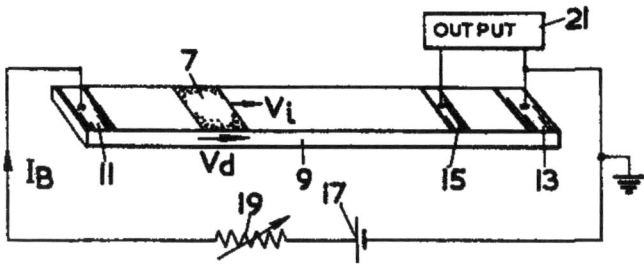

Fig. 1.2.1 (GB-A-1488258 fig. 1)

A constant bias current I_B flows between aluminium electrodes 11 and 13. An output read-out circuit 21 is connected between electrodes 13 and 15. An infrared image is projected onto the strip. The battery 17 is arranged so that minority carriers drift in the same direction as that in which an image region 7 travels along the strip. The resistor 19 is adjusted so that the bias current I_B provides an ambipolar drift having a velocity v_d which matches an image scan velocity v_i. The infrared radiation photons forming the image region 7 create electron-hole pairs in the region of the strip 9 on which they are incident, thereby increasing the carrier densities locally above their equilibrium values. Since the ambipolar carriers drift towards the electrode 15 with a drift velocity v_d which matches v_i, the minority and majority carrier densities corresponding to the image region 7 increases continuously during transit towards the electrode 15. The rate of generation of electron-hole pairs along the path of the image region 7 will depend on the photon flux in the image region 7. Thus the carrier densities at any given point along the travel path of the image region 7 increase the local conductivity by an amount which is a measure of the intensity of the image region 7. Since the bias current I_B is constant, the conductivity modulation of the material between the read-out electrodes 13 and 15 will give rise to a local electric field variation. This local field variation corresponding to the photon flux of the image region 7 and at other times of other adjacent image regions of identical size in consecutive order, is picked up as a voltage change between the electrodes 15 and 13 and is amplified and processed by the output circuit 21.

*

Apparatus for parallel-in to serial-out conversion is shown in GB-A-2007909 (The Secretary of State for Defence, GB, 23.05.79). According to this invention a stationary image is focussed on the detector and a standing distribution of ambipolar carriers is allowed to accumulate. After an accumulation period a pulsed bias current is applied to sweep the carriers to the read-out region.

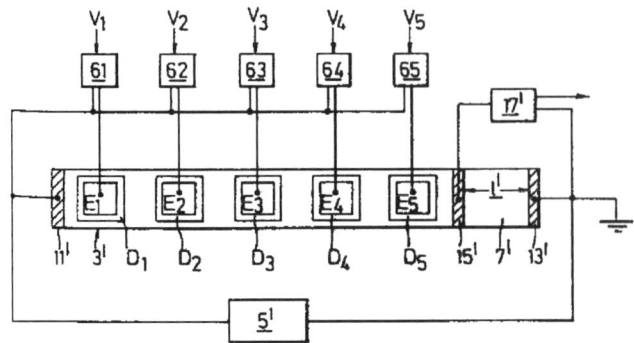

Fig. 1.2.2 (GB-A-2007909 fig. 5)

In a first embodiment, a transfer device 3' in the form of an elongate strip 7' of a photo-conductive material is provided with an array of input diodes D spaced along its length. A corresponding array of electrodes E are deposited to form ohmic contact with the diodes. Current electrodes 11' and 13' and detector electrode 15' are deposited. Electrical voltages $V_1,...,V_5$ are applied to the diodes by synchronous gates 61,...,65. Electrodes 11' and 13' are connected across a current pulse generator 5' and electrodes 13' and 15' are connected across an output amplifier circuit 17'. A distribution of ambipolar carriers is photo-electrically generated in the strip. Upon this distribution, a signal distribution which is injected via diodes $D_1,...,D_5$ may be super-imposed.

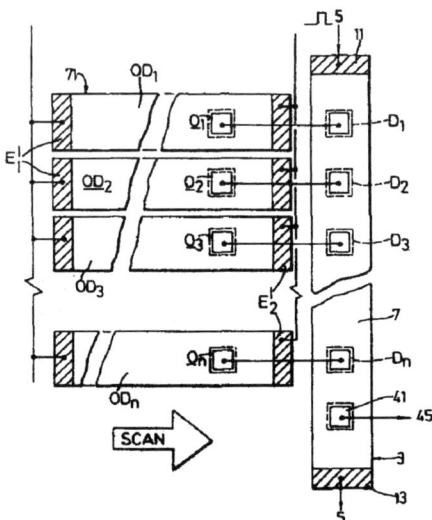

Fig. 1.2.3 (GB-A-2007909 fig. 6)

In a second embodiment an array is shown which includes a number of optical imaging detectors $OD_1,...,OD_n$, the outputs $O_1,...,O_n$ of which are electrically connected to diode inputs $D_1,...,D_n$ formed in the substrate 7 of transfer device 3. When the detectors OD are exposed to focussed optical radiation, ambipolar carriers are photo-electrically generated in the substrate material of each detector and caused to drift towards detector diodes O near the end of the detectors. As these ambipolar carriers are detected, they produce electrical variations and charge carriers are injected via the connected input diodes $D_1,...,D_n$ of the transfer device 3. For image detection in the 8-14 micron band of the infrared spectrum, 8-14 micron band type mercury cadmium telluride material (eg. $Hg_{0.79}Cd_{0.21}Te$) is used for the detectors, whilst 3-5 micron band type mercury cadmium telluride material (eg. $Hg_{0.70}Cd_{0.30}Te$) is used for the transfer device 3.

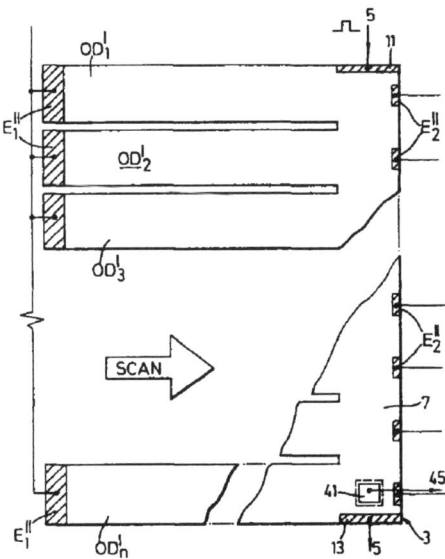

Fig. 1.2.4 (GB-A-2007909 fig. 7)

In a third embodiment the imaging detectors $OD_1',...,OD_n'$ and the transfer device 3 are fabricated from a single slice of semiconductor material. The detectors OD' are separated along their length by slots cut in the material by an etching and photolithographic process. Each detector has a current electrode E_1'' at its far end. At its rear end where it extends from the transfer device 3 spaced from the junction is another electrode E_2''. In operation, when electric potential is applied between electrode pairs E_1'', E_2'', ambipolar carriers drift along each detector towards electrodes E_2''. As the carriers approach electrodes E_2'' they drift into a common region of the semiconductor material that is shared by detectors $OD_1',...,OD_n'$ and transfer device 3. When an electric pulse is applied between device electrodes 11 and 13, the accumulated ambipolar carriers are swept towards electrode 13 and to the detecting diode 41.

*

In GB-A-1605321 (Philips Electronic and Associated Industries Limited, GB, 19.07.89) devices using a plurality of read-out means are disclosed.

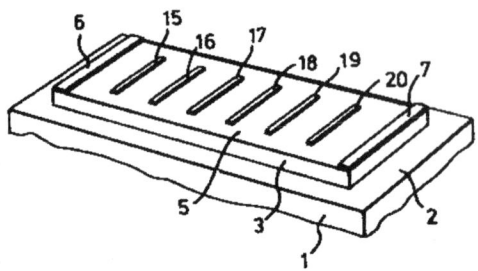

Fig. 1.2.5 (GB-A-1605321 fig. 1)

A rectangular body 3 of n-type mercury cadmium telluride, $Hg_{0.79}Cd_{0.21}Te$, is secured on a sapphire substrate 1 by an epoxy adhesive. At opposite ends of the upper suface of the body 3 there are biasing electrodes 6 and 7. P-type regions are formed and connected to read-out metal electrodes 15-20. The device is equivalent of a plurality of devices which are used sequentially. In this device a signal to noise advantage is achieved corresponding to the signal to noise integration over substantially the whole length of the semiconductor body between the biasing electrode means.

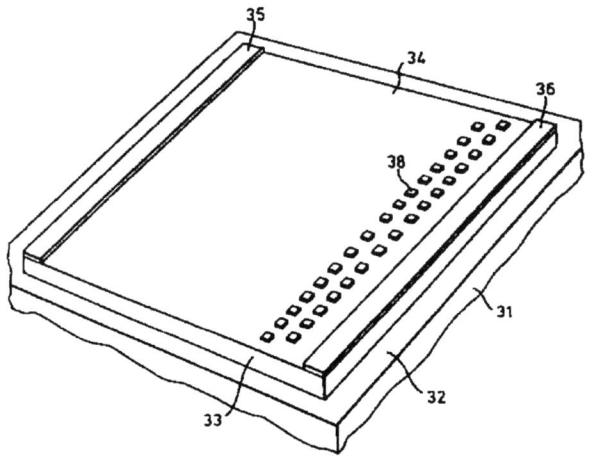

Fig. 1.2.6 (GB-A-1605321 fig. 4)

A device employing a parallel scan mode is also disclosed. An n-type semiconductor body 33 of $Hg_{0.79}Cd_{0.21}Te$ is provided with biasing electrodes 35 and 36. Read-out electrodes 38 connect to p-type regions and form a series of read-out electrodes arranged in two parallel rows.

*

The resolution is improved in GB-A-2019649 (The Secretary of State for Defence, GB, 31.10.79) where generated photocarriers are caused to drift in a flow path in which their diffusive spread in a longitudinal direction is limited to a spread substantially less than the diffusion length characteristic of the material.

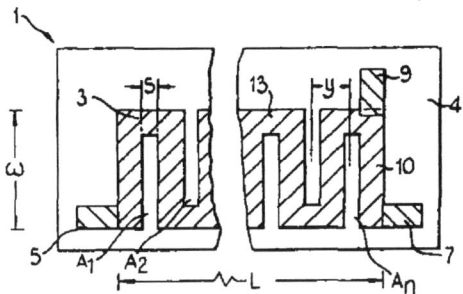

Fig. 1.2.7 (GB-A-2019649 fig. 1)

A strip detector 1 comprises a substrate 3 of mercury cadmium telluride supported on a carrier 4 of sapphire. Along the length of the strip are two spaced current electrodes 5, 7 separated a distance "L" apart. A read-out electrode 9 and the second electrode 7 define a fixed geometry sample area 10. Between the current electrodes there is a multiplicity of open-ended, fine interdigital slots A of width "s", extending alternately from opposite sides of the substrate, and periodically spaced from centre to centre a distance "y" apart, "y" being less than the diffusion length. The slots, formed for example by etching, leave a well defined serpentine structure 13, providing a drift path between input electrodes 5 and 7 of significantly increased length, compared with the electrode separation distance "L".

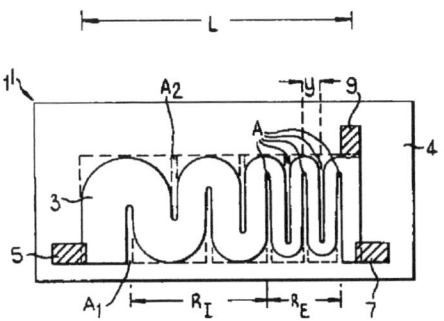

Fig. 1.2.8 (GB-A-2019649 fig. 6)

In a device 1' of a second embodiment a terminal region R_E adjacent to the electrode 7, is provided with a plurality of close-spaced slots A extending alternately from opposite sides of the strip. These slots are spaced apart each by a distance less than the characteristic diffusion length. In contrast to the construction of the first embodiment above, the close-spaced slots are only provided where the diffusion spread of the photo-carriers would reach a significant value.

The increases in the total resistance and total power dissipation resulting from an increase in flow path length are thus of reduced value compared with the increases in resistance and dissipation of the device described above. In an intermediate region R_I a further plurality of slots is introduced. The spacing between consecutive slots in this region differs from one pair of consecutive slots to the next and the magnitude of this spacing is gradually reduced as eventually match the close-spacing of the slots in the terminal region. In this way, the mean deviation of photocarrier velocities is reduced. This is further facilitated by removing the squared corners by etching.

*

A modified read-out structure is shown in GB-A-2201834 (Philips Electronic and Associated Industries Limited, GB, 07.09.88).

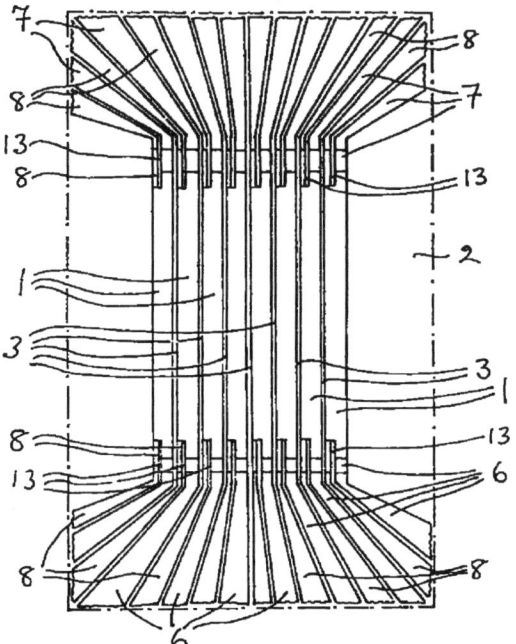

Fig. 1.2.9 (GB-A-2201834 fig. 1)

N-type $Hg_{0.79}Cd_{0.21}Te$ semiconductor strips 1 are secured to a sapphire substrate 2 by a layer of epoxy adhesive. Biasing electrodes 6 and 7 are formed using ion-etching and metal lift-off technique (see EP-A-0007667). At the area of the read-out means (8 and 6) and (8 and 7) each of the strips 1 branches into two parts separated from each other by a slot 13 which extends in a direction parallel to the strip. One part provides the continuation of the ambipolar drift path and the other part supports a metal strip connection which forms the read-out electrode 8. The metal strip may not extend beyond the inner end of the slot 13 as shown in the figure or may extend right across the strip so as to form the read-out electrode beyond the slot 13. The continuation of the ambipolar drift path is narrower than the part of the drift path before the area of the read-out means. This narrowing of the drift path results in a constriction of the bias current and so introduces a higher electric field which increases both the drift velocity and the sensitivity of the device. The device has read-out means at both ends of each strip. This permits read-out with the strips biased in either direction. If the characteristics of the device as made are better when biased in one direction rather than the other, this direction is chosen for operation.

The upper edge of each of the strips is more rounded at the opposite ends of the strip than it is along the sides of the strip. The metal layers forming the electrode connections extend onto the substrate 2 over this more rounded edge. Therefore wires can be bonded to a part of the connection remote from the sensitive area associated with the read-out means.

Ion-etching is used to form the parallel semiconductor strips 1 from a single semiconductor body and to form the separate electrodes and their connections for each strip 1 from a metal layer deposited on the semiconductor body and on the substrate 2 (see EP-A-0007668). Alternatively, the strips 1 are formed from an epitaxial layer of one conductivity type material which is deposited on an intrinsic substrate 2 or a substrate 2 of cadmium telluride.

* *

The read-out structure is further improved in EP-A-0061801 (N.V. Philips' Gloeilampenfabrieken, The Netherlands, 06.10.82).

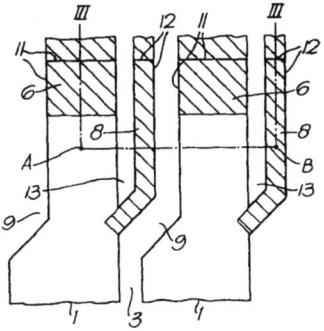

Fig. 1.2.10 (EP-A-0061801 fig. 2)

The read-out means 8 and 6 of each strip 1 comprises the electrode connection 8 which protrudes from one side of the strip at the area of the read-out means. At this area each strip comprises a recess 9 extending across a part of the width of the strip from the opposite side of the strip and towards the protruding connection 8. The protruding connection at one side of one strip extends into this recess at the opposite side of an adjacent strip. In this way a compact device geometry is achieved. Since the narrowing of the drift path is effected from the opposite side of the strip compared with the read-out connection, the resulting distortion of the electric field in this area can act in such a way as to make the effective read-out length slightly shorter than its actual physical length so improving the spatial resolution of the device.

* *

In EP-A-0061802 (Philips Electronic and Associated Industries, GB, 06.10.82) the sensitivity is increased by forming a meandering drift path in the read-out area.

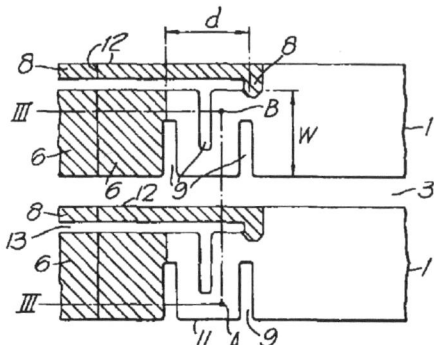

Fig. 1.2.11 (EP-A-0061802 fig. 2)

A part 11 of each strip 1 between the electrode pairs 8 and 6 provides a continuation of the ambipolar drift path through the read-out area. Spaces 9 extend locally across part of the width of the strip to narrow locally the width of the continuation of the ambipolar drift path in the read-out area thereby increasing the electric field within the read-out area. As a result a higher level of sensitivity can be obtained.

* *

A read-out structure which improves the frequency response and the spatial resolution is suggested in GB-A-2095906 (The Secretary of State for Defence, GB, 06.10.82).

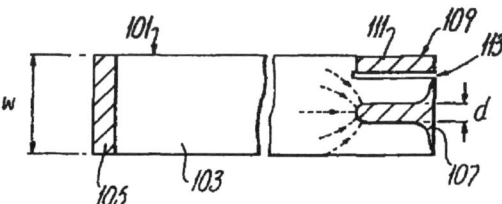

Fig. 1.2.12 (GB-A-2095906 fig. 1)

The output bias contact 107 is shaped to concentrate an electric field bias in the immediate vicinity of the contact 107. This concentrated field sweeps away minority carriers which otherwise accumulate near the output contact. It is proposed that the contact 107 is shaped by extending it towards input bias contact 105, or that the detector material 103 near this contact 107 is configured by slotting or tapering, or that an annular ring input bias contact surrounds a circular disc output contact.

*

A device providing both serial and parallel scan coverage of a scene is disclosed in GB-A-2207801 (The Secretary of State for Defence, GB, 08.02.89).

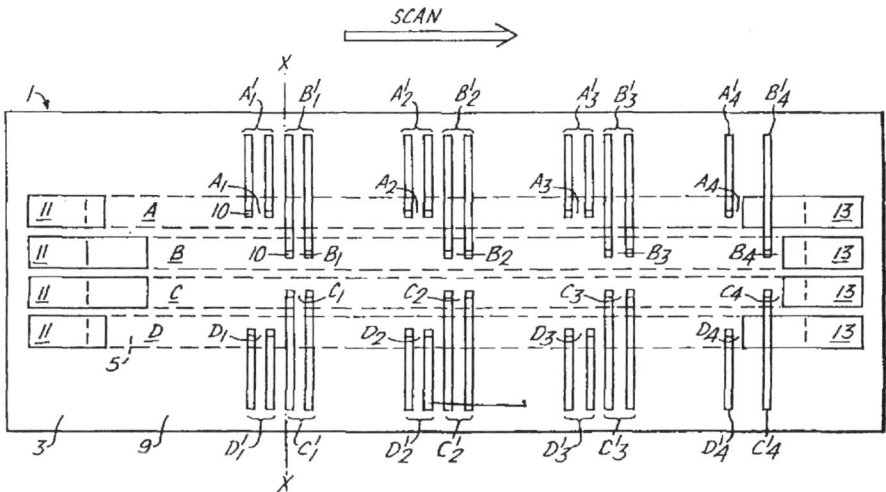

Fig. 1.2.13 (GB-A-2207801 fig. 1)

In a first embodiment strips 5 of $Hg_{0.8}Cd_{0.2}Te$ is provided on a sapphire substrate 3 by bonding an oversize slice of mercury cadmium telluride material onto the insulating substrate with an epoxy resin adhesive, defining the strip pattern photolithographically using a photo-resist medium, and cutting the slice in this pattern using an ion-beam milling technique (see EP-A-0007667, EP-A-0007668 and EP-A-0007669). Input and output bias contracts 11, 13 are provided by sputtering nichrome-gold. Read-out regions A1,A2,...,D4 are defined by paired electrical contacts also of nichrome-gold material. The read-out regions of the inner strip detectors B, C have overlaying conductor connections that also overlay strips A, D respectively. The positions or read-out regions B1 to B4 and C1 to C4 are staggered relative to regions A1 to A4 and D1 to D4, respectively, allowing the conductor connections A1' and B1', A2' and B2',...,C4' and D4' to lie in displaced side by side relationship.

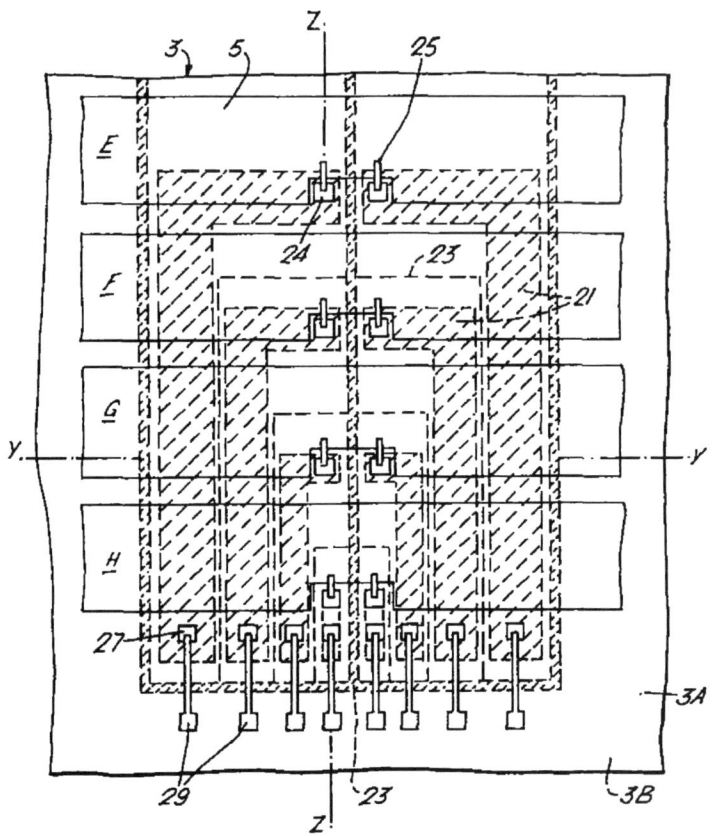

Fig. 1.2.14 (GB-A-2207801 fig. 4)

In a second embodiment the strip detectors A to H are mounted on an intrinsic p-type silicon substrate 3A covered by a silicon oxide layer 3B. A patterned arrangement of conductor tracks 21 is formed in the semiconductor base 3B. Each track is formed by diffusion or ion-implantation of an n-type dopant material, and isolated from adjacent tracks by means of a channel stop network 23. Bridging links of nichrome-gold are formed to define and connect the read-out regions to the tracks 21. The links 25 are paired and thus provide voltage detection contacts. The tracks 21 are connected to connection pads 29. Signal processing circuitry is incorporated in the semiconductor base layer 3B.

Sensitivity is improved in GB-A-2094548 (The Secretary of State for Defence, GB, 15.09.82).

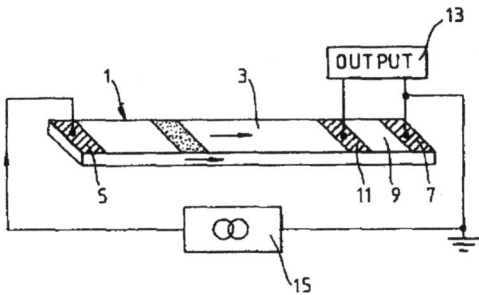

Fig. 1.2.15 (GB-A-2094548 fig. 1)

The length of the detector element 3 or the magnitudes of bias and scan velocity are selected so that the time taken to scan the detector element 3 from one end 5 to a read-out region 9 of the detector element 3 is greater than the lifetime of the photocarriers generated in the element 3. In order to avoid loss of resolution by photocarrier diffusion the photocarrier lifetime of the detector material has of necessity a relatively low value.

*

A detector involving an ambipolar drift of radiation-generated free minority carriers using the loophole technique is disclosed in GB-A-2207802 (Philips Electronic and Associated Industries Limited, GB, 08.02.89).

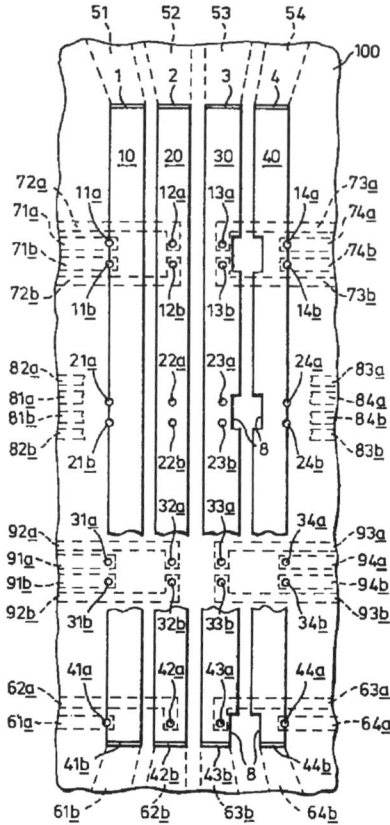

Fig. 1.2.16 (GB-A-2207802 fig. 2)

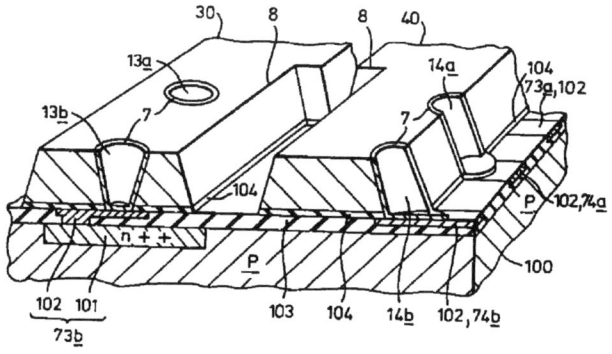

Fig. 1.2.17 (GB-A-2207802 fig. 3)

Biasing-electrode means 1 to 4 and 41b to 44b are spaced so as to cause a bias current predominantly of majority charge-carriers to flow along each semiconductor strip 10, 20, 30, 40 of n-type mercury cadmium telluride, the bias current supporting an ambipolar drift of charge carriers in the opposite direction. Metallized holes provide read-out means in the drift path of each strip. The strips are secured to a silicon substrate 100 by a layer of electrically-insulating epoxy adhesive. Conductor pattern 61-64, 71-74, 81-84, 91-94 provides electrical connections to each read-out means. The substrate comprises the read-out means and signal processing circuitry. Each read-out means is formed at holes (reference a and b) which extend through the thickness of the strips to the conductor pattern. These holes are formed with small transverse dimensions by ion-etching. Metallization 7 is deposited over substantially the whole side wall of each hole without extending on the upper surface of the strips. As a result, the read-out means and their connections do not interrupt the ambipolar drift path to any great extent. In the figures different read-out hole configurations across the width of the different strips 10, 20, 30, 40 are illustrated.

*

The detector of GB-A-2260218 (The Secretary of State for Defence, GB, 07.04.93) comprises a photo-responsive material and an array of planar antennae formed thereon which concentrate radiation in fringe fields at antenna edges and extremities interacting with the photo-responsive material. This structure allows the photo-responsive material to be formed thinly, which reduces volume-dependent generation-recombination noise with consequent increase in responsivity and detectivity. Furthermore, reduced element thickness allows higher element resistance.

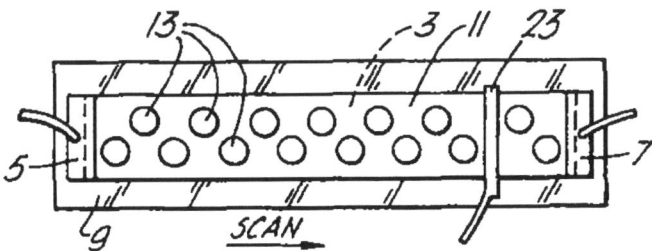

Fig. 1.2.18 (GB-A-2260218 fig. 9)

Bias contacts 5 and 7 and a read-out contact 23 are formed on an elongated strip 3. The detector is provided with an array of antennae 13 upon its surface. In operation, an image is scanned along the detector towards the read-out. A bias current is applied via contacts 5 and 7 and the current level is set so that the forced drift of photocarriers is at a velocity matched to the velocity of a moving image. Stepwise discrete summation of photocarriers takes place as photocarriers drift beneath the antennae and collect carriers generated in the fringe fields. The detector may also be used in a staring mode. Photocarriers are then allowed to accumulate and a bias pulse is applied to drive the photocarriers to the read-out after a period of integration.

*

In EP-A-0151311 (Honeywell Inc., USA, 14.08.85) a device having decreased time for read-out is disclosed.

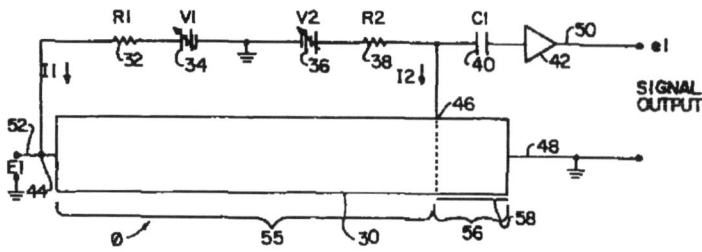

Fig. 1.2.19 (EP-A-0151311 fig. 3)

The semiconductor strip 30 includes an active region 55 and a read-out region 56. A first bias current I1 is applied through the active region of the strip thereby producing an ambipolar drift of photocarriers. A second bias current I2 is applied to the read-out region so that the drift of the photocarriers in the read-out region has a velocity which is greater than the velocity of the drift of the photocarriers in the active region of the strip.

*

A four terminal device with increased spatial frequency response is proposed in EP-A-0188241 (Honeywell Inc., USA, 23.07.86). A problem with the three terminal devices is that the use of the ground bias contact as one of the two read-out electrodes can degrade the performance of the device. This degradation manifests itself as an additional smearing out of the image being scanned along the detector. It arises because the photogenerated excess carriers remain too long in the read-out zone as a result of either blocking of minority carriers at the ground contact or an increase in the effective length of the read-out zone due to contact resistance effects.

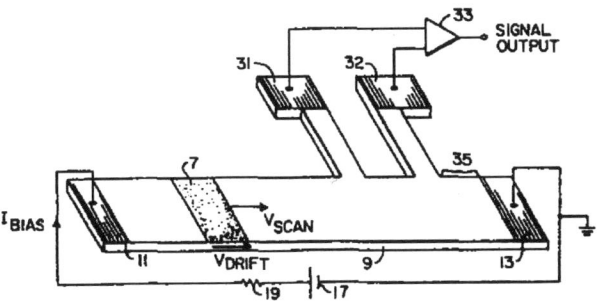

Fig. 1.2.20 (EP-A-0188241 fig. 2)

The device includes two voltage probes 31 and 32, across which the signal voltage is measured by an ac-coupled low-noise preamplifier 33. The second voltage probe 32 is located a distance 35 away from the ground bias electrode 13, such that any blocking effects at the ground bias electrode do not influence the excess carrier distribution between the voltage probe read-out electrodes.

*

Chapter 1.3

Static Induction Transistor Imagers

Summary

When light impinges on a static induction transistor, electron-hole pairs are generated within the channel region, and at least part of the holes thus produced are stored in the gate region, causing the gate potential to rise. The drain voltage varies depending on the magnitude of the gate potential produced as the result of the irradiation of light.

- A photo-electric converter which incorporates the principles of a static induction transistor is disclosed in US-A-4427990. Two dimensional imagers making use of the photo-electric converter are also presented.

- An imager using Metal-Insulator-Semiconductor gate Static Induction Transistors (MISSIT) formed in silicon, for read-out and refresh, and having a photo detecting layer of HgCdTe is presented in EP-A-0038697.

- The imager of EP-A-0038697, discussed above, has a complex structure and hence is difficult to produce. In EP-A-0042218 a simplified structure is presented. In all the embodiments the refresh transistor present in most of the embodiments of EP-A-0038697 has been eliminated.

- In EP-A-0094973 a storing capacitor is connected either to the gate, the source or the drain of the static induction transistor.

Static Induction Transistor Imagers

A photo-electric converter which incorporates the principles of a static induction transistor is disclosed in US-A-4427990 (Zaidan Hojin Handotai Kenkyu Shinkokai, Japan, 24.01.84). When light impinges on a static induction transistor, electron-hole pairs are generated within the channel region, and at least part of the holes thus produced are stored in the gate region, causing the gate potential to rise. The drain voltage varies depending on the magnitude of the gate potential produced as the result of the irradiation of light. Two dimensional imagers making use of the photo-electric converter are also presented.

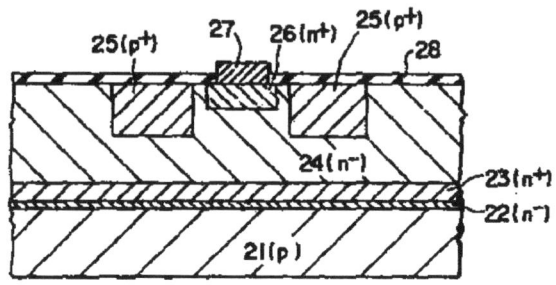

Fig. 1.3.1 (US-A-4427990 fig. 7a)

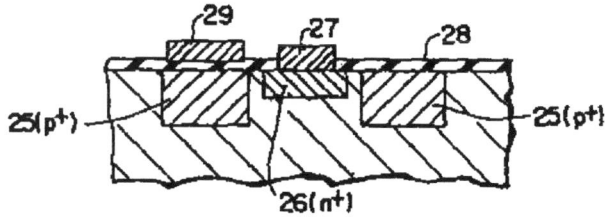

Fig. 1.3.2 (US-A-4427990 fig. 7b)

A single image pick-up element is shown in the figures above. On a p-type substrate 21, there is formed an n^+-type drain region 23 surrounded by an n^--type epitaxial layer 24. A p^+-type gate region 25 and an n^+-type source region 26 are formed within the epitaxial layer. An electrode 27 is formed on the source region. The surface of the epitaxial layer is covered with a transparent insulating layer 28. A gate electrode 29 is formed via the insulating layer above a portion of the gate region. The impurity concentration of the n^--type region 24 is selected to be sufficiently low, so that, when a reverse bias is applied between the source electrode and the gate electrode, the channel region becomes sufficiently pinched off to produce a potential barrier, and that the height of this potential barrier can be controlled also by the application of a drain voltage. The thickness of the insulating layer is selected so that the capacitance which is formed by the MIS structure is significant as compared with the capacitance between the gate region and the source region.

In operation, a positive voltage is applied to the gate electrode, while maintaining both the drain region and the source region at the same potential. The p^+-type gate region 25 is thus forwardly biased, and if an excessive amount of electrons is stored therein, they are expelled outside the gate region, so the gate region is rendered to a predetermined state. Next, a negative voltage is applied to the gate electrode 29, or a positive voltage is applied to the source electrode 27 and/or to the drain region 23, to thereby reverse bias the gate region. At this time, the gate region and the n⁻-type region 24 constitute a diode which is reverse biased. When electron-hole pairs are generated by light which impinges on the diode, the holes flow to the gate region 5 where they are stored. Owing to the stored electric charge, the potential of the gate region will vary. Next, when a positive voltage is applied to the drain electrode with respect to the source potential, a drain current flows which is dependent on the gate potential.

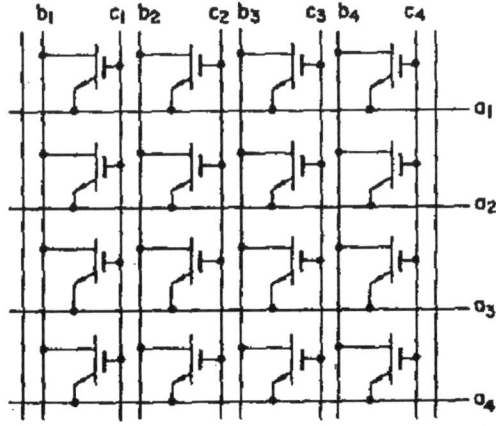

Fig. 1.3.3 (US-A-4427990 fig. 8)

The figure above shows a schematic diagram of an imager, in which image pick-up elements are arranged in a matrix. The sources are conected to row lines a_1, a_2,..., the drains are connected to column lines b_1, b_2,.... and the gate electrodes are connected to clear lines c_1,c_2,... which are disposed in parallel with the column lines. At read-out, a positive voltage is applied to the row lines a_2,a_3,..., while maintaining the row line a_1 at zero potential, and a positive voltage pulse is applied successively to the column lines b_1, b_2,... and the current flowing through the row line a_1 is detected. Next, an erasing positive voltage pulse is applied to the clear lines c_1, c_2,... to clear the elements which have been read. The operation is repeated for the following row.

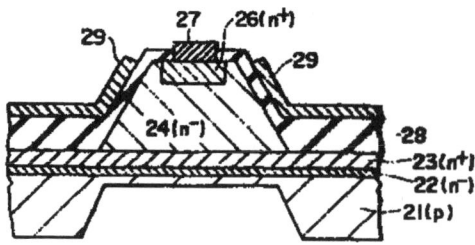

Fig. 1.3.4 (US-A-4427990 fig. 9)

Another embodiment comprises a p-type substrate 21, an n^+-type drain region 23 surrounded by an n^--type region 22, an n^--type channel region 24, n^+-type source region 26, a source electrode 27 and a gate electrode 29 which is formed on an insulating layer 28. The peripheral portions of the channel region are recessed. By causing the insulating layer located beneath the gate electrode to have a negative electric charge, an inversion layer is formed at the surface.

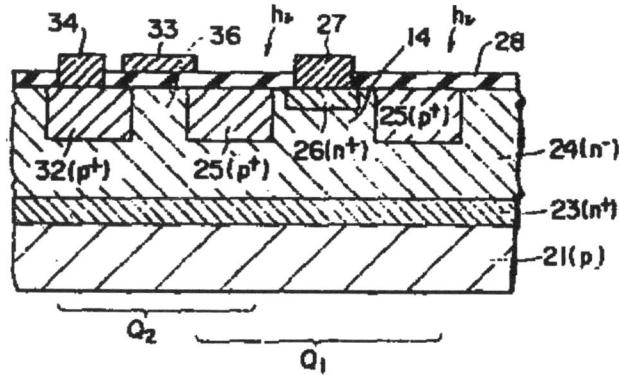

Fig. 1.3.5 (US-A-4427990 fig. 11a)

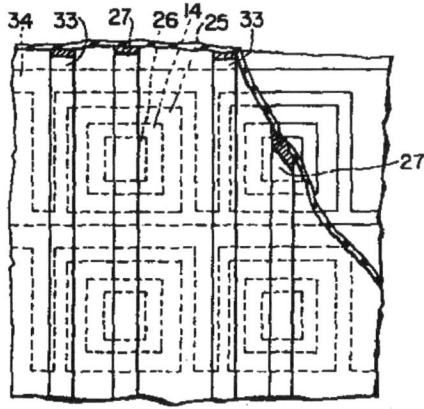

Fig. 1.3.6 (US-A-4427990 fig. 11b)

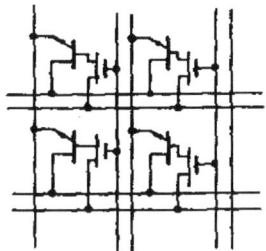

Fig. 1.3.7 (US-A-4427990 fig. 11d)

In a further embodiment, an insulated gate transistor is used as a switching transistor. A stripe shaped n^+-type drain region 23 is formed on a p-type substrate 21. An n^--type epitaxial layer 24 is grown thereon. A p^+-type gate region 25 is formed in the n^--type region and surrounds a channel region 14. Another p^+-type region 32 is formed in the n^--type region. An n^+-type source region 26 is formed in the channel region. The p^+-type regions 25 and 32 and the intervening n^--type region 36 constitute a pair of current electrode regions and a channel region 36 of an insulated gate transistor. Source electrode 27 of the photo-transistor and drain electrode 34 of the switching transistor are formed. A gate electrode 33 is formed on an insulating layer 28 above the channel region of the switching transistor. An embodiment in which a bipolar transistor is used as a switching transistor instead of an insulting gate transistor is also disclosed.

The semiconductor material used in the embodiments is silicon. However, it is mentioned that HgCdTe may be used instead.

*

An imager using Metal-Insulator-Semiconductor gate Static Induction Transistors (MISSIT) formed in silicon, for read-out and refresh, and having a photo detecting layer of HgCdTe is presented in EP-A-0038697 (Semiconductor Research Foundation, Japan, 28.10.81).

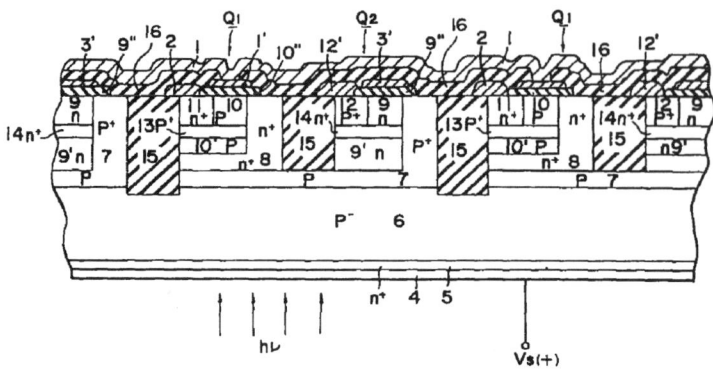

Fig. 1.3.8 (EP-A-0038697 fig. 1a)

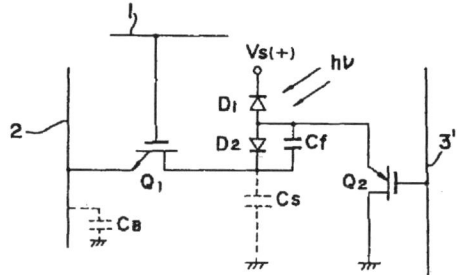

Fig. 1.3.9 (EP-A-0038697 fig. 1b)

The imager comprises a word line 1, which is connected to a gate electrode 1' of an n-channel MISSIT read-out transistor Q1, a bit line 2, which is connected to an n^+-type source region 11 of the transistor Q1, a gate electrode 3' of a p-channel MISSIT refresh transistor Q2 and a transparent electrode 4, which is biased by a power source voltage $V_S(+)$ and which is connected to an n^+-type region 5. An n^+-type region 8 serves as the drain of the transistor Q1 and a region from which electrons flow out in response to a light input and are charged positive, storing therein light information as a voltage. Holes, optically generated in a photo detecting region 6, are stored in a p region 7 which serves as a source region of the refresh transistor. A p region 10 and an n region 9, which are covered by insulating films 10" and 9", are channel regions of the transistors Q1 and Q2 respectively. A p^+-type region 13 and an n^+-type region 14 are buried regions provided for limiting the thickness of depletion layers in the channel regions, so as to avoid the variation of voltage information stored in the n^+-type region

8 and the p region 7 under the influence of alpha rays or the like, and to achieve a normally off characteristic with a short channel MOS gate structure. The transistors are separated by isolation regions 15. A storage capacitor C_S and an earth line are not illustrated in the sectional view of the imager. For detecting infrared light, HgCdTe is grown heterogeneously or deposited by sputtering, evaporation or a CVD method on the surface of the photo detecting portion.

The power source voltage $V_S(+)$ completely depletes the region 6. Optically generated electrons and holes are absorbed in the region 5 and stored in the p region 7, respectively. Accumulated holes charge the p region 7 positive. The potential barrier height for electrons in the n^+-type region 8 is lowered, permitting electrons to flow from the n^+-type region 8, across the p region 7 and towards the substrate. As a result, the n^+-type region 8, which is held floating relative to the earth line via the capacitor C_S, is depleted and its potential is increased. The potential is read out, non-destructively, to a signal output line (bit line) 2 by the read-out transistor Q1. It is shown that the potential on the signal output line is dependent on the junction capacitance between p region 7 and n^+-type region 8 but independent on the storage capacitance C_S and the capacitance of the signal output line. Furthermore, when the size of a detector element is decreased and a lens is provided in front of the optical input receiving surface, which has a resolving power large enough to sufficiently collect the incident light, the output voltage increases.

Several embodiments are disclosed in the patent application. In one of the embodiments the refresh transistor has been eliminated.

*

The imager of EP-A-0038697, shown above, has a complex structure and hence is difficult to produce. In EP-A-0042218 (Semiconductor Research Foundation, Japan, 23.12.81) a simplified structure is presented. In all the embodiments the refresh transistor present in most of the embodiments of EP-A-0038697 has been eliminated.

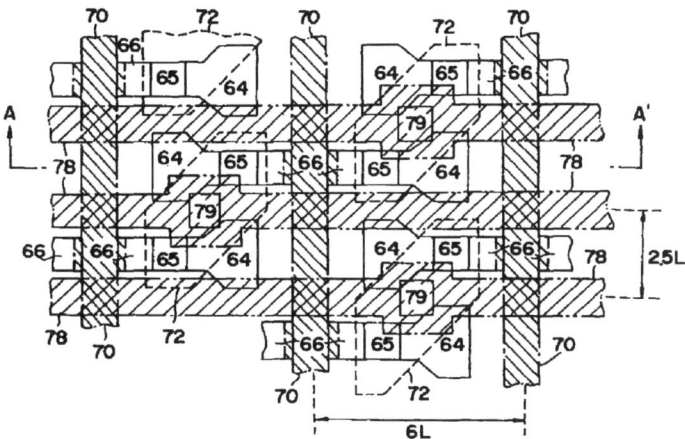

Fig. 1.3.10 (EP-A-0042218 fig. 5a)

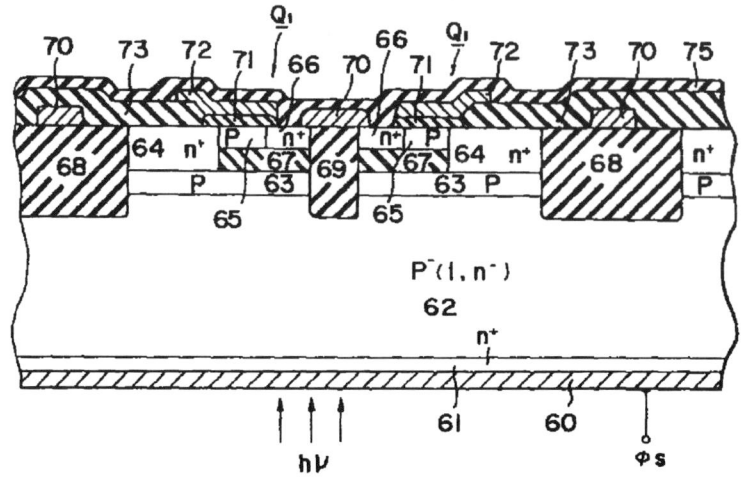

Fig. 1.3.11 (EP-A-0042218 fig. 5b)

A hook structure is formed by n^+ 64, p 63, p^- 62 and n^+ 61 junction structure. A read-out MISSIT transistor Q1 is formed having a source, a channel and a drain formed by the n^+-type region 64, a p region 65 and an n^+-type region 66. The n^+-type region 66 is connected to a signal read-out line 75. A gate region 72 of the read-out transistor is connected via a contact hole 79 to a signal address line 78. An insulating layer 67 is formed between the p region 63 and the p channel region 65 and the n^+-type drain region 66. 68, 69, 73 and 75 denote insulating regions and layers. A region 71 serves as a gate insulating film of the read-out transistor. The signal address line 78 and the via 79 are common to two adjacent cells.

By a pulse voltage Φ_S, connected to a transparent electrode 60, the substrate region 61 is supplied with the voltage $V_S(+)$ in the light integration periode. In the refresh period the voltage Φ_S becomes 0 V or slightly negative, performing the operation of drawing out holes which are excess majority carriers stored as optical information in the p region 63.

An infrared imager is formed by forming the n^+-type region 61, the p^--type region 62 and the p layer 63 of HgCdTe, forming an insulating layer of ZnS on the p layer 63 at a predetermined position, growing thereon polysilicon by a CVD technique, leaving the polysilicon at a predetermined position and adding a desired impurity by ion-implantation. A relatively good single crystal polysilicon can be made using a laser anneal technique.

*

In EP-A-0094973 (Semiconductor Research Foundation, Japan, 30.11.83) a storing capacitor is connected either to the gate, the source or the drain of the static induction transistor.

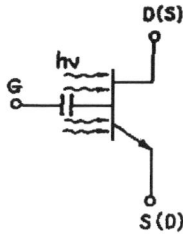

Fig. 1.3.12 (EP-A-0094973 fig. 14a)

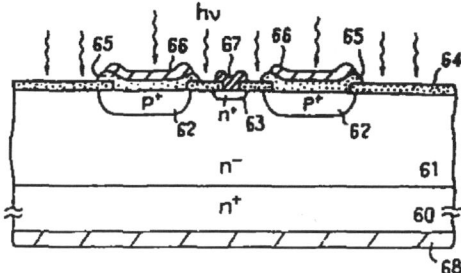

Fig. 1.3.13 (EP-A-0094973 fig. 14b)

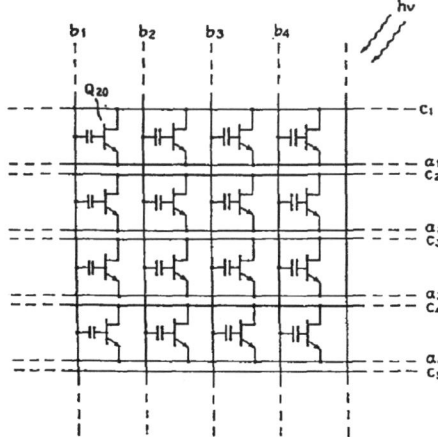

Fig. 1.3.14 (EP-A-0094973 fig. 18a)

An n$^-$-type layer 61 is formed on an n$^+$-type substrate 60. P$^+$-type gate regions 62 and n$^+$-type source region 63 are formed. The gate regions, metal electrodes 66 and an intermediate insulating layer 65 form a storage capacitor. 67 and 68 indicate metal electrodes of the source and the drain, respectively, and 66 indicates a surface protecting film. In other embodiments, the capacitor for storing the signal is connected to the source or the drain. An imager is formed by arranging a plurality of transistors having storage capacitors in two dimensions. The device may be formed of HgCdTe.

*

The structures of the imagers in EP-A-0094974 (Semiconductor Research Foundation, Japan, 30.11.83) are presented in EP-A-0094973 which has been shown above. However, atoms which form an impurity layer which is excited by light having a wavelength longer than the wavelength of the light corresponding to the energy bandgap of the channel region receiving the light are added to the channel region. The imagers may be formed of HgCdTe.

*

Chapter 1.4

Charge Injection Device Imagers

Summary

A charge injection device (CID) imager [1, 2] comprises an array of horizontal row lines and output column lines. Each row line is connected to row-MIS electrodes. Each MIS in a given row can interchange charge with only one column MIS well with each of the column MIS wells in a row being attached to a different column output line.

- The advantages of charge-injection-device (CID) and charge-coupled-device (CCD) technologies for the development of monolithic HgCdTe infrared image sensors for the 3-5 µm wavelength band are presented by Richard A. Chapman et al. in *IEEE Transactions on Electron Devices*, ED-27(1), January 1980, pp. 134-145, "Monolithic HgCdTe Charge Transfer Device Infrared Imaging Arrays".

- The invention of US-A-4259576 refers to a CID in which each detector element includes a x-electrode portion and a y-electrode portion. The rows of y-electrode portions are covered with an opaque covering. The arrangement permits read-out of individual detector elements with a reduced risk for blooming by nearby detector elements at low inputs.

- The imager of US-A-4360732 comprises integrating electrodes which are connected together in columns, transfer electrodes which are connected together in rows and drain wells. After an integration period, accumulated charge is transfered from wells under the integrating electrodes to the drain wells. The column voltages are sampled before and after

---

[1] A. J. Lewis, R. A. Chapman, E. Schallenberg, A. Simmons and C. G. Roberts, "Monolithic HgCdTe Charge Injection Devices", *IEDM Tech. Dig.*, 571-573, December 1979.
[2] C.-Y. Wei, H. H. Woodbury and S. C.-H. Wang, "A Novel CID Structure for Improved Breakdown Voltage", *IEEE Trans. Electron Devices*, 37 (3), 611-617, March 1990.

the transfer of charge and the voltage difference is proportional to the intensity of infrared radiation.

A field plate, which covers the area between active areas, is used to prevent light from creating minority carriers in the regions between detector elements and to provide channel stop by permitting accumulation of the surface between the detecting areas thereby preventing carriers from moving from one detector element to another. A MIS electrode is formed on an insulating layer which separates the MIS electrode from the field plate.

- The provision of a field plate next to a MIS electrode establishes an electric field at the edge of a column-read well next to the field plate. This electric field produces an increased tunnel current. This problem is solved in EP-A-0044610 by either having an increased thickness of an insulating layer between the field plate and the MIS electrode or by placing the field plate and the MIS electrode on the same level.

- Each pixel of the imager presented in JP-A-57010986 comprises a storage electrode, a detecting electrode and a control electrode. The control electrode allows the signal charge to be skimmed off from the total amount of charge which includes charge generated from the background.

- The elementary cells of the imager in US-A-4583108 are arranged in a staggered configuration. The arrangement permits the number of rows to be doubled without increasing the number of cells connected to a reading amplifier. Guard electrodes which are disposed below addressing and reading conductors are provided in order to prevent signals for addressing and reading the cells to interfere with charges of other cells.

- Leakage-current is reduced and the speed is improved in the imager of JP-A-59047759, in which a thin layer of n-type is formed at the surface of a p-type substrate.

- The interface between an n-type HgCdTe layer and an insulating layer of a MIS type photosensor is improved in JP-A-59061080 by forming an anodic oxide film on the surface of the n-type layer.

- In JP-A-60074568 the problem of tunnel currents, which are generated when a large voltage is applied to an electrode in order to generate a deep storage well, is reduced by forming transparent electrodes over photo-absorbing HgCdTe regions having a narrow energy bandgap and forming storage electrodes over HgCdTe regions having a large energy bandgap.

- The imager disclosed in JP-A-60103666 is similar to the imager presented in JP-A-60074568 discussed above. An HgCdTe layer having a large energy bandgap is formed on a CdTe substrate in this structure.

- The imager of JP-A-60180162 comprises active regions corresponding to detector elements which are surrounded by resistance layers buried in the substrate. The resistance layers allow stored signal charges which are injected in the direction of the substrate to recombine at high speed.

- In JP-A-60193375 HgCdTe regions having a narrow energy bandgap are formed below charge injection electrodes. When reset is performed, signal charges recombine at high speed in these HgCdTe regions.

- A thin recombination layer is employed between a substrate and a photodetective layer in the imager of JP-A-61067958. The lattice constant of the recombination layer is different from the lattice constant of the substrate and the photodetective layer. To prevent misfit dislocation, the thickness of the recombination layer is chosen to be thinner than a critical value.

- The design of the imager presented in US-A-5171994 eliminates previously used optical windows consisting of a conductive thin semi-transparent chromium layer. The elimination of the chromium optical window results in an increased pixel-to-pixel uniformity, fabrication with only four mask levels and an increase in wafer yield. The detector is claimed in this patent while the method of fabrication of the device is claimed in US-A-5130259.

- A detector element in which amplification of photon-generated charge takes place by avalanche multiplication is disclosed in US-A-4912536.

Charge Injection Device Imagers

The advantages of charge-injection-device (CID) and charge-coupled-device (CCD) technologies for the development of monolithic HgCdTe infrared image sensors for the 3-5 μm wavelength band are presented by Richard A. Chapman et al. in *IEEE Transactions on Electron Devices*, ED-27(1), January 1980, pp. 134-145, "Monolithic HgCdTe Charge Transfer Device Infrared Imaging Arrays".

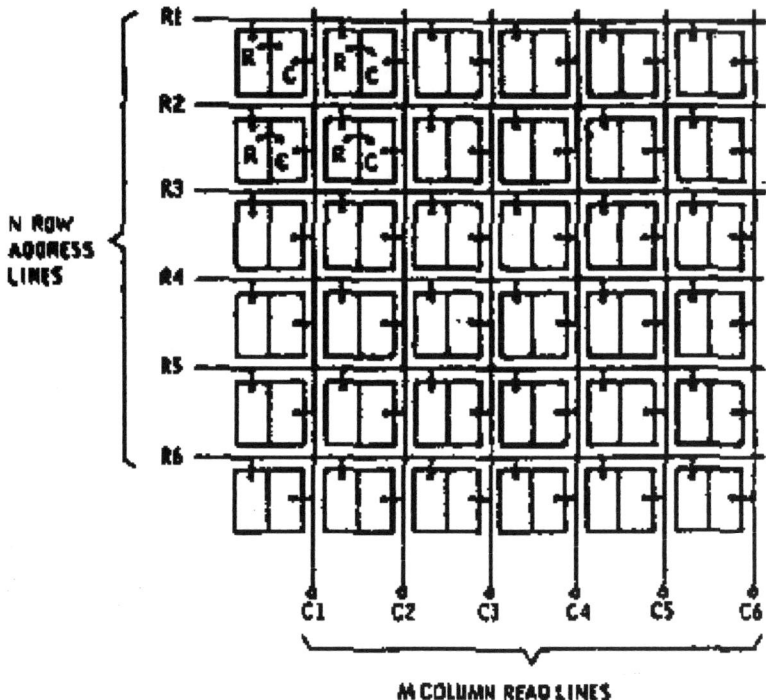

Fig. 1.4.1 (*IEEE Trans. Electron Devices*, ED-27(1), 134-145, fig. 1, © 1980 IEEE)

The figure above shows the top view of a 6 x 6 CID array addressed by six horizontal row lines (R1-R6) with six output column lines (C1-C6). Each row line is connected to six row-MIS electrodes (R). Each MIS in a given row can interchange charge (as shown by the arrows) with only one column (C) MIS well with each of the column MIS wells in a row being attached to a different column output line.

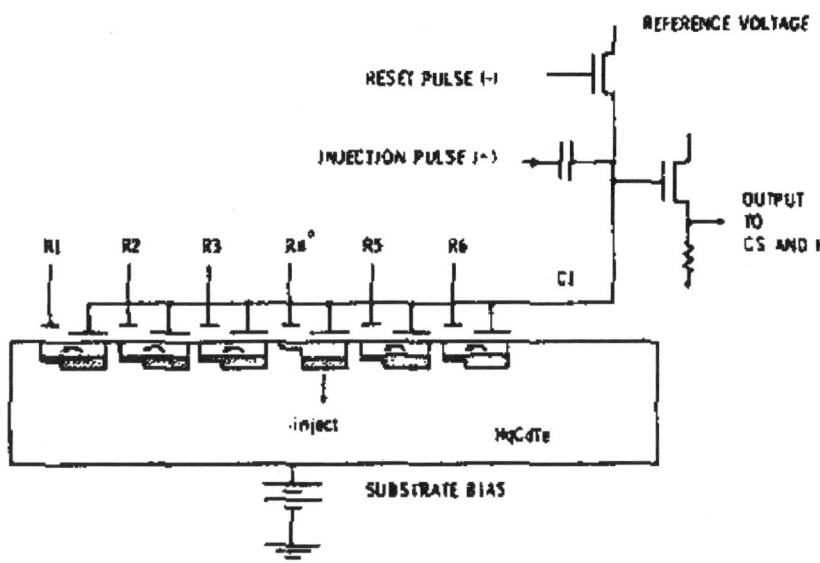

Fig. 1.4.2 (*IEEE Trans. Electron Devices*, ED-27(1), 134-145, fig. 2, © 1980 IEEE)

One way of operation has all row wells normally on (biased to produce a potential well), then one row is turned off at a time. Detection is obtained by applying an injection pulse to each column line during the time one row is turned off. The figure above shows the potential wells along the column line (C1) when row R4 has been addressed by collapsing the R4 potential wells and before an injection pulse has been applied to the column line. When an injection pulse is applied to the C1 line, all charge packets which are not to be detected are transferred to the adjoining row wells as noted by the arrows. Only at the R4 elements of the matrix is the charge injected into the substrate, because here the row well is off when the column wells are collapsed by the injection pulse. Signal is detected by sensing the column node voltage before injection, clamping the output at this value, and then sampling the change in node voltage after the injection of the charge packet by the injection pulse.

*

The invention of US-A-4259576 (The United States of America as represented by the Secretary of the Navy, USA, 31.03.81) refers to a CID in which each detector element includes a x-electrode portion and a y-electrode portion. The rows of y-electrode portions are covered with an opaque covering. The arrangement permits read-out of individual detector elements with a reduced risk for blooming by nearby detector elements at low inputs.

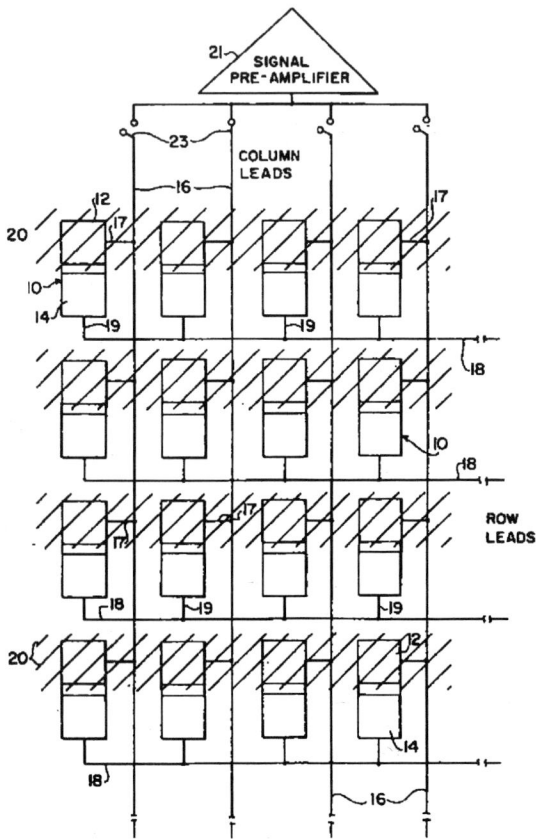

Fig. 1.4.3 (US-A-4259576 fig. 1)

The detector array comprises detector elements 10, each including a x-electrode 14 and a y-electrode 12. Each of the x-electrodes 14 of the same row is connected to a row conductor 18 by leads 19 and each of the y-electrodes of the same column is connected to a column conductor 16 by leads 17. The column conductors 16 are successively connected to a preamplifier 21 by use of switches 23. The y-electrodes and the areas nearby are covered with a material 20, represented by hash-marks, which is opaque to infrared radiation.

*

The imager of US-A-4360732 (Texas Instruments Incorporated, USA, 23.11.82) comprises integrating electrodes which are connected together in columns, transfer electrodes which are connected together in rows and drain wells. After an integration period, accumulated charge is transferred from wells under the integrating electrodes to the drain wells. The column voltages are sampled before and after the transfer of charge and the voltage difference is proportional to the intensity of infrared radiation.

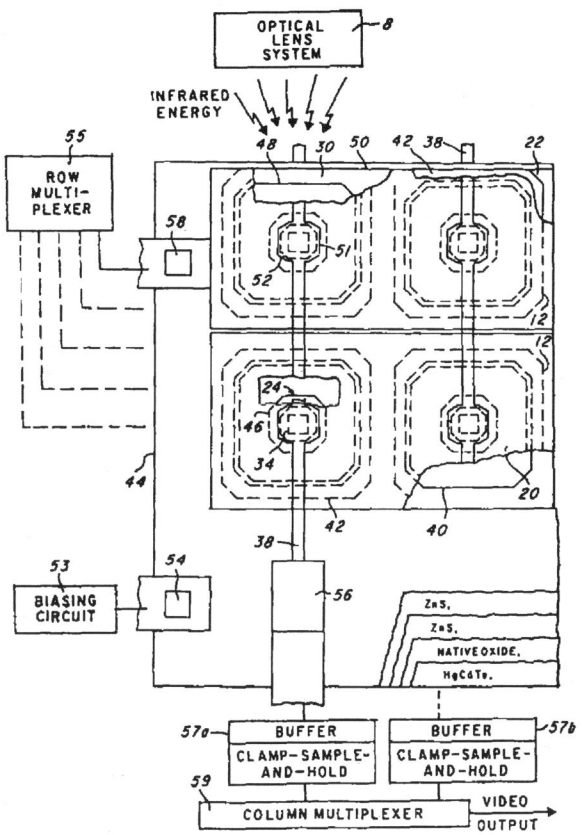

Fig. 1.4.4 (US-A-4360732 fig. 3)

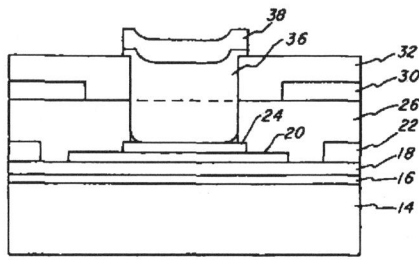

Fig. 1.4.5 (US-A-4360732 fig. 2g)

The imaging system comprises an optical lens system 8 for focusing infrared energy emanating from a scene in the field-of-view onto the imager and detector cells 12 arranged in rows and columns on an HgCdTe substrate 14. A native oxide layer 16 is formed on the HgCdTe substrate followed by a first ZnS insulating layer 18, a metal layer 20 of nickel which is transparent to infrared energy and which defines infrared sensitive areas, a thick metal layer 24 which is opaque to infrared energy and which selectively covers portions of the first insulating layer and the transparent metal layer to form a drain and integrate electrodes 24, a second ZnS insulating layer 26, opaque metal transfer gates 30 and a third ZnS insulating layer 32. Apertures 34 provide passages through insulating layers 32 and 26 to the thick metal layer 24. The apertures are filled with indium 36 and are capped with aluminium address lines 38. A via 54 connects the common drain 22 to a biasing circuit 53. Lead bond pads 56 connect the detector elements of each column to a buffer and clamp, sample and hold circuit 57. A column multiplexer 59 provides a serial video output. The transfer gates 30 are connected through vias 58 to a row multiplexer 55.

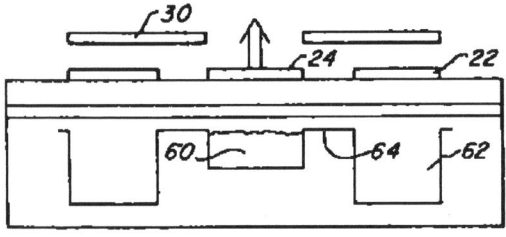

Fig. 1.4.6 (US-A-4360732 fig. 4b)

In operation, charges are first accumulated in the column wells. At the end of the integration period, the voltage levels at the column wells are measured.

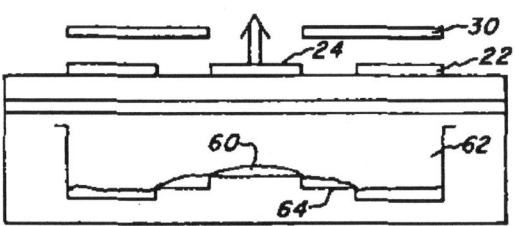

Fig. 1.4.7 (US-A-4360732 fig. 4c)

Next, the transfer gates are turned on and the charges are dumped into the drain wells 62.

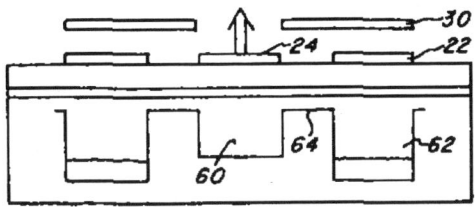

Fig. 1.4.8 (US-A-4360732 fig. 4d)

Thereafter, the transfer gates are turned off and the voltage levels at the column wells are measured again and the difference between the measured voltage levels for each detector element is established. Finally, injection pulses are applied to the drain electrodes to clear the drain wells.

In an alternative embodiment the drain 22 is replaced by a diode operative to remove charges dumped from the integrate wells by the transfer gates 30.

*

A field plate, which covers the area between active areas, is used to prevent light from creating minority carriers in the regions between detector elements and to provide channel stop by permitting accumulation of the surface between the detecting areas thereby preventing carriers from moving from one detector element to another. A MIS electrode is formed on an insulating layer which separates the MIS electrode from the field plate. However, the provision of a field plate next to a MIS electrode establishes an electric field at the edge of a column-read well next to the field plate. This electric field produces an increased tunnel current. This problem is solved in EP-A-0044610 (Texas Instruments Incorporated, USA, 04.08.82) by either having an increased thickness of an insulating layer between the field plate and the MIS electrode or by placing the field plate and the MIS electrode on the same level.

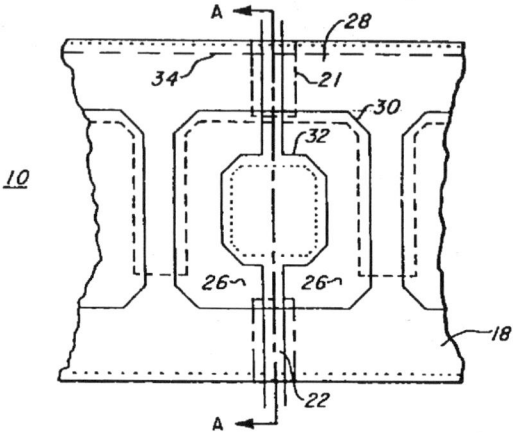

Fig. 1.4.9 (EP-A-0044610 fig. 1)

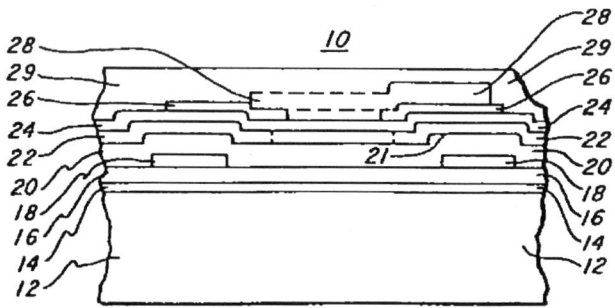

Fig. 1.4.10 (EP-A-0044610 fig. 2)

A native oxide layer 14 is formed on an HgCdTe substrate 12 followed by a first insulating layer 16 of ZnS. An opaque metal field plate 18 is formed and covered by a second insulating layer 20 of ZnS. The insulating layer 20 has an increased thickness 21 over the field plate 18. Next, the second insulating layer is masked and a thick layer 22 of metal is deposited and formed into column electrodes. A third insulating layer 24 of ZnS is formed and a thin transparent metal layer 26 of nickel is deposited and patterned to form transparent electrodes corresponding to infrared sensitive regions. Then, an opaque row address metal 28 is formed and finally a ZnS insulator layer is formed. The extra thickness 21 of insulating layer 20 is to increase the distance between the field plate 18 and the column electrode to reduce the electric field in the semiconductor underneath the field plate edge.

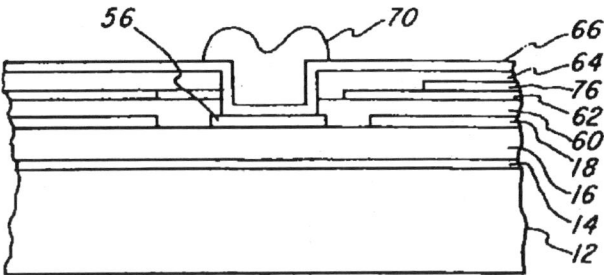

Fig. 1.4.11 (EP-A-0044610 fig. 5)

In a second embodiment, each detector element comprises an HgCdTe substrate 12, native oxide layer 14, ZnS insulator 16, a field plate 18 and a column gate electrode 56 both formed from a first level of metallization. This arrangement eliminates the overlap between the field pate and column gate and thereby the need for the field plate protect 21 described above. The detector element further comprises a ZnS insulating layer 60, a transparent row metal 62, an opaque row address line 76, ZnS insulating layer 64 and a column line 66.

*

Each pixel of the imager presented in JP-A-57010986 (Fujitsu Ltd, Japan, 20.01.82) comprises a storage electrode, a detecting electrode and a control electrode. The control electrode allows the signal charge to be skimmed off from the total amount of charge which includes charge generated from the background.

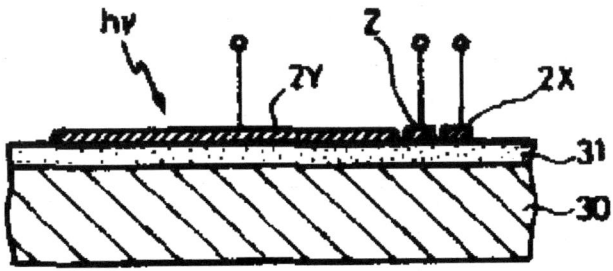

Fig. 1.4.12 (JP-A-57010986 fig. 2a)

A storage well 41 is generated under the storage electrode in which background related charge Q1 and signal charge Q2 are accumulated. The signal charge is then skimmed off by controlling the potential on the control electrode. The detecting electrode is used to detect the amount of signal charge.

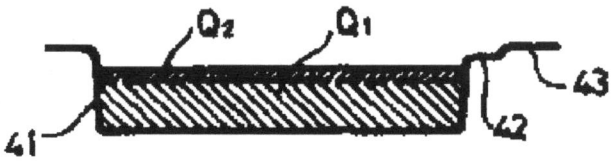

Fig. 1.4.13 (JP-A-57010986 fig. 2b)

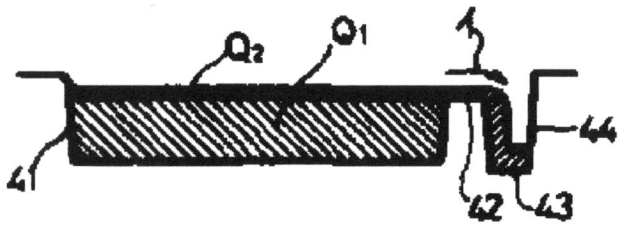

Fig. 1.4.14 (JP-A-57010986 fig. 2c)

A semiconductor substrate 30 is covered with an insulating film 31. Each pixel comprises a transparent storage electrode 2Y, an opaque control electrode Z and an opaque detecting electrode 2X.

*

The elementary cells of the imager in US-A-4583108 (Societe Anonyme de Telecommunications, France, 15.04.86) are arranged in a staggered configuration. The arrangement permits the number of rows to be doubled without increasing the number of cells connected to a reading amplifier. Guard electrodes which are disposed below addressing and reading conductors are provided in order to prevent signals for addressing and reading the cells to interfere with charges of other cells.

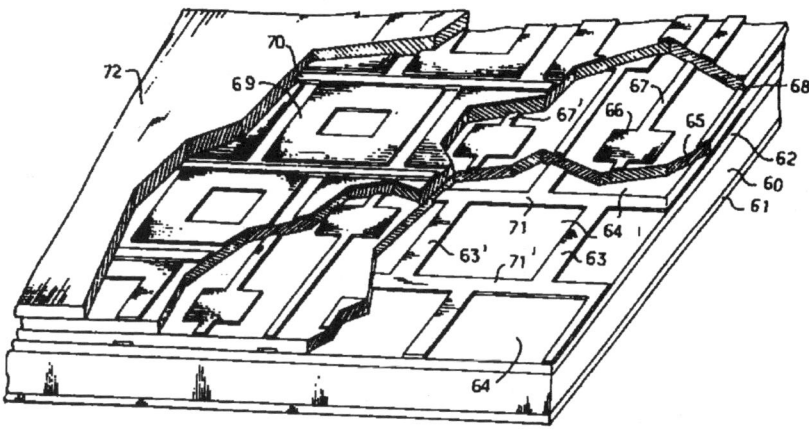

Fig. 1.4.15 (US-A-4583108 fig. 6)

The imager comprises an n-type mercury cadmium telluride substrate 60. A metal layer 61 is deposited on one of the main surfaces while a first insulating layer 62 is deposited on the opposite surface. Opaque guard electrodes 63 of chromium-gold are deposited on the first insulating layer. Those electrodes define internal surfaces 64 in the form of rectangles, which correspond to elementary cells disposed in a staggered configuration. A second insulating layer 65 covers the guard electrodes and the rectangles. Transparent reading electrodes 66 and opaque reading conductors 67 are deposited on the second insulating layer. The reading electrodes are disposed at the centre of the rectangles. The reading conductors of a column of cells are aligned with the adjacent guard electrodes on one side in a column of cells disposed in a staggered relationship. A third insulating layer 68 covers the reading electrodes, the reading conductors and the part of the second insulating layer which is not covered by the reading electrodes and the reading conductors. Transparent addressing electrodes 69 and opaque addressing conductors 70 are disposed on the third insulating layer. Each electrode for addressing a cell is disposed around the electrode for reading the same cell, and the periphery thereof coincides with the periphery of the rectangles 64. The electrodes for addressing a row of cells are connected together by an opaque addressing conductor such as that indicated at 70, and the addressing conductors are aligned with the limb portions of the subjacent guard electrodes, such as those indicated at 71 and 71'. A fourth insulating layer 72 covers the whole of the addressing electrodes, addressing conductors and intermediate spaces.

When making use of a staggered arrangement of elementary cells, it is possible to produce matrices comprising 128 rows and 200 columns. The 64 conductors of odd-number rows are connected to a first addressing circuit while the corresponding 100 conductors of odd-number columns are connected to a first reading circuit. The 64 conductors of even-number rows are connected to a second addressing circuit while the corresponding 100 conductors of even-number columns are connected to a second reading circiut. In this way, it is possible to read the odd-number rows first and then the even-number rows as in the procedure for scanning a television screen. Furthermore, four matrices can be grouped together so as to produce a matrix comprising 256 rows and 400 columns, that is to say, a matrix of television format.

In an alternative embodiment, each elementary cell is divided into two half-cells which have the same reading electrode but two addressing electrodes which are each connected to an addressing conductor.

*

Leakage-current is reduced and the speed is improved in the imager of JP-A-59047759 (Fujitsu KK, Japan, 17.03.84), in which a thin layer of n-type is formed at the surface of a p-type substrate.

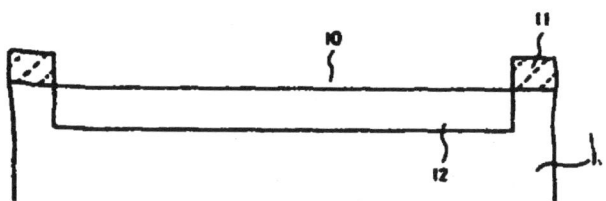

Fig. 1.4.16 (JP-A-59047759 fig. 3)

An HgCdTe substrate 1 is subjected to an heat treatment whereby Hg is evaporated turning the substrate into p-type. A SiO_2 film 11 having an opening 10 is formed. The device is subjected to a low temperature treatment at 250°C in an Hg atmosphere. The SiO_2 film acts as a mask and Hg fills vacancies in the surface layer 12 of the substrate 1 in the regions corresponding to the opening 10 turning this surface layer into n-type.

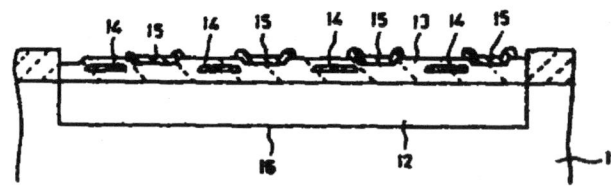

Fig. 1.4.17 (JP-A-59047759 fig. 4)

The device is completed by a SiO_2 layer 13, thin transparent read-out electrodes 13 of chrome and opaque accumulation electrodes 15.

*

The interface between an n-type HgCdTe layer and an insulating layer of a MIS type photosensor is improved in JP-A-59061080 (Fujitsu K.K., Japan, 07.04.84) by forming an anodic oxide film on the surface of the n-type layer.

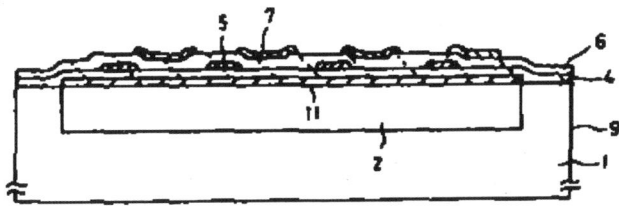

Fig. 1.4.18 (JP-A-59061080 fig. 3)

An n-type HgCdTe layer 2 is formed at a surface portion of a p-type HgCdTe substrate 1 by introducing Hg into the substrate. Anodic oxidation is performed to form an anodic oxide film 11 on the surface of the n-type layer. Next, an insulation film 4 of SiO_2 is formed over the entire surface. Read-out electrodes 5, an insulating film 6 and accumulation electrodes 7 are formed on the insulating film 4.

*

In JP-A-60074568 (Fujitsu KK, Japan, 26.04.85) the problem of tunnel currents, which are generated when a large voltage is applied to an electrode in order to generate a deep storage well, is reduced by forming transparent electrodes over photo-absorbing HgCdTe regions having a narrow energy bandgap and forming storage electrodes over HgCdTe regions having a large energy bandgap.

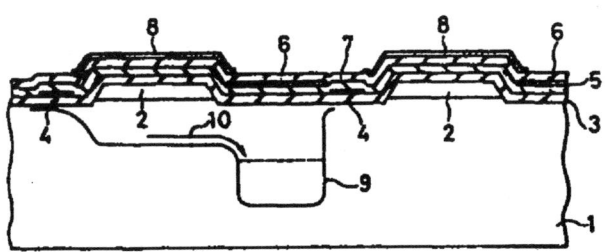

Fig. 1.4.19 (JP-A-60074568 the figure)

An HgCdTe layer having a narrow energy bandgap is epitaxially grown on an HgCdTe substrate 1 having a large energy bandgap. The epitaxially grown HgCdTe layer is etched to form mesas 2. Field plates 4 are formed to separate adjacent detector elements from each other. The structure is completed by forming ZnS insulating films 3, 5 and 7, transparent electrodes 8 and storage and reading electrodes 6.

*

The imager disclosed in JP-A-60103666 (Fujitsu KK, Japan, 07.06.85) is similar to the imager presented in JP-A-60074568 shown above. An HgCdTe layer having a large energy bandgap is formed on a CdTe substrate in this structure.

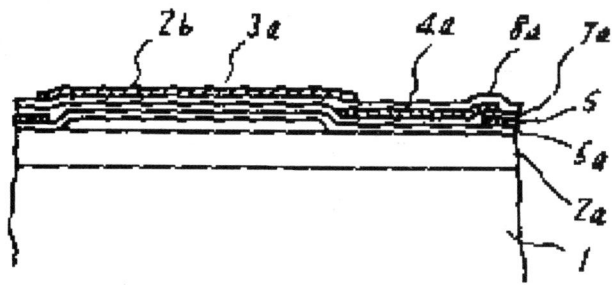

Fig. 1.4.20 (JP-A-60103666 fig. 2)

An HgCdTe layer 2a having a large energy bandgap is formed on a CdTe substrate 1. Mesa shaped HgCdTe regions having a narrow energy bandgap are formed on the layer 2a. The structure comprises also transparent electrodes 3e, storage electrodes 4e and field plates 5.

*

The imager of JP-A-60180162 (Fujitsu KK, Japan, 13.09.85) comprises active regions corresponding to detector elements which are surrounded by resistance layers buried in the substrate. The resistance layers allow stored signal charges which are injected in the direction of the substrate to recombine at high speed.

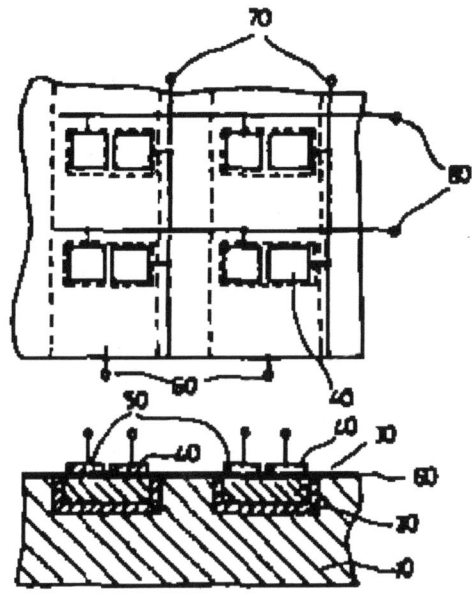

Fig. 1.4.21 (JP-A-60180162 fig. 2)

N-type HgCdTe epitaxial layers 30, surrounded by p-type CdTe resistance layers 20, are formed in an n-type CdTe substrate 10. The imager further comprises storage electrodes 40 and charge injecting electrodes 50.

*

In JP-A-60193375 (Fujitsu KK, Japan, 01.10.85) HgCdTe regions having a narrow energy bandgap are formed below charge injection electrodes. When reset is performed, signal charges recombine at high speed in these HgCdTe regions.

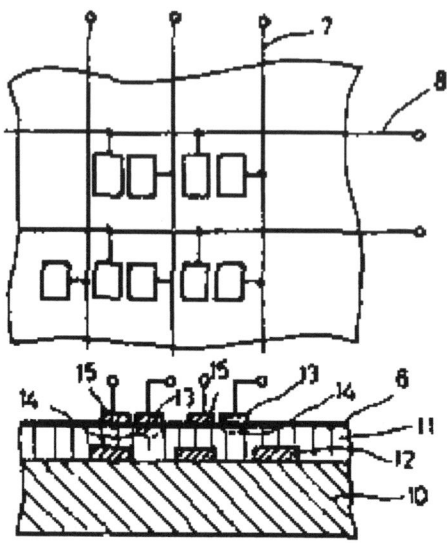

Fig. 1.4.22 (JP-A-60193375 fig. 2)

P-type HgCdTe regions 12 are formed on an n-type CdTe substrate 10. An n-type HgCdTe layer 11 is epitaxially grown on the structure. Charge accumulation electrodes 13 and charge injection electrodes 15 are formed on an insulating layer 6. The p-type HgCdTe regions 12 have a narrower energy bandgap than the energy bandgap of the n-type HgCdTe layer 11. Infrared radiation penetrates through the substrate and generates charge which is accumulated in potential wells 14.

*

A thin recombination layer is employed between a substrate and a photodetective layer in the imager of JP-A-61067958 (Fujitsu Ltd, Japan, 08.04.86). The lattice constant of the recombination layer is different from the lattice constant of the substrate and the photodetective layer. To prevent misfit dislocation, the thickness of the recombination layer is chosen to be thinner than a critical value.

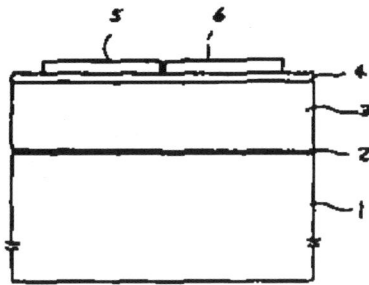

Fig. 1.4.23 (JP-A-61067958 fig. 1)

An $Hg_{0.9}Cd_{0.1}Te$ recombination layer 2 is formed between a $CdTe_{0.96}Se_{0.04}$ substrate 1 and an n-type photodetective layer 3 of $Hg_{0.7}Cd_{0.3}Te$. A storage electrode and a read electrode 6 are formed on an insulating layer 4. The lattice constants of the substrate and the photodetecting layer are both about 6.45 Å which gives a lattice mismatch for the recombination layer of about 0.06 %. The generation of misfit dislocations can be prevented if the thickness of the recombination layer is chosen to be less than 0.2 µm.

*

The design of the imager presented in US-A-5171994 (Northrop Corporation, USA, 15.12.92) eliminates previously used optical windows consisting of a conductive thin semi-transparent chromium layer. The elimination of the chromium optical window results in an increased pixel-to-pixel uniformity, fabrication with only four mask levels and an increase in wafer yield. The detector is claimed in this patent while the method of fabrication of the device is claimed in US-A-5130259 (Northrop Corporation, USA, 14.07.92).

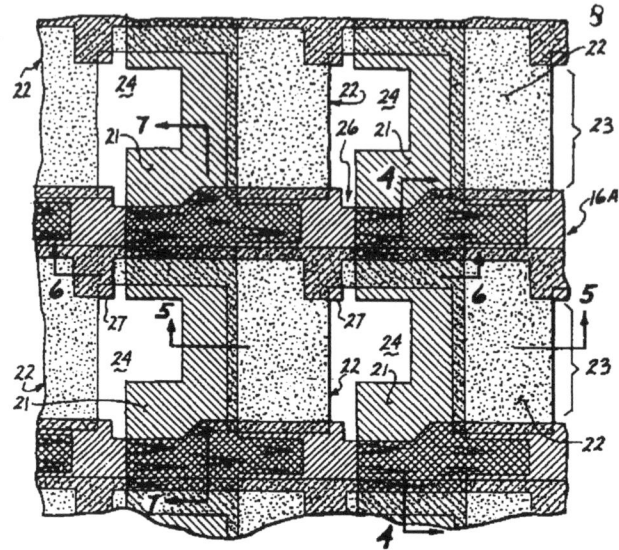

Fig. 1.4.24 (US-A-5171994 fig. 2)

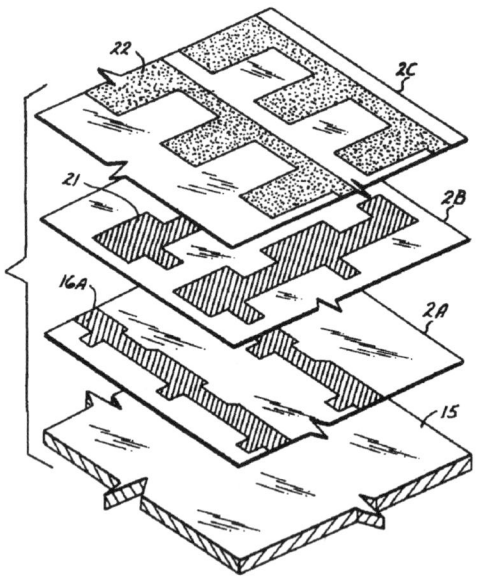

Fig. 1.4.25 (US-A-5171994 fig. 2d)

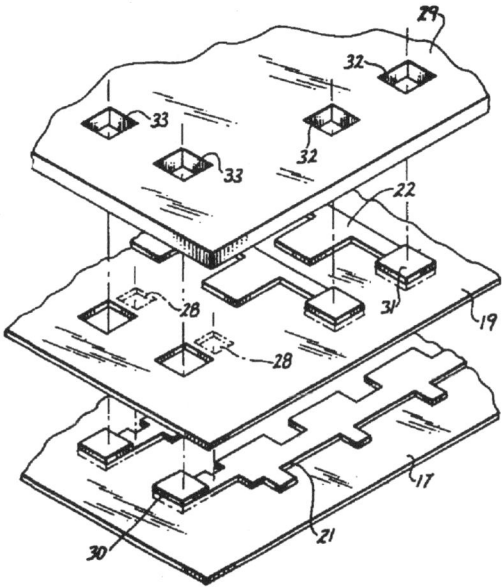

Fig. 1.4.26 (US-A-5171994 fig. 15)

The devices presented have been realized in InSb material, but they can also be implemented in HgCdTe. A substrate 15 is coated with a field oxide layer 16. The first masking step provides for the selective removal of portions of the field oxide to define the active area 23. This is followed by deposition of a first gate oxide 17. A first metal layer 18 is deposited and patterned to form columns 21. A second gate oxide layer 19 is then deposited. The second gate oxide layer is patterned by masking and etching to provide contact holes 28. A second metal layer 20 is subsequently deposited over the second gate oxide layer. The second metal layer is patterned by masking and etching to selectively remove portions thereof to provide rows 22.

In a first embodiment in which windows 28 are etched through the second gate oxide layer 19, column bonding pads 30, shown in solid line, are derived from the second metal layer 20 and contact is made to the columns 21 via extensions from the solid portion of the bonding pad 30 through the windows 28 to the columns 21.

In a second embodiment in which the column bonding pads 30 are derived from the first metal layer 18, the bonding pads 30 correspond to the lower portion, shown in a broken line, of the column bonding pads which are integral with the remainder of the columns 21. Thus, in this embodiment the columns, the column bonding pads and the interconnect between the column bonding pads and the columns are formed and patterned from the first metal layer.

With use of a passivation layer 29 a further embodiment allows for utilization of thin metal layers for both the first metal layer 18 and the second metal layer 20 to optimize yield and desired characteristics for the rows 21 and columns 22. The desired thickness of the bonding pads 30 and 31 being built up by a subsequent metal deposition. This is achieved by patterning the passivation layer 29 with a mask to form column bonding pad vias 32 and row bonding pad vias 33 therein. These are etched into the passivation layer and for the column bonding pads also through the second gate oxide 19. The passivation layer can serve as an anti-reflection coating.

*

A detector element in which amplification of photon-generated charge takes place by avalanche multiplication is disclosed in US-A-4912536 (Northrop Corporation, USA, 27.03.90).

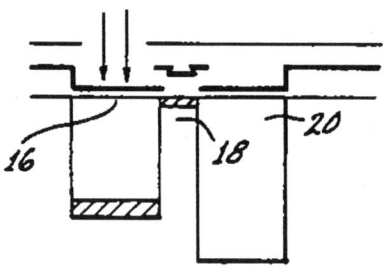

Fig. 1.4.27 (US-A-4912536 fig. 3b)

A first gate 16 is utilized as a photon-detector and charge-accumulator. In response to a photon flux, charges are generated in the gate 16. A second gate 18 is utilized to control charge transfer of accumulated charge from gate 16 to a third gate 20, the avalanche multiplication gate. The avalanche multiplication gate is utilized to provide one component of the internal gain of the detector whereas the time of transfer between the gates 16 and 20, under the control of the transfer gate 18, provides a further component of the gain of the detector.

*

Chapter 1.5

Charge Imaging Matrices

Summary

In an unit cell of a Charge Imaging Matrix (CIM) [1,2], charge carriers are created by photoabsorption and are collected in a detector well. Upon activation of a charge transfer electrode the charge passes over a dc transfer gate and down a read line to an output charge collector means for processing by a signal processor.

- A charge imaging matrix is presented in US-A-4429330.

- The charge imaging matrix of JP-A-60182766 has read-out diodes made of HgCdTe with a larger bandgap than the bandgap of the HgCdTe accumulation regions. The structure improves the reverse bias characteristics of the read-out diodes.

- A stepped insulator is used in US-A-4658277 to allow the functions of a transfer gate and a field gate to be combined into one gate. The design also allows the detector gates, under which photo-generated charge accumulate, to be placed close to the semiconductor surface thereby allowing a large charge storage capacity.

- A two colour infrared focal plane array with the two colours' detectors interleaved and with a single set of read-out lines for both colours is presented in US-A-4956686.

---

[1] S. R. Borrello, H. B. Morris, R. A. Schiebel and C. G. Roberts, "Charge Imaging Matrix for Infrared Scanning", *Proc. Soc. Photo-Opt. Instrum. Eng.*, 409, 69-75, April 1983.
[2] J. P. Omaggio, "Monolithic HgCdTe MIS IR Detectors with a Floating-Diode Sense Mechanism", *IEEE Trans. Electron Devices*, ED-33 (10), 1494-1501, October 1986.

- In order to suppress dark current, the read-out structure of JP-A-2272766 is formed on a semiconductor layer having a larger energy bandgap than the energy bandgap of a semiconductor layer in which photodiodes are formed.

- The imager of US-A-5144138 comprises photocapacitors which include a heterojunction. The structure has the advantages of a large potential well capacity, small amount of dark current and detection of two colours. Other features of this document are presented in chapter 2.5.

- In GB-A-2261323 an imager is presented in which both ends of a pn-junction of a light receiving layer and both ends of a pn-junction of an impurity diffusion region constituting a MIS switch are all formed in a semiconductor layer having a large energy bandgap. This structure shows small amount of recombination of signal charges in the light receiving layer, leakage current at the surface of the light receiving layer and dark current in the MIS switch.

Charge Imaging Matrices

In an unit cell of the imager of US-A-4429330 (Texas Instruments Incorporated, USA, 31.01.84), charge carriers are created by photoabsorption and are collected in a detector well. Upon activation of a charge transfer electrode the charge passes over a dc transfer gate and down a read line to an output charge collector means for processing by a signal processor.

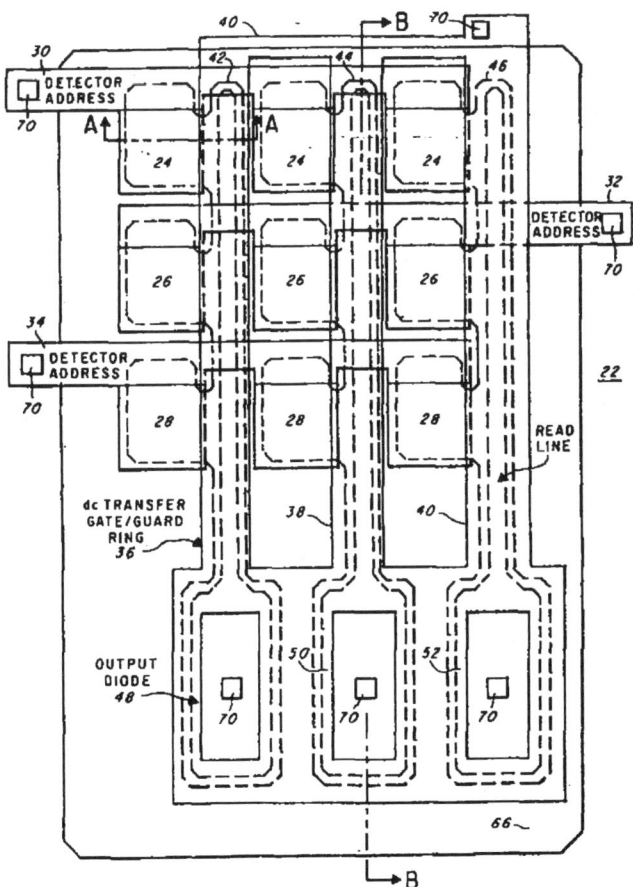

Fig. 1.5.1 (US-A-4429330 fig. 2a)

A random access imager 22 comprising detector elements, which are arranged in rows and columns, is shown. Each row of detector elements 24, 26 and 28, are addressed, respectively, through detector address buses 30, 32 and 34. The columns of the detector elements are connected by transfer gates 36, 38 and 40, respectively, to read lines 42, 44 and 46. The read lines are connected, respectively, to output collectors 48, 50 and 52. The next figure shows a cross-section of one detector and adjoining read line taken along line A-A of the figure above.

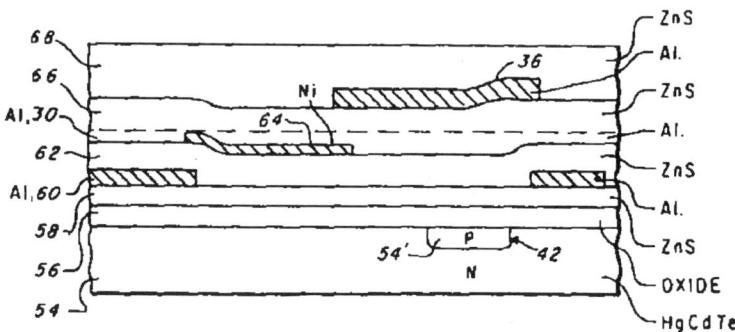

Fig. 1.5.2 (US-A-4429330 fig. 3a)

The imager comprises a substrate 54 of HgCdTe, a native oxide layer 56, pn-junction read lines 42, 44 and 46 formed by ion-implantation, impurity diffusion or mercury in-or-out diffusion, a first insulating layer 58 of ZnS, field plate/channel stop MIS electrode 60 of aluminium or nickel, a second insulating layer 62 of ZnS, thin semitransparent electrodes 64 of nickel, opaque address lines 30, 32 and 34 of aluminium, a third insulating layer 66 of ZnS, opaque transfer gates 36, 38 and 40 of aluminium and a fourth insulating layer 68 of ZnS. The aluminium address lines overlap the thin nickel electrodes to electrically connect the rows of the detector elements to the address lines. Vias 70 are etched and indium is used to form bonding pads.

In the initial state, the row of detector elements 24 is off and they have already dumped their charges into the collector diodes 48, 50, or 52. A negative preset voltage pulse is applied to the charge collecting diodes. The reset pulse sets the diode to a fixed voltage prior to dumping charge onto the diode from row 26 detectors. Next, the reset pulse is turned off and the voltage on the charge collecting diodes are measured and amplified. Then, the row of detector elements 26 is turned off and the charges of their potential wells dumped through read lines 42, 44, and 46 into the collector diodes. The collector diodes voltage outputs are measured and subtracted form the first measured voltages. The difference represents the intensity of the infrared photon flux impinging on the row of detector elements 26. Next, the same operational steps are performed to read the third and the first row of detector elements.

If the gate voltage of the detector well is pulsed from an initial low value to a large value, thereby creating a potential well in the semiconductor, charge very rapidly tunnels to the well to partially fill it. After this initial tunnel current, the infrared generated minority carriers slowly begin to fill the well. Finally, the potential well overfills because the tunnel current has reduced the charge capacity available for the storage of infrared generated minority carriers. Signal charge cannot be collected for a time longer than that for which the tunnel and background generated carriers completely fill the well. When the gate voltage is ramped, the large charge generation rate due to infrared background can be used to suppress the creation of tunnel generated charge. Large increases in maximum integration time can be obtained for closely balanced ramps.

The ramp effect may be improved through device design by utilizing an MIS electrode which has a small opaque region, and which has an inversion threshold slightly smaller than that for the larger area transparent region. In this design, the first effect of the ramp is to create a small empty well under the opaque electrode which is insensitive to infrared. When the gate voltage becomes larger than inversion threshold for the large area transparent gate region of the MIS, the infrared sensitive potential well starts out partially empty due to the diversion of photogenerated carriers to the opaque portion of the well. The ramp rate is chosen so that the surface potential does not increase after the gate voltage is large enough to turn on both portions of the well. The difference in flatband voltages between the transparent and the opaque portions of the MIS must be smaller than that difference which would cause the potential well underneath the opaque portion from reaching the tunnel threshold before the potential well underneath the transparent electrode portion turns on.

In a second embodiment the imager is a metal-insulator-semiconductor, MIS, structure throughout. That is, the read lines 42, 44 and 46 and output collectors 48, 50 and 52 are MIS structures. The contiguous read lines 42, 44 and 46 and output collectors 48, 50 and 52, in another embodiment, are formed in a spaced relationship by gaps and a pulsed barrier switch is utilized to control charge flow through the gaps. The use of the pulsed barrier switch makes possible a pn-junction read line and MIS output diode, or an MIS read line and pn-junction output diode.

*

The charge imaging matrix of JP-A-60182766 (Fujitsu KK, Japan, 18.09.85) has read-out diodes made of HgCdTe with a larger bandgap than the bandgap of the HgCdTe accumulation regions. The structure improves the reverse bias characteristics of the read-out diodes.

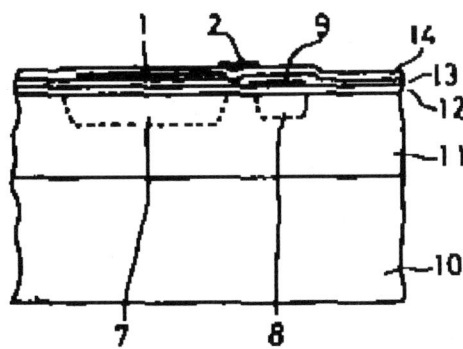

Fig. 1.5.3 (JP-A-60182766 fig. 2b)

An epitaxial layer 11 of $Hg_{0.8}Cd_{0.2}Te$ is grown on a CdTe substrate 10. Charge accumulation regions 7 are provided in the epitaxial layer. Read-out diode regions 8 are formed by selectively growing $Hg_{0.7}Cd_{0.3}Te$. Accumulation electrodes 1, transfer electrodes 2 and read-out electrodes 9 complete the structure.

*

A stepped insulator is used in US-A-4658277 (Texas Instruments Incorporated, USA, 14.04.87) to allow the functions of a transfer gate and a field gate to be combined into one gate. The design also allows the detector gates, under which photo-generated charge accumulate, to be placed close to the semiconductor surface thereby allowing a large charge storage capacity.

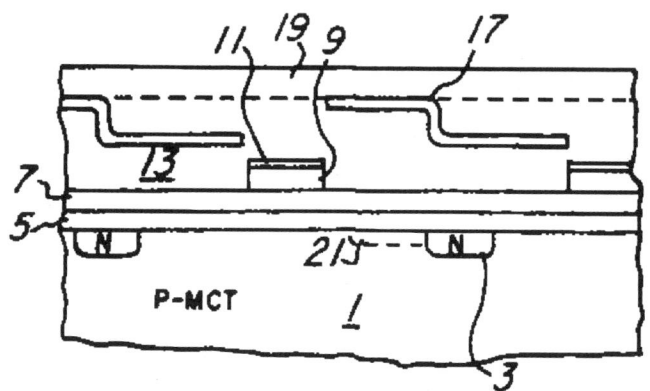

Fig. 1.5.4 (US-A-4658277 fig. 1e)

A p-type HgCdTe substrate 1 has formed therein ion-implanted n-type regions 3. A ZnS insulating layer 7 is deposited over a native oxide layer 5. An infrared transparent nickel layer 9 and an aluminium layer 11 are deposited and patterned to form detector gates. An insulating layer 13 of ZnS is formed over the entire surface of the substrate. The insulating layer is patterned with a photo-resist to provide a pattern thereon which overlaps the edge of each detector gate and the edge of each diode 3. The insulating layer is then milled in the region of overlap with the diode and further over the diode to overlap the adjacent detector gate to provide a two level insulator layer. The photo-resist is then removed and an opaque conductor 17 of aluminium is deposited and patterned into a combined transfer and field gate. A further insulating layer 19 of ZnS is deposited.

In operation a negative bias is placed on the detector gate to provide a depletion area thereunder. The combined transfer and field gate is biased to provide a flat band condition in the substrate under the lower portion and to provide an n-channel 21 beneath the higher portion. When the bias on the detector gate is made sufficient negative, the electrons trapped in the depletion area move out and proceed through the n-channel and are collected in the diode 3.

*

A two colour infrared focal plane array with the two colours' detectors interleaved and with a single set of read-out lines for both colours is presented in US-A-4956686 (Texas Instruments Incorporated, USA, 11.09.90).

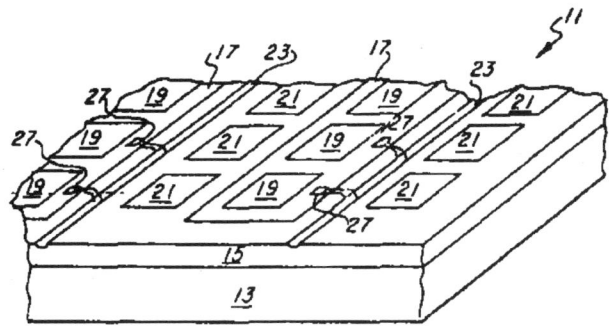

Fig. 1.5.5 (US-A-4956686 fig. 1)

A layer 15 of $Hg_{0.73}Cd_{0.27}Te$ is epitaxially grown on a CdTe substrate 13. Regions 17 of $Hg_{0.78}Cd_{0.22}Te$ are epitaxially grown in slots in the layer 15. The $Hg_{0.73}Cd_{0.27}Te$ layer absorbs photons with wavelengths of less than or approximately equal to five microns and the $Hg_{0.78}Cd_{0.22}Te$ regions absorb photons of wavelength less than or approximately equal to ten microns at an operating temperature of 77 K. Individual detector gates 19 and 21, of thin nickel so as to be transparent to infrared, and corresponding diodes 27 and diode read lines 23 are formed. Each diode read line 23 serves both a column of detector gates 21 and a column of detector gates 19. Each of detector gates 19 and 21 forms an MIS capacitor with the underlying HgCdTe and, when biased, accumulates charge generated by absorption of infrared photons in the HgCdTe near the detector gate. This charge is then read off by diode lines 23.

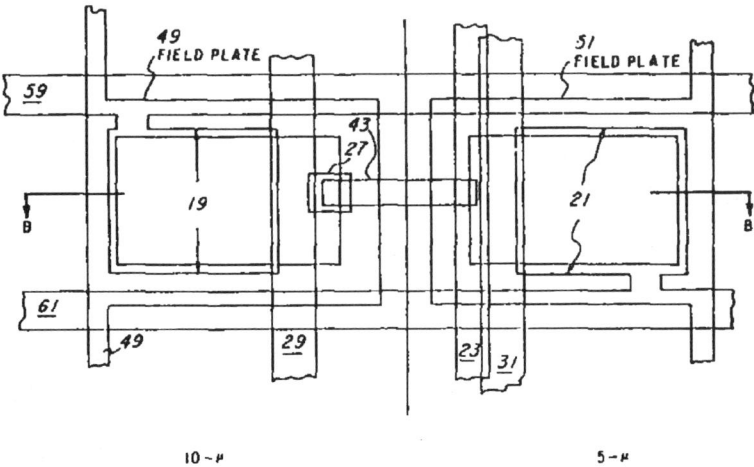

Fig. 1.5.6 (US-A-4956686 fig. 6a)

A pair of detector gates 19 and 21 connected to a common diode read line 23 are shown above. A cross-section along line B-B is shown below.

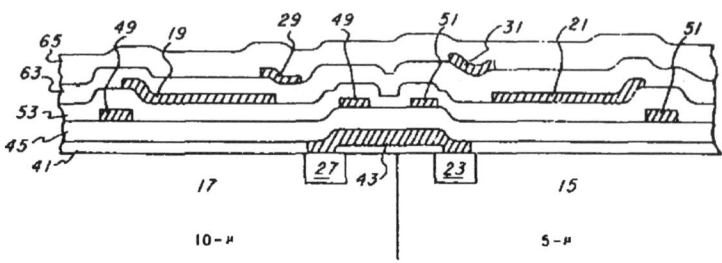

Fig. 1.5.7 (US-A-4956686 fig. 6b)

The diodes 27 in the ten micron HgCdTe regions 17 are small and connected, by metallization 43, to diode lines 23 which are formed in the five micron HgCdTe. This structure gives a small leakage-current in the ten micron HgCdTe. 41 indicates a layer of anodic oxide and 45, 53, 63 and 65 layers of ZnS. The imager also comprises opaque field plates 49 and 51, which are to compensate for trapped positive charge caused by the anodic oxidation. The detector gates 19 and 21 are row connected by buses 59 and 61. Opaque transfer gates 29 and 31 are formed on the insulating layer 63.

*

In order to suppress dark current, the read-out structure of JP-A-2272766 (Fujitsu Ltd, Japan, 07.11.90) is formed on a semiconductor layer having a larger energy bandgap than the energy bandgap of a semiconductor layer in which photodiodes are formed.

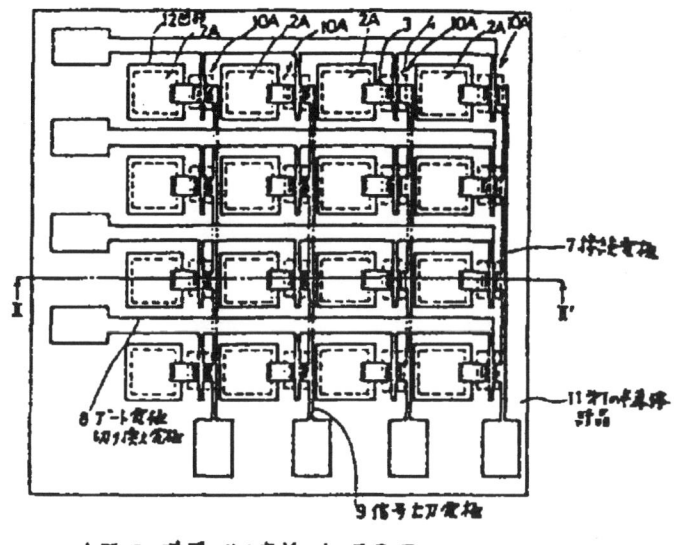

Fig. 1.5.8 (JP-A-2272766 fig. 2)

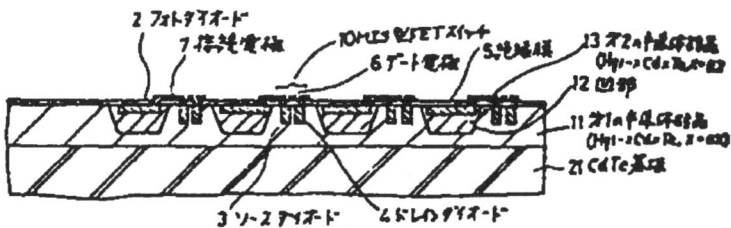

Fig. 1.5.9 (JP-A-2272766 fig. 3)

An HgCdTe layer 11 having a large energy bandgap is disposed on a CdTe substrate 21. HgCdTe layers 12 having a small energy bandgap are buried in the layer 11 and photodiodes are formed in the layers 12. Drain diodes 3 and source diodes 4 are formed in the layer 11. Connection electrodes 7 connect the photodiodes with the drain diodes, and gate electrodes 6 are formed on an insulating layer between the source diodes and the drain diodes.

*

The imager of US-A-5144138 (Texas Instruments Incorporated, USA, 01.09.92) comprises photocapacitors which include a heterojunction. The structure has the advantages of a large potential well capacity, small amount of dark current and detection of two colours. Other features of this document are presented in chapter 2.5.

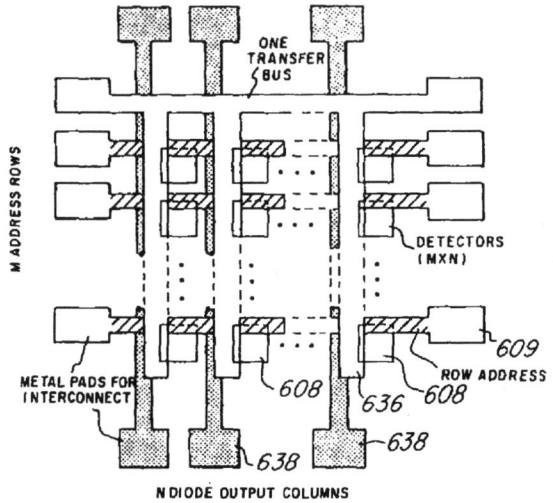

Fig. 1.5.10 (US-A-5144138 fig. 6a)

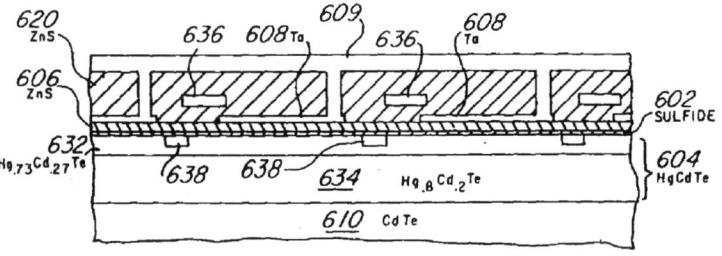

Fig. 1.5.11 (US-A-5144138 fig. 6b)

The charge imaging matrix comprises a CdTe substrate 610, HgCdTe layer 604, anodic sulphide passivation layer 602, ZnS gate dielectric layer 606, 10 nm thick tantalum gates 608, ZnS layer 620, gate row buses 609 and transfer gates 636. The layer 604 is a heterostructure of 1-4 μm thich n-type $Hg_{0.73}Cd_{0.27}Te$ layer 632 on 20 μm thick n-type $Hg_{0.8}Cd_{0.2}Te$ layer 634. The layer 632 includes p-type diode lines 638.

The imager may operate to detect two colours either by varying the gate bias voltage as is described in chapter 2.5 or by using alternating rows of detectors at high and low gate biases.

*

In GB-A-2261323 (Mitsubishi Denki Kabushiki Kaisha, Japan, 12.05.93) an imager is presented in which both ends of a pn-junction of a light receiving layer and both ends of a pn-junction of an impurity diffusion region constituting a MIS switch are all formed in a semiconductor layer having a large energy bandgap. This structure shows small amount of recombination of signal charges in the light receiving layer, leakage current at the surface of the light receiving layer and dark current in the MIS switch.

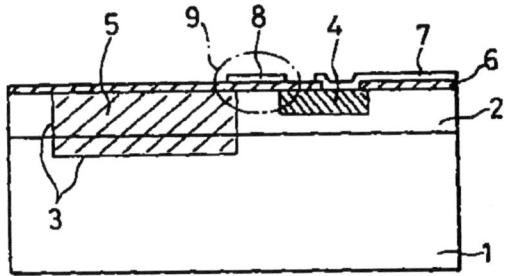

Fig. 1.5.12 (GB-A-2261323 fig. 1)

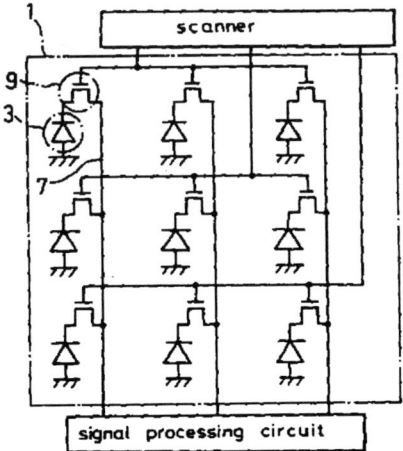

Fig. 1.5.13 (GB-A-2261323 fig. 4)

A p-type $Hg_{1-x}Cd_xTe$ ($x > 0.3$) layer 2 having a large energy bandgap is disposed on a p-type substrate 1 of $Hg_{0.8}Cd_{0.2}Te$ with a small energy bandgap. An n-type light receiving layer 5 penetrates the layer 2 to reach into the substrate. A high concentration n-type region 4 is formed in the layer 2. An insulating film 6 is disposed on the layer 2, and a MIS electrode 8 is disposed on the insulating film. An electrode wiring 7 is connected to a signal processing circuit.

In a second embodiment, the n-type light receiving layer 5 and the high concentration n-type region 4 are formed by ion-implantation in the same step. In this case the n-type region 4 reaches into the substrate 1.

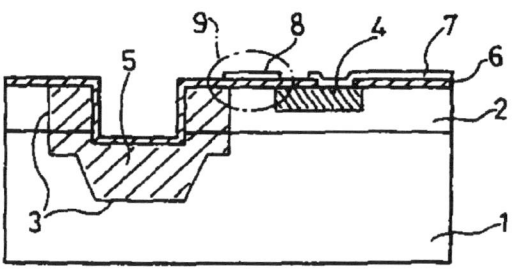

Fig. 1.5.14 (GB-A-2261323 fig. 3)

In a third embodiment, an aperture is formed penetrating the p-type layer 2, and the light receiving layer 5 surrounds the aperture. Since it is not necessary to precisely control the depth of the pn-junction of the light receiving layer 5, the production yield can be improved.

*

PART TWO

HYBRID ARRAYS

Imagers having detector elements and read-out means provided in separate semiconductor bodies are presented in this part.

Chapter 2.1

Detector Arrays with Individual Detector Element Read-Out Leads

Imagers having individual detector elements provided with individual read-out leads formed in or on a non-active supporting substrate are presented in this chapter.

Summary

- An interconnection pattern is used instead of wire interconnects in US-A-3965568 to increase the practical limit of the number of elements in an array.

- A method to provide individual detector elements, which are to be assemblied and contacted in later steps, rather than a monolithic assembly, is given by GB-A-1559473. This approach allows a smaller spacing between detector elements of the imager.

- A method to assemble individual detector elements, such as described in GB-A-1559473 above, to an imager is shown in GB-A-1559474.

- The method of providing detector elements as disclosed in GB-A-1559473, shown above, is further improved in GB-A-1568958 in which an anodic surface treatment is used to form a passivation layer. This layer is found to have a protective or passivating effect which allows the device to be subjected to elevated temperatures (90°C) without severe degradation in the detectivity of the detector elements.

- In EP-A-0007667 detector elements of an imager comprise mesas to which side-wall electrodes are formed. This structure permits a significant proportion of the current flows to pass across the bulk of the mesas between their side-walls and not adjacent their top surfaces where the carrier recombination velocity may be higher.

There is a loss in responsivity in a detector element comprising a HgCdTe layer having an active area between two separate electrodes due to recombination of radiation-generated charge carriers at one of the electrodes. This loss in responsivity results from a phenomenon called "sweep-out". The responsivity is proportional to the time the radiation-generated charge carriers spend in the HgCdTe layer. The electrodes act as regions of intense recombination of these carriers. The carrier lifetime is a measure of the time a minority carrier can spend in the HgCdTe layer before recombining therein with majority carriers. If, before this recombination occurs, the minority carriers reach the electrode a part of their useful life in radiation detection is lost.

- In EP-A-0007669 ion-etching is used to provide meandering current path in the active area. This increases the current path and the charge-carrier transit time thereby improving the responsivity.

- An imager having conductive patterns formed on both sides of a HgCdTe layer is shown in US-A-4025793. This structure allows wires to be bonded without exerting pressure on the HgCdTe layer.

- US-A-4162507 refers to a linear imager with a comb-shaped HgCdTe layer. Each of the finger-shaped tongues of the HgCdTe layer comprises a detector element. Wires are bonded to metal regions which are formed on each of the finger-shaped tongues. The spacing between two detector elements is affected by the need to provide space for the bonding of the wires. A structure which allows closer spacing of the detector elements is disclosed.

- A conversion treatment comprising electrolytical anodising of portions of a surface of an HgCdTe body is used in GB-A-1600599 to form imagers having photo-conductors or photodiodes as detector elements.

- In JP-A-57091557 row and column conductors are connected to detector elements which have been formed in a matrix.

- When semiconductive layers of photo-conductors are grown onto or adhesively bonded to an insulating chip their formation often includes a step to establish passivation between the semiconductor and the insulating chip. This passivation often results in a slight charge accumulation at the interface of the semiconductor with the insulating chip. This accumulation may adversely or beneficially affect the detector sensitivity but has the disadvantage that it cannot be controlled or corrected after the detector is made. In US-A-4445269 a structure is disclosed which allows a bias to control the accumulation of charge.

- When indium, gold, platinum or alloy combinations of these metals are used to contact HgCdTe detectors, they tend to cause stress at cryogenic temperatures due to the difference of thermal expansion between these metals and HgCdTe. If the connecting leads are formed of a molybdenum layer covered by a gold-germanium or a gold layer, as disclosed in US-A-4439912, the amount of stress is reduced because the thermal coefficients of expansion of these metals matches the thermal coefficient of expansion of HgCdTe. The detector arrays using these conductors instead of conductors of materials such as indium, gold or platinum have a greater detectivity D^*, and the noise level is reduced by a factor of approximatey three. The definition of the detectivity D^* is explained in, for example, Proceedings of the IEEE, Vol. 79, No. 1, January 1991, pp. 66-85, Dean A. Scribner et al., "Infrared Focal Plane Array Technology".

- In US-A-4549195 an imager having pn-junctions formed by two layers of HgCdTe with different energy bandgap forming heterojunction diodes is shown. A reduction in surface leakage-currents is achieved by making the layer having the greatest energy bandgap fully cover the boundaries of the layer having a lesser energy bandgap.

- Electrical contact between electrodes and detector elements is improved in JP-A-60028266 by the use of a conductive bonding agent.

- The peak power of infrared radiation at 300 K has a wavelength of 10 microns. This presents the problem that an infrared radiation detector operating in this range will quickly be saturated because many surrounding objects will be at 300 K. The invention of US-A-4807007 refers to metal-insulator-semiconductor, MIS, detectors, each having a potential well of which only a small portion is exposed to infrared radiation.

- In US-A-4570329 a method for fabricating a backside contacted mosaic detector array is claimed. A cold weldable metal is used both as a means for fastening the array to a circuit board and as a means for providing electrical contacts. This arrangement allows closer detector packing compared with arrays using over the edge contacts. The detector array is claimed separately in US-A-4695861.

- A thin metal layer of nickel is often used as a transparent gate of MIS photo-detectors in the infrared range. To reduce the absorbtion of photons in the gate, it is made as thin as possible. However, when the layer thickness is less than 6 nm the device yields drop off dramatically. A gate of nickel having a thickness of 6 nm absorbs from about 30 to 40 percent of the radiation. A solution to this problem is found in US-A-4654686 where a portion of the metal layer forming the gate is eliminated.

- A method to form closely spaced detector elements having slanted edges onto which electrodes are deposited is taught in JP-A-61220459.

- The photodiodes presented in GB-A-2241605 have their pn-junctions located in planes which intersect two opposite major faces as well as two side faces of the diode bodies. The dimensions of the diode bodies can be made small to avoid the provision of excess body material around the pn-junction planes thereby reducing thermal generation of carriers which would contribute significant noise to the photodiode signals.

- In GB-A-2241377 an imager is presented which is hardened against ionizing radiation by forming each detector element as a very small separate body of the kind disclosed in GB-A-2241605 discussed above.

- A high resistance photo-conductive detector structure is disclosed in US-A-4731640. An n-type photosensitive layer connected via two n^+-type regions is disposed between two blocking regions which confine electrons and holes within the photosensitive layer. The increase of the resistance of the photosensitve layer is accomplished by electrically depleting majority carriers, electrons, out of the photosensitive layer.

- An interconnection layer is used in JP-A-63070454 to allow all external connections to be performed on one side of the detector elements, thereby facilitating the assembly of a plurality of imagers.

- An imager is disclosed in JP-A-3104278 in which detector elements are separated by grooves and a connection pad of a common electrode is provided in line with the connection pads of the detector elements.

- Detector elements having three layer bonding pads are disclosed in JP-A-63148677. This structure allows bonding of the elements without damaging the HgCdTe detector material.

- EP-A-0293216 refers to manufacturing methods of an imager having separated detector elements formed on an insulating substrate.

- The detector disclosed in US-A-5006711 refers to an HgCdTe multi-element detector array formed on a sapphire substrate. The elements have individual electrodes and a common electrode. When the electrodes are formed by a lift-off process, the maximum thickness of the metal layer forming the electrodes is limited due to the operational characteristics of the lift-off process. This results in a non-negligible resistance of the common electrode, which gives the device a low sensitivity. To reduce the resistance of the common electrode, an auxiliary electrode is formed on an aperture plate such that when the aperture plate is assembled onto the detector, the auxiliary electrode is pressed onto the common electrode.

- Manufacturing an imager, such as the imager disclosed in US-A-4197633, presented in chapter 2.5, comprises the steps of bonding a mercury cadmium telluride body via a first surface by an adhesive layer to a substrate on which a conductor pattern is present, removing portions of the body throughout its thickness and providing electrical connections from the substrate to regions of the body. A problem is that the conductor pattern is exposed in the removal step. Furthermore, pn-junctions formed in the body adjacent to a second surface, opposite to the first surface, can be degraded by heat treatments. In GB-A-2231199 the body is adhered via the second surface to the substrate thereby protecting the pn-junctions against material degradation during heat treatments.

- The invention of US-A-4989067 relates to interconnections from an HgCdTe detector chip to a silicon read-out chip. A beam lead interconnect formed on a silicon chip is used to provide connections with a finer pitch than is achievable when wire-bonding is used.

- US-A-5146303 refers to the connection of an HgCdTe imager chip with Si read-out chips by the use of an interconnect structure. The structure facilitates testing and repair of the sensor.

Detector Arrays with Individual Detector Element Read-Out Leads

An interconnection pattern is used instead of wire interconnects in US-A-3965568 (Texas Instruments Incorporated, USA, 29.06.76) to increase the practical limit of the number of elements in an array.

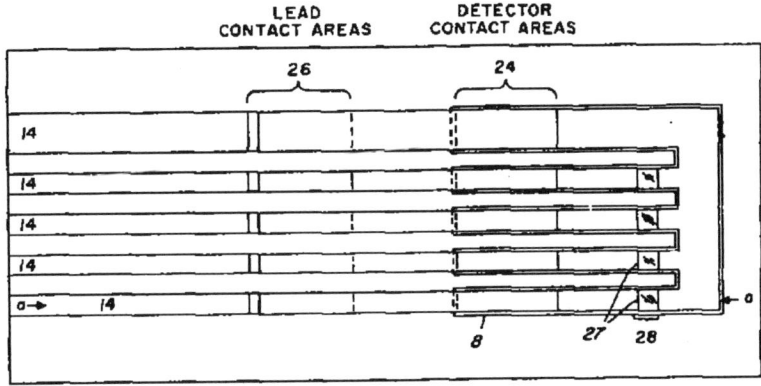

Fig. 2.1.1 (US-A-3965568 fig. 3)

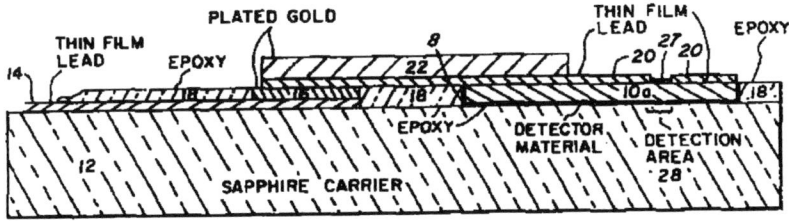

Fig. 2.1.2 (US-A-3965568 fig. 4)

Mesas in the mirror image of a desired array pattern are formed on the surface of a bar of HgCdTe by etching to a depth greater than the desired final detector array thickness. A lead pattern 14 is formed on a sapphire substrate 12 with contact pads 16 plated to a thickness greater than the final detector thickness. The bar is turned over and the etched surface bonded to the substrate by an epoxy resin 18 to form an unit. The unit is lapped to final detector thickness to form a coplanar surface of the lead pattern contact pads, detector contact areas, and the epoxy resin. An interconnect pattern 20 is then formed by vacuum depositing indium on the coplanar surface to connect the detector array to the support member lead pattern. The conductivity and the mechanical strength of the interconnect is improved by plating gold 22 on the portion of the interconnect. Alternatively, the interconnect pattern connects the detector array with a semiconductor device placed adjacent to and on the same level as the bar of HgCdTe.

*

A method to provide individual detector elements, which are to be assemblied and contacted in later steps, rather than a monolithic assembly, is given by GB-A-1559473 (Mullard Limited, GB, 16.01.80). This approach allows a smaller spacing between detector elements of the imager.

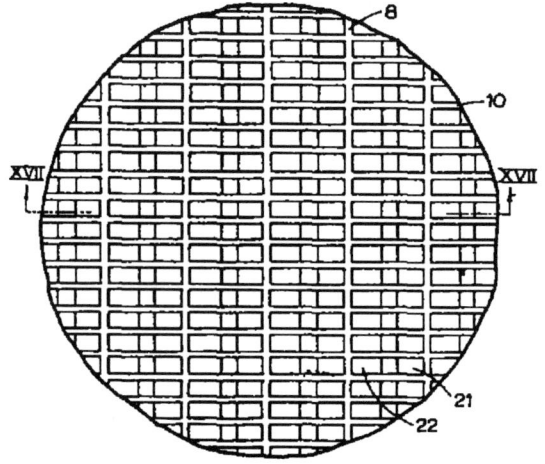

Fig. 2.1.3 (GB-A-1559473 fig. 18)

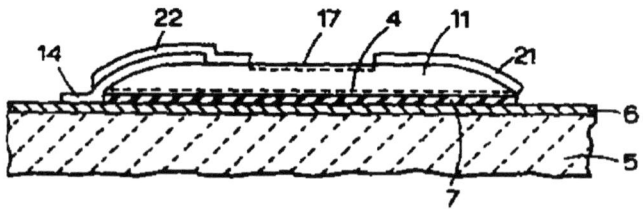

Fig. 2.1.4 (GB-A-1559473 fig. 20)

A disc shaped wafer of HgCdTe having a passivation layer 4 is thinned in a multi-step process. During one of the steps the wafer is adhered by a wax layer 7 to a tantalum layer 6 which covers a polishing block 5. A first plurality of parallel extending channels 8 which defines strip portions are formed in the wafer by etching. Thereafter the thickness is further reduced to 8 µm by a combined polishing and etching step and at the same time the upper longitudinal edges of the strips are rounded. Next a plurality of parallel extending channels 10 are formed by etching thereby providing a plurality of elemental body portions 11. Gold layer strips 14 are formed over a first of the rounded edges of the strips. The gold layer is applied by sputtering and the strips are formed by using a lift-off technique. Passivation layers 17 are formed after portions of the surface have been subjected to an etch treatment. A gold layer is deposited by sputtering and contacts 21 and 22 are formed by a lift-off technique. The contact 21 extend over a second of the rounded edges of the strips. The elemental body portions 11 are individually removed from the polishing block mechanically by lifting from the wax with the aid of a fine tool, the detector element are tested and assembled as described in GB-A-1559474 which is shown below.

*

A method to assemble individual detector elements, such as described in GB-A-1559473 above, to an imager is shown in GB-A-1559474 (Mullard Limited, GB, 16.01.80).

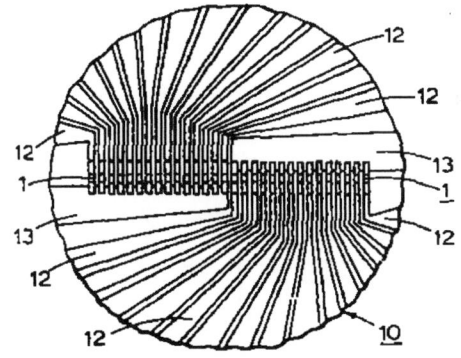

Fig. 2.1.5 (GB-A-1559474 fig. 4)

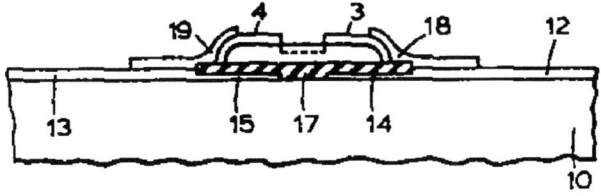

Fig. 2.1.6 (GB-A-1559474 fig. 10)

A conductive pattern of lead-out conductors 12 and 13 with parts 14 and 15 of reduced thickness is formed on an upper surface of an insulating substrate 10 of alumina, silicon, sapphire or beryllia. Detector elements 1 having electrodes 3 and 4 are adhered to the substrate by an epoxy resin 17, and the electrodes are connected to the lead-out conductors by a gold interconnection layer which is formed by the use of a lift-off technique. Due to the reduced thickness of the lead-out conductors and the curved edges of the detector elements no large steps or discontinuities occur in the interconnection layer. The imager is tested and defective detector elements are substituted for new ones.

* *

The method of providing detector elements as disclosed in GB-A-1559473, shown above, is further improved in GB-A-1568958 (Mullard Limited, GB, 11.06.80) in which an anodic surface treatment is used to form a passivation layer. This layer is found to have a protective or passivating effect which allows the device to be subjected to elevated temperatures (90°C) without severe degradation in the detectivity of the detector elements.

* *

In EP-A-0007667 (Philips Electronic and Associated Industries Limited, GB, 06.02.80) detector elements of an imager comprise mesas to which side-wall electrodes are formed. This structure permits a significant proportion of the current flows to pass across the bulk of the mesas between their side-walls and not adjacent their top surfaces where the carrier recombination velocity may be higher.

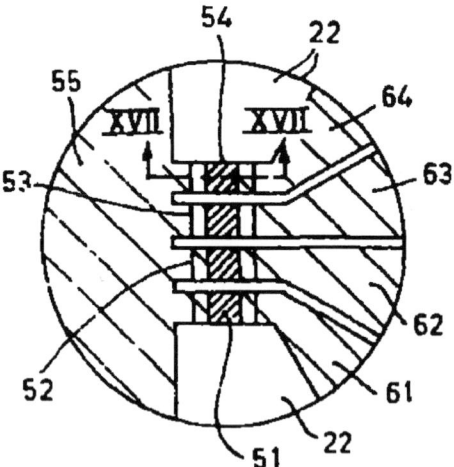

Fig. 2.1.7 (EP-A-0007667 fig. 16)

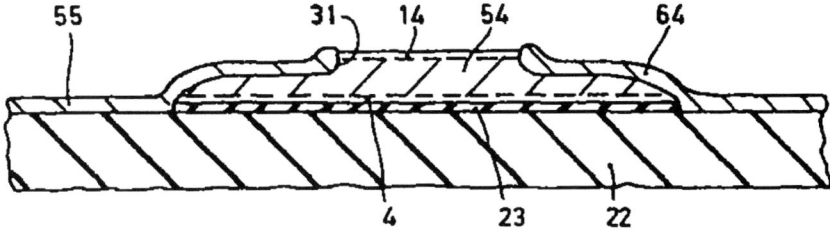

Fig. 2.1.8 (EP-A-0007667 fig. 17)

Elemental body portions are formed in a similar way as disclosed in GB-A-1559473, shown above, and are mounted on an insulating substrate 22 of sapphire. A first masking layer is provided and mesas 31 are formed at the body surfaces by ion-beam etching. Metal layers are deposited by evaporation of chromium followed by a layer of gold. Good adhesion is achieved to the topographically rough surfaces produced by the ion-beam etching on both the HgCdTe body and the sapphire substrate. The first masking layer is removed to lift away the metal thereon, thereby separating electrode 55 from electrodes 61-64. A second masking layer is provided and argon ions are used to separate the detector elements 51-54 by ion-etching through the metal layer and through the thickness of the body portion. This fabrication process using two masking layers is emphasized in EP-A-0007668 (Philips Electronic and Associated Industries Limited, GB, 06.02.80). An imager having two rows of detector elements formed back-to-back is also shown.

* *

There is a loss in responsivity in a detector element comprising a HgCdTe layer having an active area between two separate electrodes due to recombination of radiation-generated charge carriers at one of the electrodes. This loss in responsivity results from a phenomenon called "sweep-out". The responsivity is proportional to the time the radiation-generated charge carriers spend in the HgCdTe layer. The electrodes act as regions of intense recombination of these carriers. The carrier lifetime is a measure of the time a minority carrier can spend in the HgCdTe layer before recombining therein with majority carriers. If, before this recombination occurs, the minority carriers reach the electrode a part of their useful life in radiation detection is lost. In EP-A-0007669 (Philips Electronic and Associated Industries Limited, GB, 06.02.80) ion-etching is used to provide meandering current path in the active area. This increases the current path and the charge-carrier transit time thereby improving the responsivity.

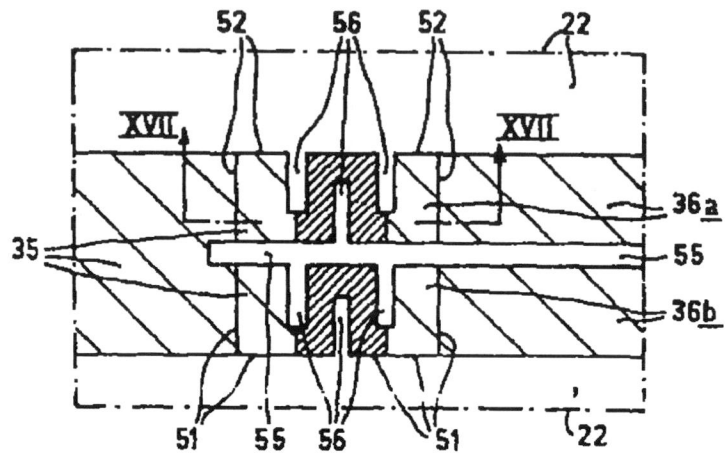

Fig. 2.1.9 (EP-A-0007669 fig. 16)

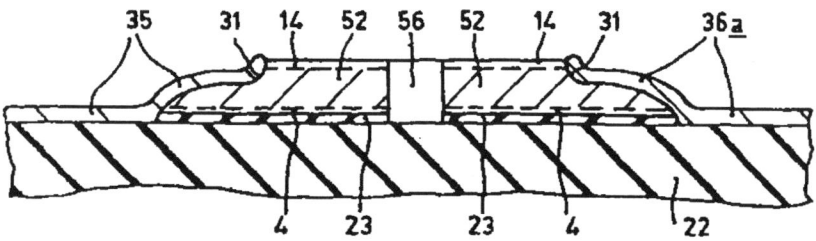

Fig. 2.1.10 (EP-A-0007669 fig. 17)

Elemental body portions are formed and placed on an insulating substrate and provided with metal layers as described in EP-A-0007667, shown above. Then, argon ions are used to ion-etch throughout the thickness of the HgCdTe body portion and through the metal layers to form slots 56 and a separation 55, thereby shaping the electrodes 36a and 36b and providing meandering current paths in the detector elements 51 and 52.

*

An imager having conductive patterns formed on both sides of a HgCdTe layer is shown in US-A-4025793 (Santa Barbara Research Center, USA, 24.05.77). This structure allows wires to be bonded without exerting pressure on the HgCdTe layer.

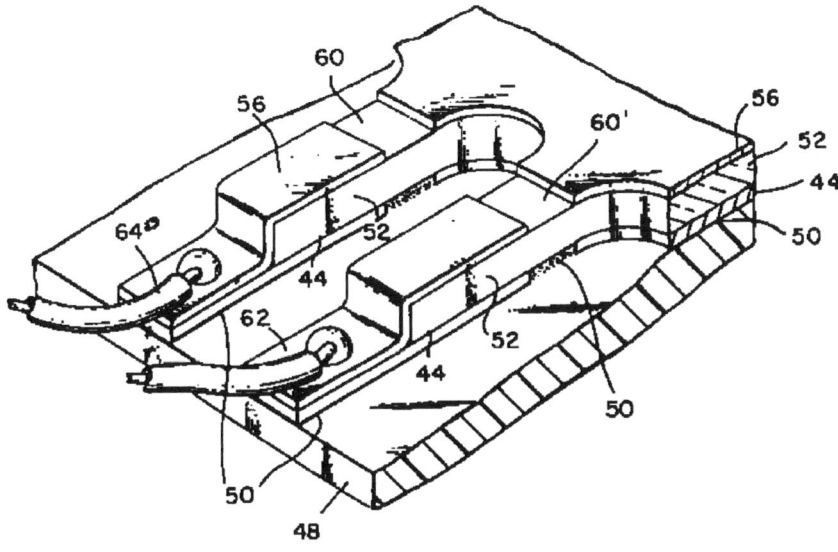

Fig. 2.1.11 (US-A-4025793 fig. 4)

A first metallization pattern 44 is deposited on an HgCdTe wafer. The pattern includes openings which will define active regions. The wafer is bonded by means of an epoxy resin 50 to a substrate 48 of sapphire with the first metallization pattern facing the substrate. The wafer is thinned and patterned so as to leave bodies 52 which do not cover contact pad regions 62 of the pattern 44. A second metallization pattern 56 is deposited and configured to have openings corresponding to the openings in the first metallization pattern, thereby defining active regions 60. Electrical wires 64 are bonded to the contact pad regions remote from the HgCdTe bodies 52.

*

US-A-4162507 (Licentia Patent-Verwaltungs G.m.b.H, FRG, 24.07.79) refers to a linear imager with a comb-shaped HgCdTe layer. Each of the finger-shaped tongues of the HgCdTe layer comprises a detector element. Wires are bonded to metal regions which are formed on each of the finger-shaped tongues. The spacing between two detector elements is affected by the need to provide space for the bonding of the wires. A structure which allows closer spacing of the detector elements is disclosed.

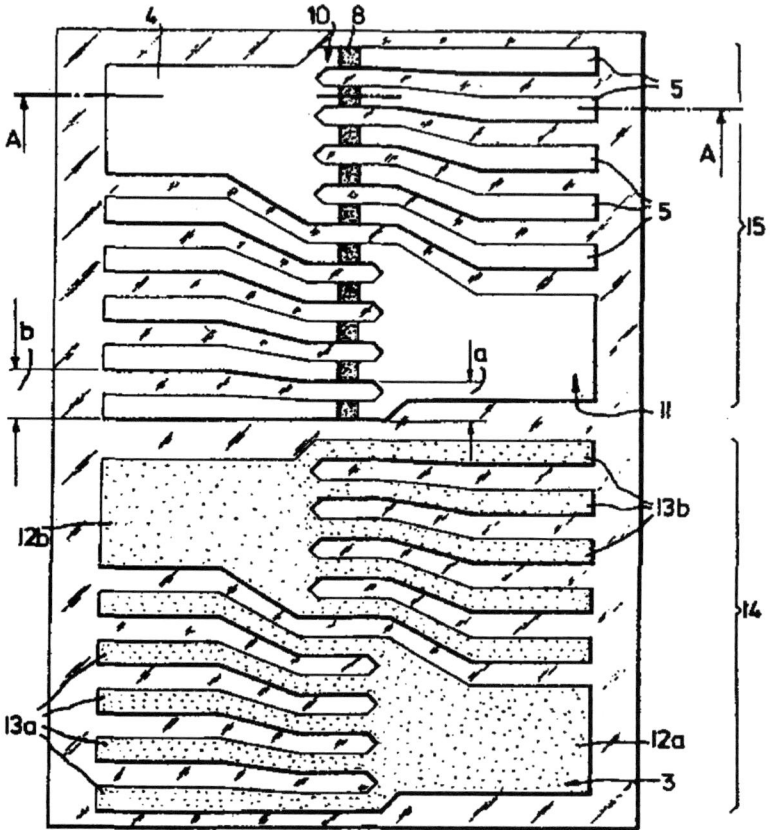

Fig. 2.1.12 (US-A-4162507 fig. 2)

Comb-shaped HgCdTe layers are applied to a carrier body of sapphire. Adjacent comb-shaped layers are rotated 180° relative to each other. Detector elements 8 are connected by a common metal contact 4 and individual metal contacts 5 which are formed on the finger-shaped HgCdTe layers. The metal contacts 5 fan-out and merge into a common metal contact of an adjacent comb-shaped HgCdTe layer thereby allowing wider spacing between bonding pads compared to the spacing between the detector elements.

*

A conversion treatment comprising electrolytical anodising of portions of a surface of an HgCdTe body is used in GB-A-1600599 (Philips Electronic and Associated Industries Limited, GB, 21.10.81) to form imagers having photo-conductors or photodiodes as detector elements.

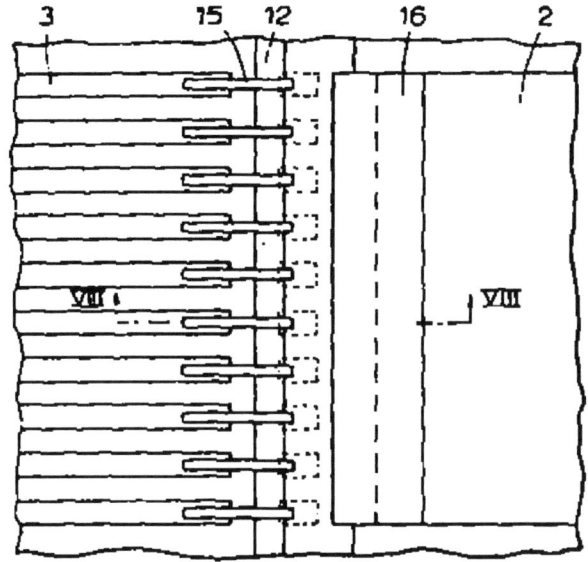

Fig. 2.1.13 (GB-A-1600599 fig. 7)

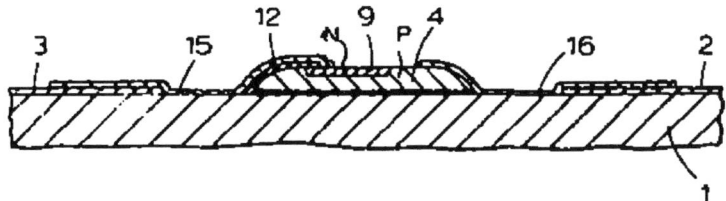

Fig. 2.1.14 (GB-A-1600599 fig. 8)

A p-type $Hg_{0.8}Cd_{0.2}Te$ body 4 is thinned in a multi-stage polishing process and mounted on a substrate 1 of alumina having lead-out contacts 2 and 3. A photo-resist layer is applied and a photomasking and developing process is carried out to define openings corresponding to the active regions of the detector elements. An electrolytic anodising treatment is effected, which produces a surface layer of approximately 200 nm. The photo-resist layer is removed and a heating treatment is carried out which results in the conversion of surface adjoining regions of the body creating photodiodes. The anodically produced surface layer is removed by etching and a further etching treatment removes 0.5 micron from the surface of the body. The active regions are protected by a new anodic surface layer. A dielectric layer 12 is formed and a layer of gold is deposited. Electrodes 15 and 16 are then formed by a lift-off technique.

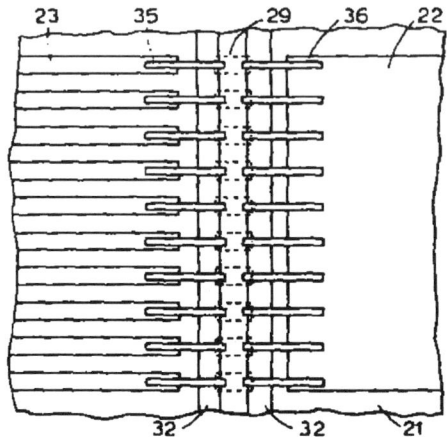

Fig. 2.1.15 (GB-A-1600599 fig. 9)

The method to convert a surface layer of a p-type HgCdTe body can also be used to form photo-conductive detector elements 29 of n-type separated from each other by pn-junctions. Electrodes 35 and 36 connect the detector elements to lead-out contacts 22 and 23 over insulating layers 32.

*

In JP-A-57091557 (Fujitsu Ltd, Japan, 07.06.82) row and column conductors are connected to detector elements which have been formed in a matrix.

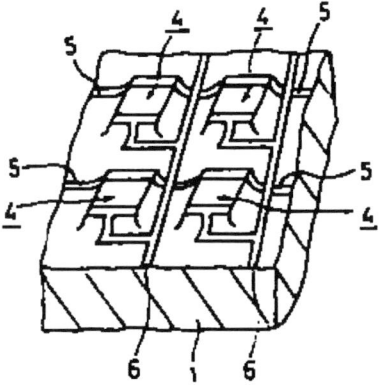

Fig. 2.1.16 (JP-A-57091557 fig. 4)

An HgCdTe layer is epitaxially grown on an insulating substrate 1 of CdTe. Electrodes are formed and individual detector elements 4 are separated by etching. An insulating layer is provided having openings corresponding to electrodes 3a of the detector elements. A conductive film 5 is patterned, thereby providing row conductors. Each row conductor connects the electrodes 3a of the detector elements of the same row. Next, column conductors are formed in the same way as the row conductors.

*

Semiconductive layers of photo-conductors are grown onto or adhesively bonded to an insulating chip. Their formation often includes a step to establish passivation between the semiconductor and the insulating chip. This passivation often results in a slight charge accumulation at the interface of the semiconductor with the insulating chip. This accumulation may adversely or beneficially affect the detector sensitivity but has the disadvantage that it cannot be controlled or corrected after the detector is made. In US-A-4445269 (The United States of America as represented by the Secretary of the Army, USA, 01.05.84) a structure is disclosed which allows a bias to control the accumulation of charge.

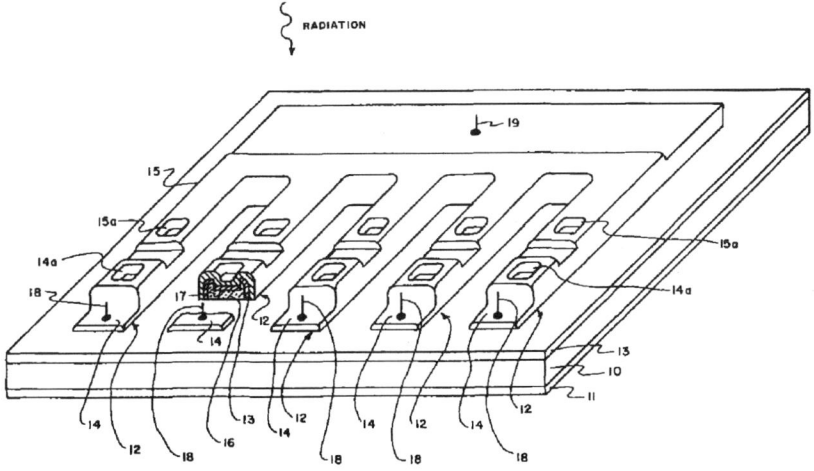

Fig. 2.1.17 (US-A-4445269 fig. 1)

A semiconductor substrate 10 of CdTe is doped slightly conductive, and is coated with an ohmic contact layer 11. An array of photo-conductive detectors 12 is formed on the substrate in a comb shape. The array is covered by a thin layer 13 of CdTe, except for perforations made to allow ohmic connections. Conductive leads 14 and 15 are formed on the layer 13. Each detector includes an HgCdTe bar 16 with two cavities formed therein. The bottom of each cavity is covered with an ohmic contact metal of indium. On top of the bar is a continuation of the CdTe layer 13 except for the area of the cavity in 16. Lead 14 is formed over layer 13 and forms a dimple 14a where it overlies the cavity and covers a contact 17. The region under dimple 15a has the same construction as the region under 14a. Layer 11 and contact 17 form ohmic connections with the substrate and the bar respectively. The substrate and layer 13 form heterojunctions with the bar and the leads 14, and 15 are Schottky barrier metals to the layer 13 and the bar 16. Layer 13 is connected to the contact 11 and an adjustable bias is applied between the detector and the layer 13/contact 11 to control the accumulation between the detector and the substrate.

*

When indium, gold, platinum or alloy combinations of these metals are used to contact HgCdTe detectors, they tend to cause stress at cryogenic temperatures due to the difference of thermal expansion between these metals and HgCdTe. If the connecting leads are formed of a molybdenum layer covered by a gold-germanium or a gold layer, as disclosed in US-A-4439912 (The United States of America as represented by the Secretary of the Army, USA, 03.04.84), the amount of stress is reduced because the thermal coefficients of expansion of these metals matches the thermal coefficient of expansion of HgCdTe. The detector arrays using these conductors instead of conductors of materials such as indium, gold or platinum have a greater detectivity D^*, and the noise level is reduced by a factor of approximatey three. The definition of the detectivity D^* is explained in, for example, Proceedings of the IEEE, Vol. 79, No. 1, January 1991, pp. 66-85, Dean A. Scribner et al., "Infrared Focal Plane Array Technology".

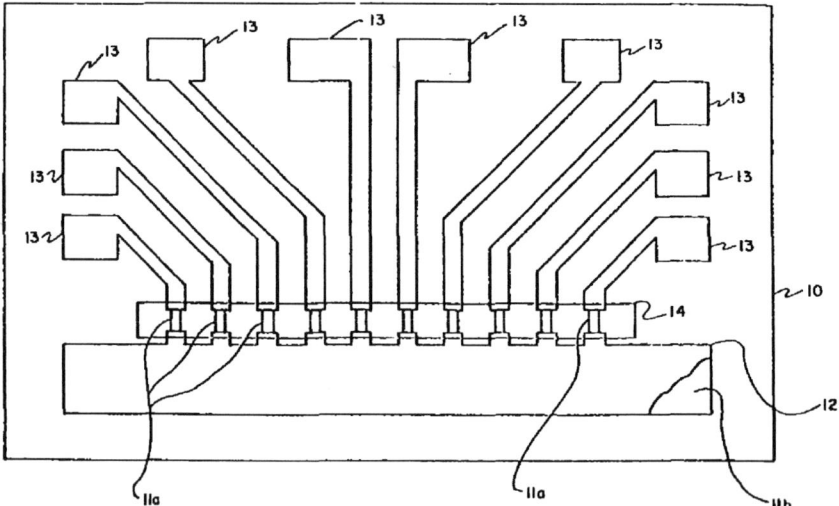

Fig. 2.1.18 (US-A-4439912 fig. 1)

A comb-shaped region of HgCdTe is formed on a CdTe substrate 10. This region includes detecting regions 11a which are teeth of the comb and regions 11b which are the back of the comb. Overlaying 11b is a common conductor 12. Individual detector conductors 13 are connected to the detector regions. Passivation and anti-reflection layer 14 covers the detector regions. Each of the conductors 12 and 13 consists of two layers, a molybdenum layer and a gold-germanium or gold layer. The gold-germanium or gold layer on top of the molybdenum layer provides a good electrical connection while the coefficient of thermal expansion of the molybdenum layer has a good match to that of HgCdTe.

*

In US-A-4549195 (Westinghouse Electric Corp., USA, 22.10.85) an imager having pn-junctions formed by two layers of HgCdTe with different energy bandgap forming heterojunction diodes is shown. A reduction in surface leakage-currents is achieved by making the layer having the greatest energy bandgap fully cover the boundaries of the layer having a lesser energy bandgap.

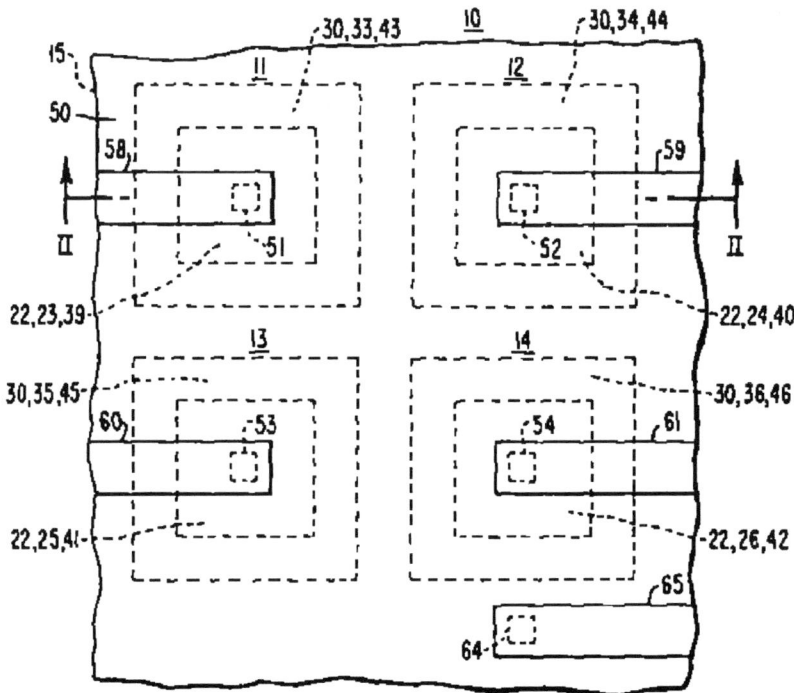

Fig. 2.1.19 (US-A-4549195 fig. 1)

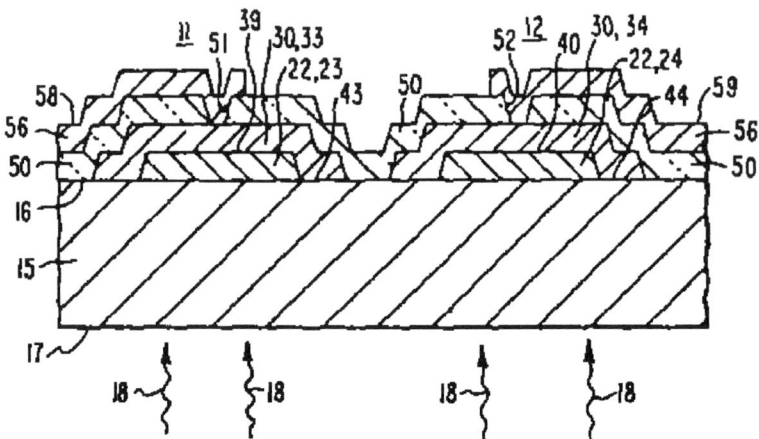

Fig. 2.1.20 (US-A-4549195 fig. 2)

An array 10 of diodes 11-14 are formed on a semiconductor substrate 15 of n-type CdTe. An HgCdTe layer 22 of n-type is formed on the substrate. The substrate may be formed of two layers and function as an optical spectral filter. Mesa islands 23-26 are formed by etching and an HgCdTe layer 30 having greater energy bandgap than layer 22 is deposited and regions 33-36 are formed by etching. These regions, which extend beyond the perimeter of islands 23-26, form the anodes of the diodes. Heterojunctions 39-42 of layer 22 are buried below regions 33-36 and are not exposed to any other layer except the substrate. A passivation layer 50 is deposited and conductive paths 58-61 are formed. By utilizing regions 33-36 with a larger energy bandgap than regions 23-26, leakage-currents on the surface of regions 23-26 and across pn-junctions 39-42 are reduced, since with a higher bandgap material, thermal generation and/or diffusion currents are smaller because of the larger activation energy associated with a wider bandgap material. Heterojunction diodes having a planar structure are shown in a second embodiment.

*

Electrical contact between electrodes and detector elements is improved in JP-A-60028266 (Fujitsu KK, Japan, 13.02.85) by the use of a conductive bonding agent.

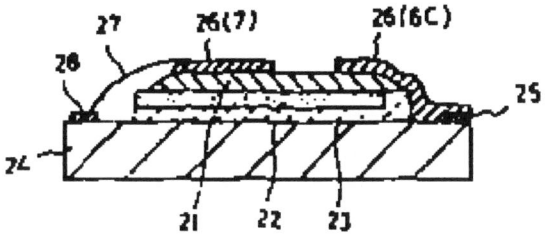

Fig. 2.1.21 (JP-A-60028266 fig. 9)

An anodic oxide film 22 is formed on selected parts of a surface of a HgCdTe substrate 21. A conductive bonding agent 23 is used to attach the substrate 21 to a sapphire substrate 24. The substrate 21 is thinned and the edges are rounded. The HgCdTe layer and the conductive bonding agent are shaped in an etching step into a comb shape. Gold is vapor-deposited to form electrodes 26. A common electrode 7 is connected by a wire 27 to pad 28 and individual electrodes 6 are connected to pads 25 by the aid of the conductive bonding agent 23.

*

The peak power of infrared radiation at 300 K has a wavelength of 10 microns. This presents the problem that an infrared radiation detector operating in this range will quickly be saturated because many surrounding objects will be at 300 K. The invention of US-A-4807007 (Texas Instruments Corporation, USA, 21.02.89) refers to metal-insulator-semiconductor, MIS, detectors, each having a potential well of which only a small portion is exposed to infrared radiation.

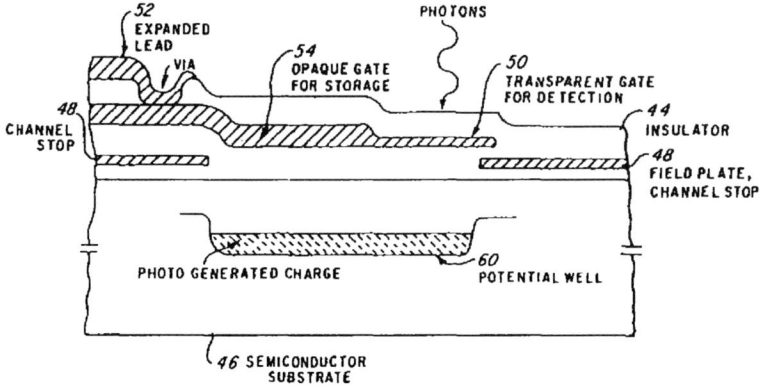

Fig. 2.1.22 (US-A-4807007 fig. 3a)

A field plate 48 which acts as a channel stop is formed inside an insulating layer 44 placed on an HgCdTe substrate 46. A transparent detection gate 50 of thin metal so as to be semi-transparent to infrared radiation is physically connected to an opaque storage gate 54. A potential well 60 is created below the gates to store charge generated by infrared radiation which has penetrated the transparent gate.

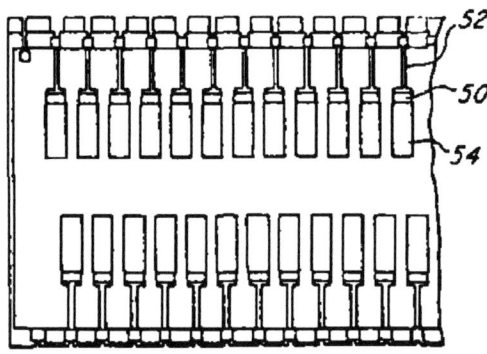

Fig. 2.1.23 (US-A-4807007 fig. 5b)

An array of such devices is shown above. The storage gate 54 and the detector gate of each device is connected to an electrode by a lead 52.

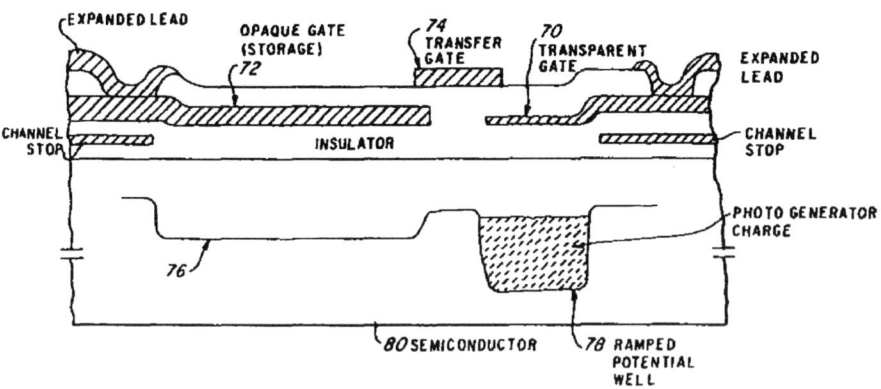

Fig. 2.1.24 (US-A-4807007 fig. 4a)

In an alternative embodiment the detection gate 70 is physically separated from the storage gate 72 and a transfer gate 74 is formed there between. This permits individual control of the gates. Further a ramped gate voltage may be applied to the detector gate.

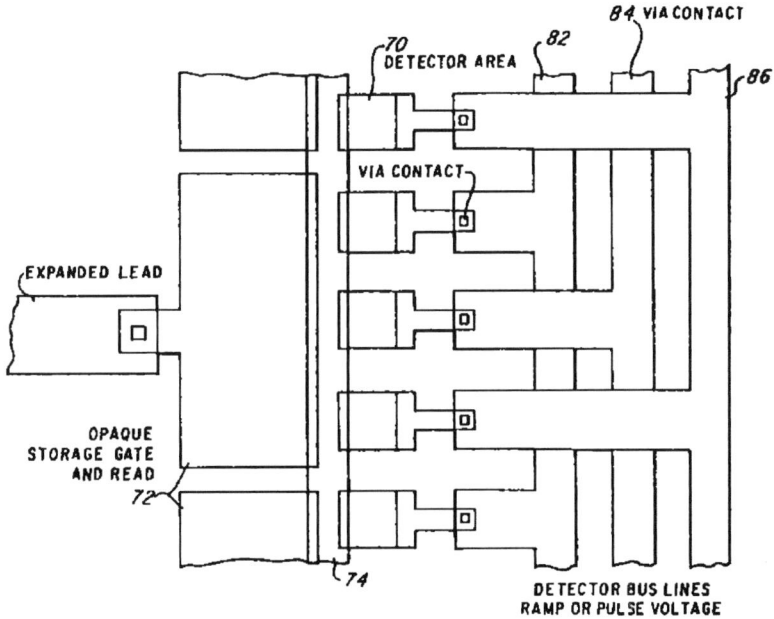

Fig. 2.1.25 (US-A-4807007 fig. 7)

An embodiment comprising a 3 to 1 multiplexer is shown above. Detector buses 82, 84 and 86 are pulsed at different times so that each storage and read gate 72 receives charge from one of the three adjacent detectors at a time.

*

In US-A-4570329 (Honeywell Inc., USA, 18.02.86) a method for fabricating a backside contacted mosaic detector array is claimed. A cold weldable metal is used both as a means for fastening the array to a circuit board and as a means for providing electrical contacts. This arrangement allows closer detector packing compared with arrays using over the edge contacts. The detector array is claimed separately in US-A-4695861 (Honeywell Inc., USA, 22.09.87).

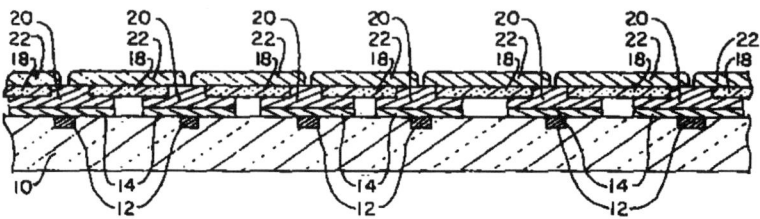

Fig. 2.1.26 (US-A-4570329 fig. 1)

A lead-out wire matrix substrate 10 is prepared having contact pads 12. Deformable contact pads 14 of indium are formed over the contact pads 12. Insulator passivations 18 are formed on an HgCdTe substrate in regions which will be between photo-conductor electrodes 20. Deformable metal contacts 20 which match the contact pads 14 in size and location are deposited. The contact pads 20 are pressed into contact with the contact pads 14, thus cold welding or reflow soldering the HgCdTe substrate and the substrate 10. Some heat may be used to cause adherence in order to minimize the possibility of damaging the HgCdTe substrate. The top surface of the HgCdTe substrate is lapped and/or etched to provide a desired thickness of detector elements 22. Individual detector elements are formed by etching or ion-milling.

*

A thin metal layer of nickel is often used as a transparent gate of MIS photo-detectors in the infrared range. To reduce the absorbtion of photons in the gate, it is made as thin as possible. However, when the layer thickness is less than 6 nm the device yields drop off dramatically. A gate of nickel having a thickness of 6 nm absorbs from about 30 to 40 percent of the radiation. A solution to this problem is found in US-A-4654686 (Texas Instruments Incorporated, USA, 31.03.87) where a portion of the metal layer forming the gate is eliminated.

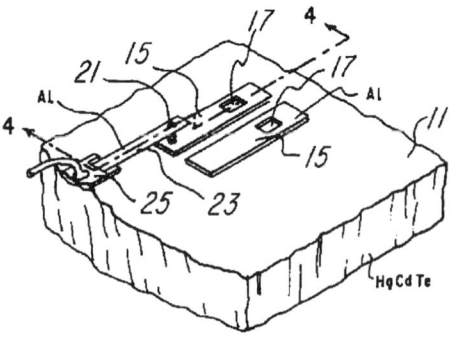

Fig. 2.1.27 (US-A-4654686 fig. 3)

MIS photdetctors of an imager include a p-type HgCdTe substrate 11 having a first insulating layer of ZnS thereon. Aluminium gate members 15 having apertures 17 are formed followed by a second insulating layer also of ZnS. Conductors 23 connect the connection pads 25 with the gates through vias 21. The system operates by establishing a fixed charge density at the semiconductor-insulator interface to create a near flat band voltage in the aperture area. No potential well is formed under the aperture. The metal gate formed around the aperture controls the surface potential and carriers generated in the apperture area diffuse and drift to the potential well created beneath the metal gate which surrounds the aperture.

*

A method to form closely spaced detector elements having slanted edges onto which electrodes are deposited is taught in JP-A-61220459 (Fujitsu Ltd, Japan, 30.09.86).

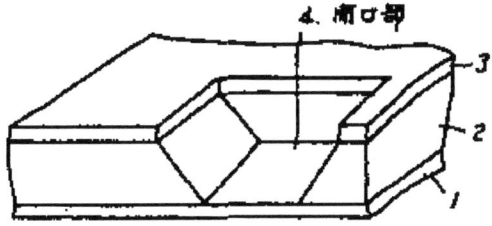

Fig. 2.1.28 (JP-A-61220459 fig. 1(2))

An HgCdTe layer 2 is formed on a sapphire substrate 1. A resist layer 3 having windows is shaped on the HgCdTe layer and an etching treatment is carried out to form through holes 4 in the HgCdTe layer. At the same time two opposite side-walls are tapered. An indium layer is deposited and electrodes 5 are formed on the slanted side-walls and on the exposed regions of the sapphire substrate when the resist layer is removed.

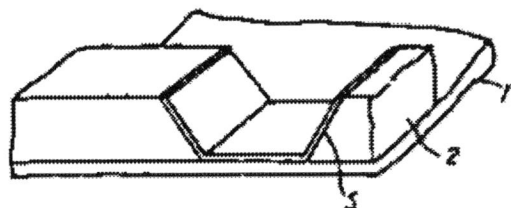

Fig. 2.1.29 (JP-A-61220459 fig. 1(4))

Individual detector elements are then formed by removing HgCdTe material between the detector elements.

*

The photodiodes presented in GB-A-2241605 (Philips Electronic and Associated Industries Limited, GB, 04.09.91) have their pn-junctions located in planes which intersect two opposite major faces as well as two side faces of the diode bodies. The dimensions of the diode bodies can be made small to avoid the provision of excess body material around the pn-junction planes thereby reducing thermal generation of carriers which would contribute significant noise to the photodiode signals.

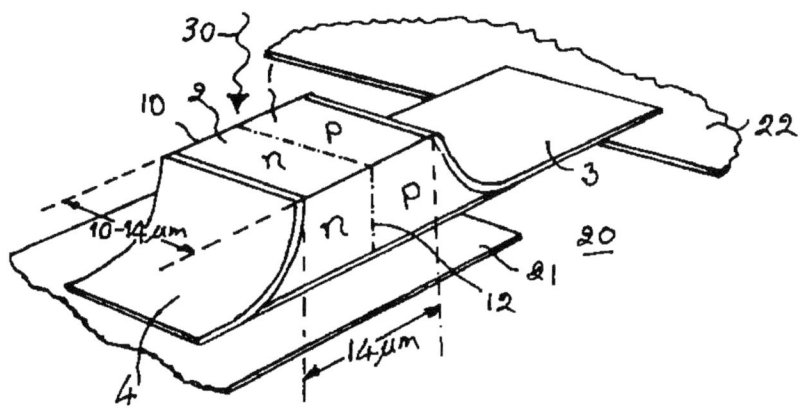

Fig. 2.1.30 (GB-A-2241605 fig. 1)

Each photodiode has a pn-junction 12 formed in a body 10 of mercury cadmium telluride between a p-type region 1 and an n-type region 2 of the body. Metal layers 3 and 4 form electrode connections to the p-type and n-type regions. The body has a thickness of less than 5 µm between two opposite major faces and a width in the range of 10 to 14 µm between two side faces. The body is divided throughout its thickness and width into the p-type region at one end and the n-type region at an opposite end. An array of photodiodes are mounted on an insulating substrate of sapphire. An infrared-reflective layer 21 of gold is present on the substrate below the optical area of the body. This permits the bodies to be as small as 1 to 2 µm thick while obtaining adequate absorption of incident infrared radiation 30 to be detected. Metal layer 3 forms an electrode connection to a common conductive track 22 and metal layer 4 forms an individual signal electrode for each photodiode. A lens body may be positioned over the array of photodiodes to provide each photodiode with an optical immersion lens. The small dimensions of the photodiodes and the use of lenses permits the attainment of adequate performance at temperatures at or near room temperature.

* *

In GB-A-2241377 (Philips Electronics and Associated Industries Limited, GB, 28.08.91) an imager is presented which is hardened against ionizing radiation by forming each detector element as a very small separate body of the kind disclosed in GB-A-2241605 presented above.

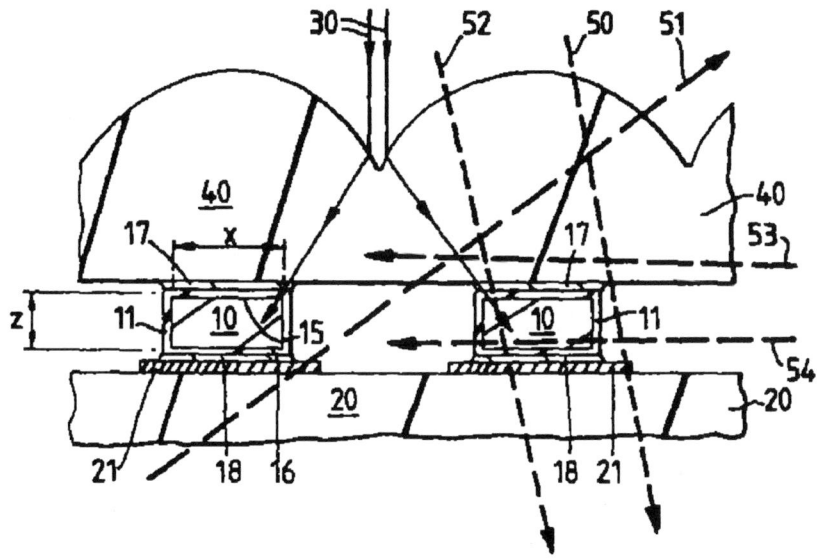

Fig. 2.1.31 (GB-A-2241377 fig. 1)

Detector element bodies 10 each having a thickness of 5 µm and a width and length of 14 µm are mounted on an insulating substrate 20 of sapphire by an adhesive layer 18. The thickness can be reduced even further by the use of rear optical reflectors 21 to give two or more traverses of the optical radiation 30. An array of optical concentrator elements 40 is attached to the detector elements by an adhesive layer 17. Because the detector elements have a small thickness they present a very small target area for ionizing radiation 53, 54 incident either directly or obliquely from the sides of the array, as well as for intermediate trajectories 51. For radiation 52, 54 which passes through the detector elements, the distance travelled in the detector element is so small that there is only a very low probability that radiation 52, 54 causes an ionizing encounter in a detector element.

*

A high resistance photo-conductive detector structure is disclosed in US-A-4731640 (Westinghouse Electric Corp., USA, 15.03.88). An n-type photosensitive layer connected via two n^+-type regions is disposed between two blocking regions which confine electrons and holes within the photosensitive layer. The increase of the resistance of the photosensitve layer is accomplished by electrically depleting majority carriers, electrons, out of the photosensitive layer.

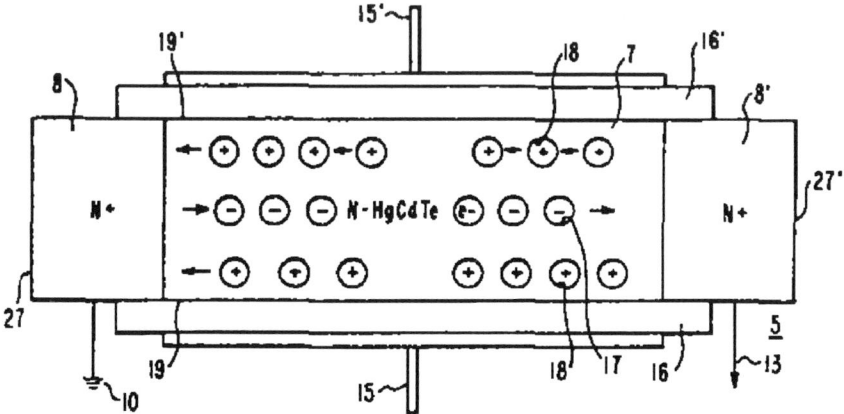

Fig. 2.1.32 (US-A-4731640 fig. 1)

A thin photosensitive layer of n-type HgCdTe 7 is connected via n^+-type regions 8 and 8'. Region 8 is grounded 10 and 8' is biased positive 13. Two gates 15 and 15' biased at negative potential are located opposite to one another with the photosensitive layer in between. Two blocking layers 16 and 16' prevent electrons 17 and holes 18 from being removed from layer 7. These blocking layers are made of an insulating material or semiconductor materials such as combinations of n-doped and p-doped HgCdTe layers. During operation, the electric field produced by the gates 15, 15' will partially deplete the photosensitive layer 7 of majority carriers. Every photogenerated electron-hole pair will be subjected to the external fields produced by the biases applied to the gates 15, 15' and regions 8, 8'. Electrons will move away from the two interfaces 19, 19' toward the center of the layer 7 while concurrently moving away from the negatively biased region 8 towards the region 8'. Holes will of course move in the opposite direction. Because the electron mobility is higher than the hole mobility, electrons will be removed more quickly than holes. Thus, every electron-hole pair will undergo transport where the electrons move faster than the holes. Once an electron is removed a charge imbalance is created and this imbalance will cause another electron from the region 8 to enter the layer 7. Charge balance will occur within the relaxation time. Replacement of the removed electrons continue until the hole is also removed. The number of electrons removed for each absorbed photon is called the photo-conductive gain. This is improved by the electric field produced by the gates 15, 15' which separates electrons and holes and prevents electron hole recombination and also because the electrons are transported within the bulk of layer 7 while holes are transported at the interfaces 19, 19' where the mobility is lower. The resistance of the photosensitive layer, and thereby the current flowing in it, will be modulated by the population of mobile carriers which will change with absorption of incident photons. The amount of current increase depends on the quantum efficiency and the photo-conductive gain of the detector.

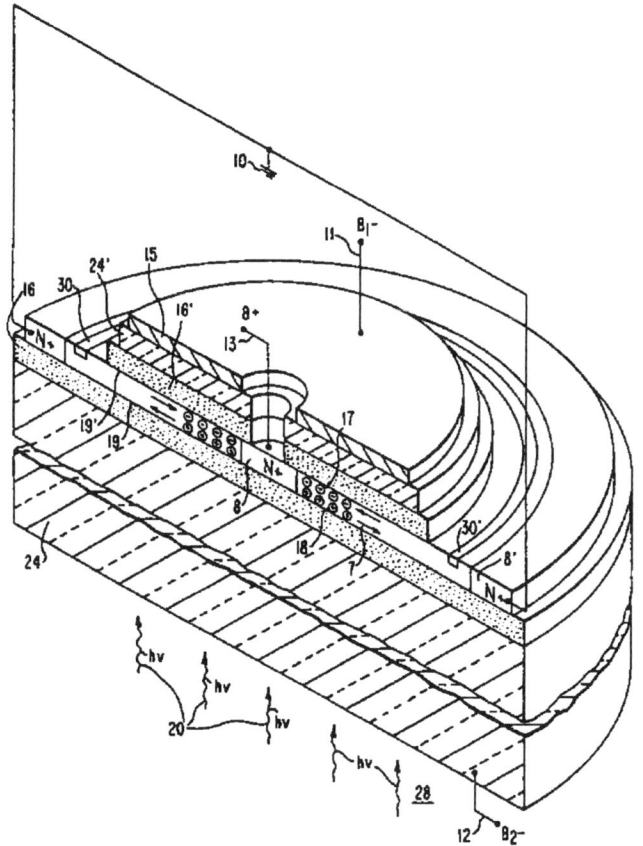

Fig. 2.1.33 (US-A-4731640 fig. 5)

An embodiment having a cylindrical structure is shown above. The detector element is built on a substrate 24 of p-doped CdTe, CdMnTe, InSb or ZnTe. To facilitate the removal of holes, p^+-type regions 30, 30' are embedded in the layer 7. A resonant cavity structure is also provided. The width of the layers 7, 16' and 24' combine to equal one-quarter of the wavelength of the frequency of energy the detector is designed to receive. These layers form a resonating function not unlike the concept of resonating cavities used in optics.

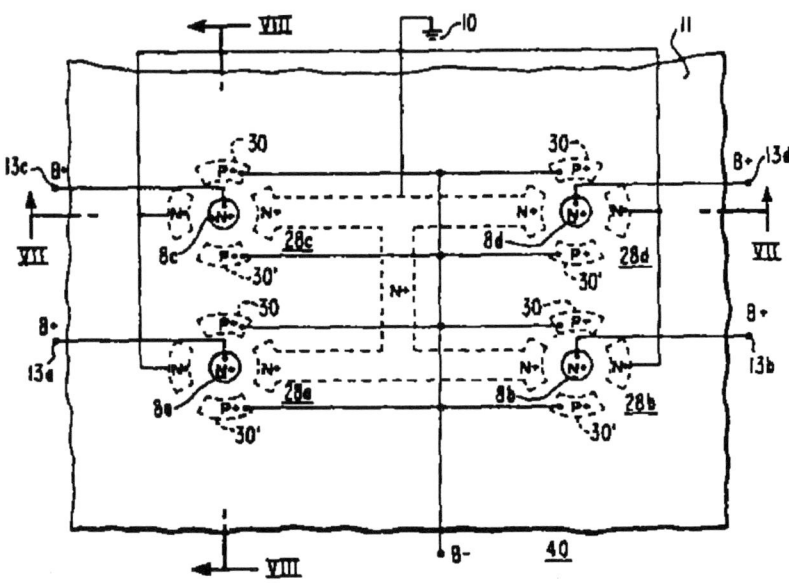

Fig. 2.1.34 (US-A-4731640 fig. 6)

A plan view of an array comprising cylindrical detectors 28a-28d is shown above.

*

An interconnection layer is used in JP-A-63070454 (Fujitsu Ltd, Japan, 30.03.88) to allow all external connections to be performed on one side of the detector elements, thereby facilitating the assembly of a plurality of imagers.

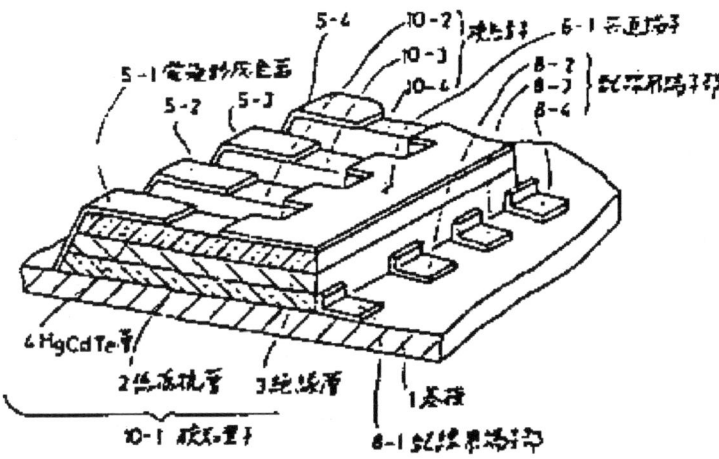

Fig. 2.1.35 (JP-A-63070454 fig. 2)

A low resistance layer 2 is formed on an HgTe or HgCdTe substrate 1. The layer is etched to form interconnection strips. An insulating layer 3 of CdTe is formed followed by an epitaxially formed HgCdTe layer 4. Detector elements, 10-1 to 10-4, are formed between individual electrodes, 5-1 to 5-4, and a common electrode 6-1. The individual electrodes are formed over the edge of the HgCdTe layer and connect to the interconnection strips. Connection pads, 8-1 to 8-4, are connected to the interconnection strips and allow external connection on the same side of the detector elements as the external connection to the common electrode.

* *

An imager is disclosed in JP-A-3104278 (Mitsubishi Electric Corp, Japan, 01.05.91) in which detector elements are separated by grooves and a connection pad of a common electrode is provided in line with the connection pads of the detector elements.

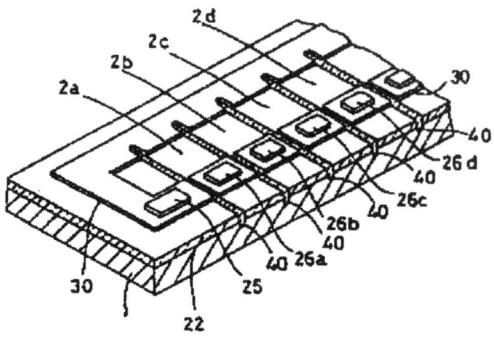

Fig. 2.1.36 (JP-A-3104278 fig. 1)

An HgTe layer 30 is formed on an HgCdTe layer 22 which is epitaxially grown on an insulating substrate 1. Connection pads 25 and 26a-26d are formed in a row. Individual detector elements 2a-2d are formed by etching grooves 40 through the HgTe and HgCdTe layers. The connection pad 25 forms a common electrode while connection pads 26a-26d form individual electrodes.

*

Detector elements having three layer bonding pads are disclosed in JP-A-63148677 (Fujitsu Ltd, Japan, 21.06.88). This structure allows bonding of the elements without damaging the HgCdTe detector material.

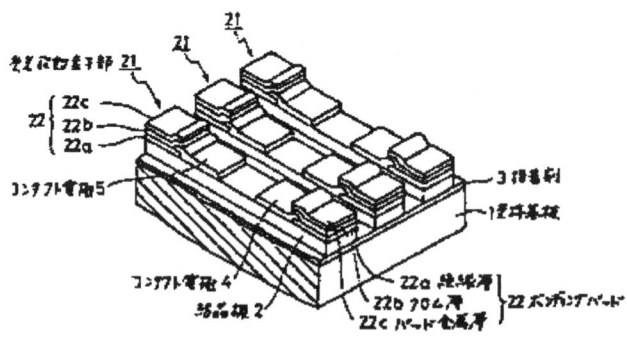

Fig. 2.1.37 (JP-A-63148677 fig. 1)

An HgCdTe layer 2 comprising detector elements, which are connected to bonding pads 22, is formed on a substrate 1. Each bonding pad comprises an insulating layer 22a formed between the HgCdTe layer and the end part of a detector electrode 4, a Cr layer 22b and an Al layer 22c. Lead lines of Al can be bonded to the pads without damaging the HgCdTe material.

*

EP-A-0293216 (The Marconi Company Limited, GB, 30.11.88) refers to manufacturing methods of an imager having separated detector elements formed on an insulating substrate.

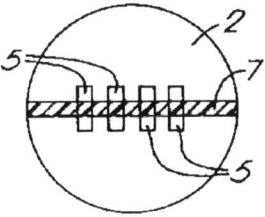

Fig. 2.1.38 (EP-A-0293216 fig. 2)

A body of HgCdTe is first mounted on a sapphire substrate 2. A first etchant mask is formed and individual detector elements 5 are formed by etching. Thereafter a second etchant mask 7 is formed, a metal layer is deposited and metal regions are formed by a lift-off process.

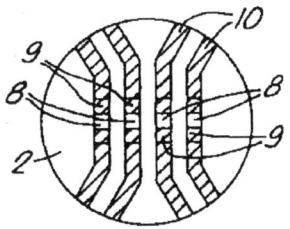

Fig. 2.1.39 (EP-A-0293216 fig. 3)

The metal regions are formed into electrodes 10 by the use of a third masking layer and subsequent etching. The second and third masks can be combined. The detector elements 8 are connected to the electrodes 10 at the contact regions 9. Alternatively, the metal electrodes may be formed on the substrate before the HgCdTe body is attached thereto. In this case individual detector elements are then formed by etching followed by the deposition of a metal layer which connects the detector elements to the electrodes.

*

The detector disclosed in US-A-5006711 (Fujitsu Limited, Japan, 09.04.91) refers to an HgCdTe multi-element detector array formed on a sapphire substrate. The elements have individual electrodes and a common electrode. When the electrodes are formed by a lift-off process, the maximum thickness of the metal layer forming the electrodes is limited due to the operational characteristics of the lift-off process. This results in a non-negligible resistance of the common electrode, which gives the device a low sensitivity. To reduce the resistance of the common electrode, an auxiliary electrode is formed on an aperture plate such that when the aperture plate is assembled onto the detector, the auxiliary electrode is pressed onto the common electrode.

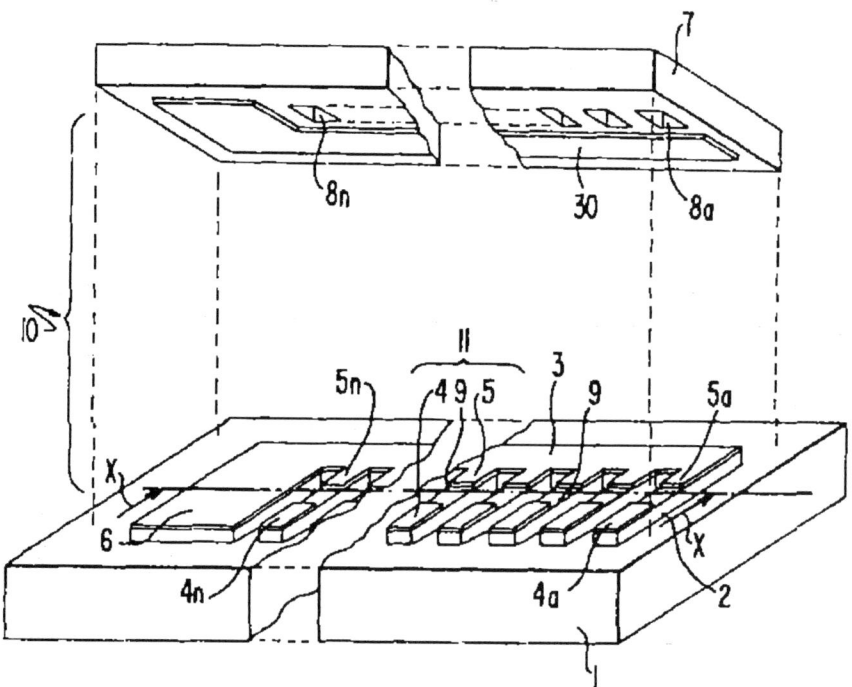

Fig. 2.1.40 (US-A-5006711 fig. 2)

Detector elements 11 are formed on a ceramic substrate 1. Each detector element includes a photosensitive zone 9, an output terminal 4 and a common terminal 5. The detector elements are arranged in an array protruding from a common metal line 3, which is connected to a terminal pad 6. The terminal electrodes and the common metal line are formed on a comb-like patterned photo-conductive layer 2. An aperture plate 7 of silicon or ZnS having apertures 8 formed therein by an anisotropic etching method is prepared. The purpose of the aperture plate is to restrict the field of view of the photosensitive zones. An auxiliary electrode 30 is formed on the aperture plate. When the aperture plate is assembled with the substrate using an adhesive, the auxiliary electrode is pressed against the common metal line and the common terminal, which together reduce the electrical resistance.

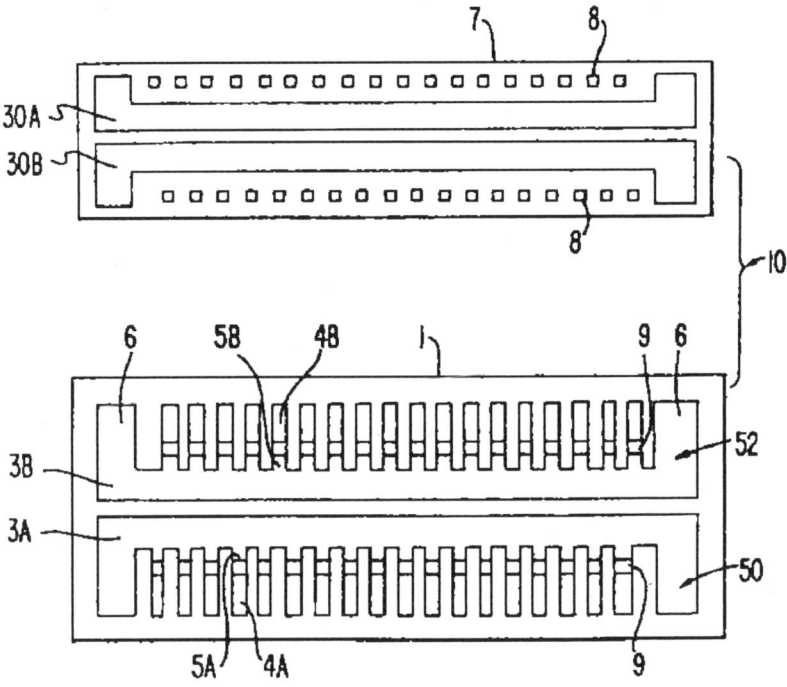

Fig. 2.1.41 (US-A-5006711 fig. 7)

In alternative embodiments the auxiliary electrode 30 and the corresponding structures on the substrate are not L-shaped but U-shaped. Two such arrays of detector elements 50 and 52 can be arranged in parallel with each other and with the positions of the detector elements of the first array shifted relative to the second array by a half pitch of the element spacing.

*

Manufacturing an imager, such as the imager disclosed in US-A-4197633, presented in chapter 2.5, comprises the steps of bonding a mercury cadmium telluride body via a first surface by an adhesive layer to a substrate on which a conductor pattern is present, removing portions of the body throughout its thickness and providing electrical connections from the substrate to regions of the body. A problem is that the conductor pattern is exposed in the removal step. Furthermore, pn-junctions formed in the body adjacent to a second surface, opposite to the first surface, can be degraded by heat treatments. In GB-A-2231199 (Philips Electronic and Associated Industries Limited, GB, 07.11.90) the body is adhered via the second surface to the substrate thereby protecting the pn-junctions against material degradation during heat treatments.

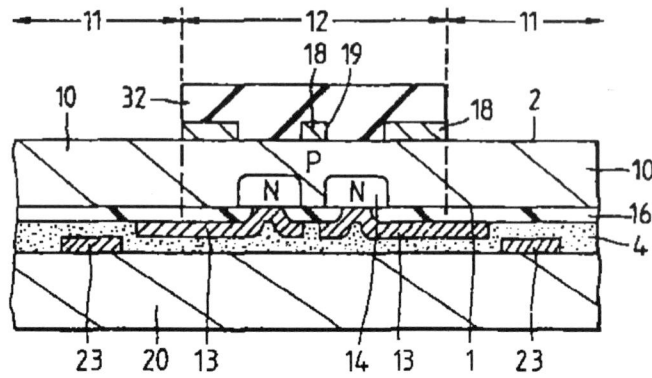

Fig. 2.1.42 (GB-A-2231199 fig. 3)

The manufacturing method comprises the steps of forming n-type regions 14 in a p-type mercury cadmium telluride body 10, forming electrode leads 13 on a passivation layer of ZnS or cadmium telluride, bonding the body 10 by an insulating adhesive layer 4 to a single-crystal sapphire substrate 20 which comprises contact pads 23, forming an infrared transmission array 19 comprising windows in an infrared mask layer 18, removing portions 11 by spray-etching of the body throughout its thickness, and removing exposed adhesive 4 from contact pads 23 by plasma etching. The body 10 may be formed by growing a mercury cadmium telluride layer on a base of cadmium telluride. The n-type regions are either formed by first ion-milling the passivation layer and continuing slightly into the mercury cadmium telluride surface, by localised mercury in-diffusion or by donor dopant implantation.

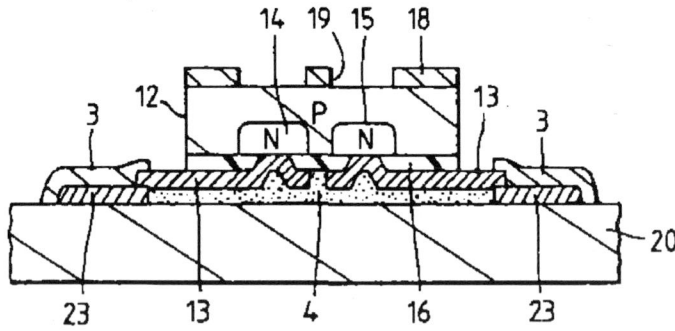

Fig. 2.1.43 (GB-A-2231199 fig. 5)

The next step is to provide electrical connections 3 between the electrode leads 13 and the contact pads 23. Because the active regions are present adjacent the passivated lower surface during the heat treatments, for example when the adhesive layer 4 is cured, they do not suffer significantly from degradation of the material properties of the mercury cadmium telluride which occur most easily and rapidly at the top surface and exposed sides of the body 10 due to out-diffusion of mercury. Note that the protruding parts of the electrode leads 13 are supported throughout their length first by the body 10 and after the bonding by the substrate 20. An imager having two rows of detector elements is shown above, but the number of rows may be different.

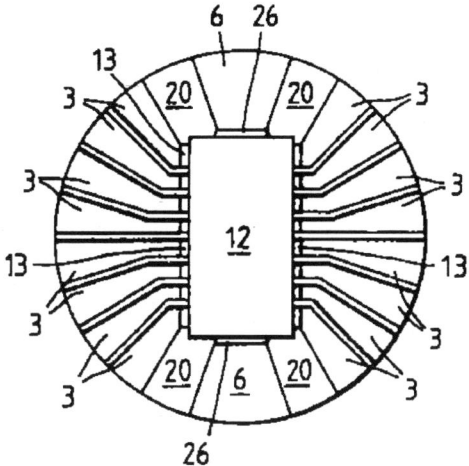

Fig. 2.1.44 (GB-A-2231199 fig. 6)

This figure shows the conductor pattern 3 overlapping exposed parts of the electrode leads 13 and fanning out on the substrate 20 to form wider terminal areas for wire-bonding. Common connections 6 and 26 are provided.

Several embodiments of the invention are disclosed. For example, an imager having an array of lenses corresponding to the array of detector elements and an imager provided with a cold-shield are shown. One embodiment is shown in chapter 2.5.

*

The invention of US-A-4989067 (General Electric Company, USA, 29.01.91) relates to interconnections from an HgCdTe detector chip to a silicon read-out chip. A beam lead interconnect formed on a silicon chip is used to provide connections with a finer pitch than is achievable when wire-bonding is used.

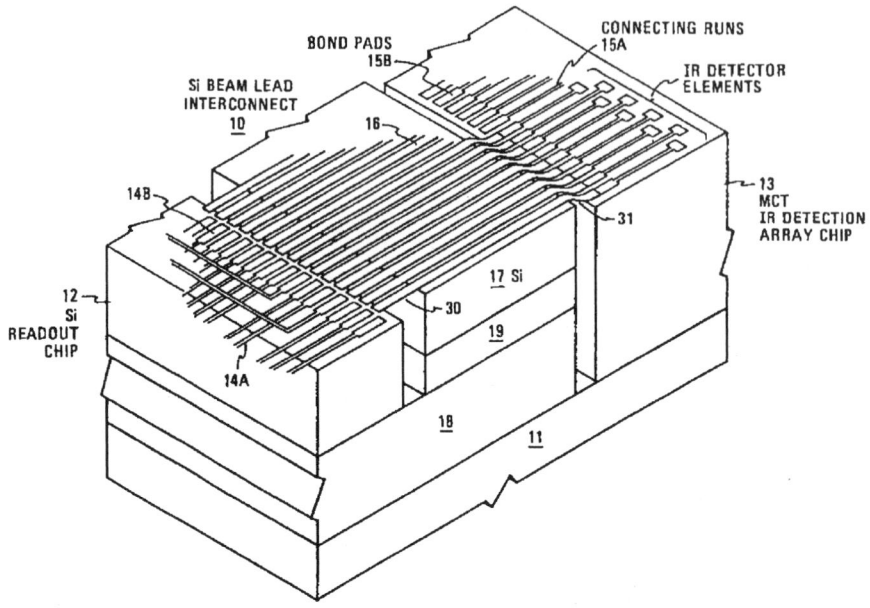

Fig. 2.1.45 (US-A-4989067 fig. 1)

A silicon beam lead strip interconnect 10 is supported upon a substrate 11 of alumina via shims 18. The substrate 11 also supports a silicon read-out chip 12 having conductor runs 14A and bonding pads 14B and an HgCdTe chip 13 having conductor runs 15A and bonding pads 15B. The chip 13 is spaced from the chip 12 by a distance required for insertion of the interconnect 10 between the two chips. A silicon chip 17 of the interconnect is inserted in a position closely spaced between the two chips 12 and 13 and rigidly attached via shims 18 and 19 to the substrate 11. Conductor runs 16 are provided on the chip 17. The runs extend between the vertical boundaries of the chip 17 and they become self supporting beam leads 30 and 31 as they extend over the left and right chip boundaries respectively. The top surface of the chip 17 is slightly above the top surfaces of the chips 12 and 13 to allow insertion of the interconnect into position between these chips without damaging the beam leads. The beam leads 31 are in a depressed position having been bonded to the bonding pads 15B while the beam leads 30 are undepressed and in a position above the bonding pads 14B preparatory to bonding.

*

US-A-5146303 (General Electric Company, USA, 08.09.92) refers to the connection of an HgCdTe imager chip with Si read-out chips by the use of an interconnect structure. The structure facilitates testing and repair of the sensor.

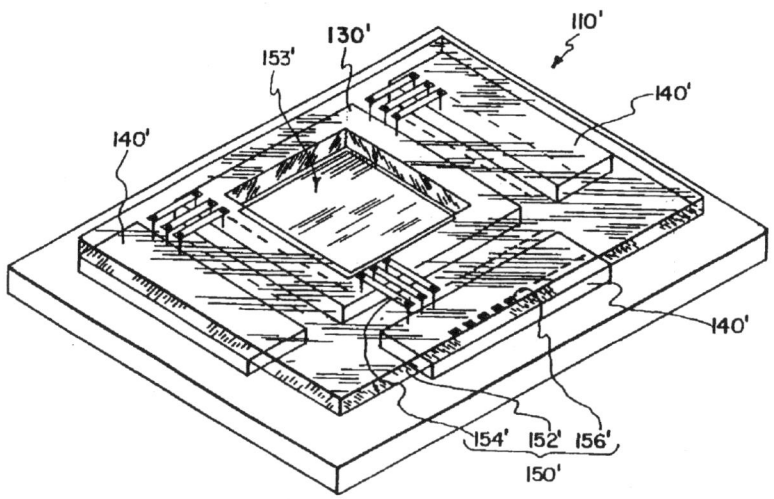

Fig. 2.1.46 (US-A-5146303 fig. 5)

An HgCdTe imager chip 130' and Si read-out chips 140' are mounted on a substrate. The imager chip is connected to the read-out chips by an interconnect structure 150' in which a layer 152' of dielectric material is bonded to the chips and has interconnecting conductors disposed thereon and extending through via holes therein into ohmic contact with contact pads of the chips. A flexible portion in the structure enables the read-out and the imager chips to be disposed in different planes to provide a compact structure.

*

Chapter 2.2

Detector Arrays without Individual Detector Element Read-Out Leads

Imagers having individual detector elements provided with bump connectors or with no connectors at all are presented in this chapter. Cross-talk preventing measures and passivation and leakage current preventing measures are presented in separate sections.

Summary

An imager having a mesa structure comprises a single n-type semiconductor block on which there are located p-type regions in the form of projections, each projection having its own connection and each projection corresponding to one detector element. An electrode common to all the detector elements is connected to the n-type semiconductor block.

- US-A-3930161 points out that the individual connections of the detector elements are fixed in the plane of the sensitive surface which influence the effective surface of the sensitive face of the imager negatively. Therefore, it is proposed to receive the radiation through the n-type semiconductor block.

At the time US-A-3808435 was filed, a conventional infrared detector system had an array of infrared detectors and each detector had an amplifier for amplifying its output. The amplified output was connected to a multiplexer scan converter. Such a system required a large number of amplifiers. It is acknowledged in US-A-3808435 that an improved system would be obtained if an imager with a charge transfer device (CTD) could be used. Two problems, however, were encountered in this technique. The first problem was the fabrication of a CTD on an infrared sensitive material and the second the fact that the thermal background radiation at room temperature is many orders of magnitude greater in the infrared than in the visible (41 orders of magnitude greater at 10 µm than at 500 nm).

- The problems stated above are overcome in US-A-3808435 by a.c. coupling detector signals of an infrared detector array to a CTD array. The detector array and the CTD array are formed in an HgCdTe chip and a silicon chip respectively and are connected by placing the chips on top of each other with intermediate conductor elements. This technique of connecting two chips is generally called the flip-chip technique.

- The invention of US-A-3902924 is concerned with low temperature growth of mercury cadmium telluride layers on insulating substrates by liquid phase epitaxy. Infrared detectors are fabricated to assist with the evaluation of the grown layers.

- An imager with photo-conductive detectors formed in an HgCdTe body is presented in US-A-3949223. In this approach the detectors have the mechanical strength of the entire semiconductor body.

- In US-A-3980915 Schottky diodes constitute the detector elements of an imager.

- The smallest distance between two adjacent detector elements of a linear imager comprising detector elements separated by grooves etched through an HgCdTe chip is limited because the engraved walls are not completely perpendicular to the chip surface but slope obliquely toward the base of the grooves. In GB-A-1597581 the distance between two detector elements is reduced by etching the chip from two opposite surfaces.

- A silicone rubber adhesive layer is used in US-A-4081819 to bond an HgCdTe substrate, which includes an epitaxial layer, to a second substrate. The silicone rubber adhesive reduces the risk that the substrate cracks when cooled to cryogenic temperatures.

- Methods to form image detectors by ion-implanting Hg ions into a CdTe substrate and thereafter annealing the substrate with a high-energy laser are disclosed in US-A-4242149.

- In JP-A-57024580 detector elements are formed in an HgCdTe layer which has been grown on a CdTe substrate.

- A method to convert a region of a p-type HgCdTe body to n-type by the use of ion-etching is presented in EP-A-0062367. A high concentration of mercury is produced from an etched-away part of the body as to act as a dopant source. The method is used to form a detector device.

- The performace of a photodiode is affected by the resistance and the capacitance of the photodiode. A large resistance results in little thermal noise; a small capacitance allows a high operating frequency. A method to increase the resistance and to reduce the capacitance of photodiodes formed as mesas is to drill holes in the mesa as disclosed in EP-A-0064918.

- In EP-A-0068652 photodiodes are formed by diffusing indium into a layer of semi-insulating CdTe which covers an HgCdTe substrate.

- Neither a silicon multiplexer input cell nor a hybrid interconnect can be reduced in size to match the smallest size of a detector. The imager presented in US-A-4727406 allows several detectors to be connected to each input cell of a multiplexer. This is achieved by disposing a plurality of conducting plates around a common junction region. The conducting plates are controled, one at a time, to create an induced junction there under.

- The photodiodes of the imagers presented in US-A-4566024 are formed in a p-type HgCdTe substrate by diffusing n-type impurities from two opposite faces, a front face and a rear face, of the substrate. The photodiodes can be connected to a read-out device by a flip-chip bonding process on the rear face of the substrate and still be illuminated from its front face.

- An imager having detector elements supported by a resin is disclosed in JP-A-58164261.

- An imager using photocapacitive detector elements is presented in FR-A-2526227. Each detector element comprises a gate with an opening and an implanted region where the amount of generated charge is measured. In this design the potential applied to the gate is not crucial and the output impedance is relatively high.

- Two arrays of HgCdTe photovoltaic detectors with pn-junctions of different sizes are combined in US-A-4517464 to form a device which has an increased functionality when exposed to ionizing radiation.

- An arrangement for multi-spectral imaging of objects is presented in US-A-4596930. Groups of detectors are disposed such that an optical system, due to its scatter circle, images each object dot on at least one detector of each group.

- A broad band detector array with homogeneous sensitivity is achieved in JP-A-60102530 by providing a thin single crystal film of $Hg_{1-x}Cd_xTe$ having a concentration gradient of the composition ratio x.

- The invention of EP-A-0167305 refers to a photo-detector which includes a photosensitive region adjacent to a minority carrier extraction region arranged when biased to depress the photosensitive region minority carrier concentration. Depression of the minority carrier concentration produces low noise and high responsivity properties as obtained by cooling, but without the need for cooling equipment. Another structure which simulates cooling is presented in WO-A-9006597. The structure, which comprises three successively disposed sections each containing three layers, a mirror and a pair of electrodes, makes up a multiple heterostructure photodetector.

- A method for processing a backside illuminated detector assembly, comprising the use of a temporary substrate on which a detector wafer is attached, is presented in EP-A-0171801.

- An imager having HgCdTe detectors made of different compositions to be responsive to two different infrared frequency windows, and having the detectors formed in the same focal plane is presented in US-A-4620209.

- A shield comprising apertures corresponding to detector elements are attached to an HgCdTe detector layer in JP-A-61147118 to determine the field of view of the detector elements.

- An imager having high resolution is presented in JP-A-61214462. An HgCdTe layer is grown on a CdTe substrate. The HgCdTe layer is connected through a hole which is formed through the CdTe substrate.

- Insulation between detector elements is automatically achieved in JP-A-61222161 by forming detector elements of epitaxially grown HgCdTe in recessed portions of a sapphire substrate.

- A comb-shaped imager having double layer metal electrodes is shown in JP-A-61251167.

- N-type regions corresponding to photodiodes, are formed in JP-A-62011265 by diffusing indium into a p-type HgCdTe layer. The indium is diffused from straps which also connect the formed n-type regions to a read-out device comprised in a silicon substrate. No alignment betweeen the detector layer and the read-out device is needed which would be the case if the n-type regions were formed before the bonding of the detector array and the read-out device takes place.

The conversion efficiency of individual photodiode elements in an imaging array is a function of the area of the pn-junction. As the surface area of an n-type material in contact with a p-type material increases so does the conversion efficiency of infrared radiation to electrical current. However, by increasing the area of the n-type region, the amount of thermal leakage-current, or dark current, generated by the diode also increases, with a consequent reduction in the diode signal-to-noise ratio. Furthermore, the n-type regions within the diode array are typically formed by ion implantation techniques. This implantation process causes damage to the crystalline lattice thereby increasing the thermally induced diode leakage-current.

- The problems stated above are addressed in the invention of US-A-4751560. A field plate and a guard plate are formed which, when correctly biased, generate an inversion layer at a semiconductor surface surrounding each photodiode.

- In US-A-4868622 isolation regions between detector elements are formed by irradiating the regions with a laser beam.

- A SiO_xN_y reflection preventive film is introduced on the imager presented in JP-A-63150976. This film is formed on a first side of a CdTe substrate with an HgCdTe detector layer formed on an opposite side.

- A process to manufacture an imager is disclosed in JP-A-63260171. Photodiodes, formed in HgCdTe chips, are tested before they are mounted on a temporary substrate. The chips are further treated before they are bonded to a CCD with a flip-chip bonding process.

- An imager having a close spacing between adjacent detector elements is disclosed in JP-A-1050557.

- The method of fabrication of an imager presented in WO-A-8910007 provides for virtually all of the critical diode fabrication processes to be performed with one photomask layer in a single processing chamber. The problem of alignment of successively applied photomask layers is therefore eliminated. Furthermore, the use of one photomask layer provides for the fabrication of pn-junctions having small areas. The small pn-junction area gives a small junction capacitance and a small leakage-current. Another advantage is that repeated surface cleaning procedures are not required.

- In many applications, photodiodes operate at zero polarization. They must therefore have high dynamic resistance to ensure a high detectivity. This is achieved in US-A-4972244 by separating detector elements by trenches. Furthermore, a common electrode is provided at the bottom of the trenches thereby reducing the interconnection resistance of the photodiodes.

- In JP-A-2009180 two layers of HgCdTe having a different energy bandgap are used to form an imager which performs wavelength separation. The imager is utilized in an image tracking apparatus.

- In GB-A-2238165 wavelength separation is carried out by two sets of detectors having different optical path lengths.

- A multicolour focal plane array is presented in US-A-4956555. A first and a second group of filters each includes a dielectric/thin metal/dielectric layer combination. The filters of the two groups differ in that the thickness of the dielectric layers are adjusted to selected wavelength bands.

- The detector array of US-A-5293036 achieves a center-surround type output from individual detectors without post-processing the detector signals externally to the array. Each detector of the detector array comprises for this purpose a central radiation absorbing area which is surrounded by a peripheral radiation absorbing area. The center-surround type output signals from the detectors are readily coupled to a neural network signal processor.

- In US-A-5113076 a two-terminal radiation detector which is sensitive to two spectral bands of infrared radiation is presented. Two heterojunctions are coupled in series and function electrically as two back-to-back diodes.

- A method of manufacturing a detector that is sensitive to two spectral bands of infrared radiation is shown in US-A-5149956.

- A hybrid detector responsive to visible wavelength radiation and also to at least two different wavelengths of infrared radiation is presented in US-A-5373182.

- The invention of GB-A-2246662 refers to testing of an imager by the use of a test element group. Elements of the test element group are connected by indium bumps to connection pads on a read-out substrate in the same flip-chip bonding process which connects detector elements to the read-out substrate.

- From the measured I-V characteristics of the test element group, presented in GB-A-2246662 above, only the presence or absence of the pn-junction in the light responsive region is detected. Even if a photo-detector has insufficient spatial resolution caused by the excessive diffusion length of the minority charge carrier in the semiconductor layer in which the pn-junctions are formed, the insufficient space resolution is not detected from the I-V characteristics of the test element group, and this photo-detector is selected as a non-defective. In GB-A-2274739 a photo-detector is presented which includes a test element group that detects the spatial resolution of the light responsive region.

- The invention of JP-A-4133363 refers to an imager comprising photodiodes. The sizes of n-type regions, which are formed in a p-type layer, are made equal by making holes in the p-type layer and forming the n-type regions in the holes.

- The patterning accuracy of detection elements is improved in JP-A-4253344 by forming the detection elements in recessed regions of a substrate.

- An imager which can select a mode of either high resolution or high sensitivity is presented in EP-A-0497326.

- A photo-detector having photodiodes which are each formed of a main and an auxiliary region, is disclosed in EP-A-0543537. The auxiliary regions reduce the blind areas of the photo-detector.

- A method to diffuse Hg into a p-type HgCdTe substrate and thereby converting regions of the substrate to n-type is disclosed in GB-A-2261767. A Hg/Cd amalgam is formed on the substrate followed by a protective film. Hg vapor which is generated from the amalgam during an annealing step diffuses into the substrate while the protective film prevents the Hg vapor from diffusing into the atmosphere.

- A two colour infrared detector is described in US-A-5300777 which for each detector element comprises a heterojunction diode and a metal-insulator-semiconductor device.

- A method of manufacturing an imager having mesa structures is presented in JP-A-6013642. The base layer of the mesas comprises a shallow n^-/n^+ structure.

- A method of fabricating photodiodes having a planar topside surface is presented in US-A-5279974. The structure exhibits less retro-reflection and allows the photodiodes to have a smaller physical size than the size of the indium bumps which are connected to the photodiodes.

- An imager comprising photo-conductors each having a charge generation region coupled to a confinement region is presented in US-A-5254850. The geometric shape of the photo-conductors increase the responsivity.

When detecting incident energy transmitted from a remote source, it is often desirable that the remote source is not aware that the energy it is transmitting is being detected. A problem in this regard is that focal plane arrays tend to reflect the received energy along the incident energy path. To minimize the reflection, focal plane arrays are provided with a slot shield which operates on the theory that most of the energy impinging upon the detector is travelling at an acute or obtuse angle relative thereto.

- A slot shield which is fabricated as part of the focal plane array fabrication procedure is disclosed in US-A-5298733.

- An imager having HgCdTe mesa structures is presented in JP-A-6209096. The mesas are formed by dry etching using a mask. The etching ratio of the mask to HgCdTe is such that an uniform layer of type converted HgCdTe is formed at the surface of the mesas.

Radiation incident to a detector array, which is not absorbed but reflected back into space by the detector array, is referred to as the "light signature", LS. In order to minimize the light signature of a detector array having trench walls or mesa structures, these wall-sides must be reduced to be much smaller than an optical blur diameter. The optical blur diameter is given by 1.22 times the wavelength, divided by the numerical aperture.

- In US-A-5414294 the width of the mesa and trench walls are reduced to a fraction of the optical blur diameter. Each mesa comprises a photodiode and the effective area of a pixel is made larger than the area of one mesa region by coupling photodiodes in parallel.

- An imager having a mesa structure is presented in JP-A-7079008. The mesas are provided with an n^+-type surface layer.

- In JP-A-7094693 an RC-network is formed between each detector element and a corresponding indium bump which connects the element to a read-out circuit.

Summary - Cross-Talk Preventing Measures

To improve the resolution of an imager the distance between adjacent photodiodes is reduced. A problem occurs when the distance reaches the same dimensions as the diffusion length of photogenerated charge carriers.

- In EP-A-0024970 an insulating opaque film is introduced between the photodiodes allowing a close spacing of the photodiodes.

- The structure discussed above in EP-A-0024970 is further improved in US-A-4665609 in which an anodic oxide is applied on the surface of the detector before an opaque metal film is formed at regions surrounding the photodiodes. This anodic oxide stabilizes the pn-junctions.

- An imager using photocapacitive detector elements is presented in FR-A-2526227. Each detector element comprises a gate with an opening and an implanted region where the amount of generated charge is measured. In this design the potential applied to the gate is not crucial and the output impedance is relatively high.

- The detector elements of GB-A-2132017 are separated by polycrystalline regions of HgCdTe.

- The problem of cross-talk between two adjacent detector elements is approached in JP-A-61198787 in which a p^+-type substrate is used and each photodiode is formed in a p-type region.

- In US-A-4646120 an ohmic contact layer is introduced which allows individual detector elements to be completely separated thereby reducing the problem of cross-talk.

- Insulation between detector elements is automatically achieved in JP-A-61222161 by forming detector elements of epitaxially grown HgCdTe in recessed portions of a sapphire substrate.

- In JP-A-62036858 reflectors are provided in V-grooves which separate individual detector elements. The amount of cross-talk between adjacent detector elements is thereby reduced.

- Light shields are provided between a CdTe substrate and an HgCdTe film in JP-A-62104163 to reduce the problem of cross-talk.

- Absorption regions of HgCdTe are formed at regions between detector elements in the invention of JP-A-63043366. The design reduces the problem of cross-talk.

- The structure of JP-A-63043366, discussed above, is further improved in GB-A-2229036. An absorption layer is combined with n-type photodiode regions which penetrate through a p-type layer.

- A structure of an imager having a low amount of cross-talk between adjacent photodiodes is disclosed in JP-A-63133580. The structure comprises islands of p-type material formed in a p^+-type HgCdTe substrate. A photodiode is provided in each island by forming an n^+-type region in each p-type island.

- Cross-talk may occur in an imager comprising an HgCdTe layer formed on a first surface of a transparent substrate when photons which are not absorbed in a particular detector element in the HgCdTe layer are reflected at a second surface of the substrate, opposite to the first surface, and thereafter absorbed in an adjacent detector element. In JP-A-63170960 this kind of cross-talk is prevented by the use of a substrate which absorbes infrared radiation.

- The problem of cross-talk, presented in JP-A-63170960 above, is solved in JP-A-63229751 where an infrared light absorbing layer is provided on a substrate, opposite the surface holding an HgCdTe detector layer.

- A backside illuminated imager is presented in JP-A-63273365. The amount of cross-talk between adjacent photodiodes, which have been separated by grooves, is reduced by forming an infrared blocking film in the groove regions.

- The problem of cross-talk between adjacent photodiodes and the problem of different thermal expansion coefficients between a silicon substrate and an HgCdTe substrate are approached in JP-A-63281460. A detector is made up of HgCdTe wells, comprising photodiodes, formed in a CdTe substrate. The detector is bonded to a silicon substrate by a flip-chip process, and finally the CdTe substrate is etched away.

- A similar structure to the structure discussed in JP-A-63281460 above is disclosed in JP-A-63296272. In the latter, the HgCdTe wells are not connected by indium bumps but by Au films supported by an epoxy resin.

- The problem of difference in thermal expansion coefficients between an HgCdTe substrate and a silicon read-out substrate is addressed in JP-A-2214159 where a design similar to the design of JP-A-63296272 is presented.

- Cross-talk is reduced in US-A-5376558 by forming a three layered light absorption structure in the areas between the photodiodes of the array. The second layer has a refractive index which is greater than that of the first layer and the third layer is a metal layer.

- The problem of difference in thermal expansion coefficients between a silicon read-out substrate and an HgCdTe detector substrate is approached in US-A-4783594 by filling the space between the two substrates with a resilient electrically insulating polymeric material and thereafter separating the detector elements from each other by removing a layer of the HgCdTe detector substrate.

- An infrared light absorbing layer is formed at regions between detector elements in the imager presented in JP-A-1201971.

- Photo-detector regions are formed in JP-A-1205476 by laser annealing a multi-layer structure of HgTe and CdTe. Radiation which has generated charge carriers between two detector elements will be absorbed in the multi-layer structure thereby reducing cross-talk.

- A structure which reduces the amount of cross-talk and which is similar to the structure proposed in JP-A-1205476, discussed above, is presented in JP-A-1228180.

- In JP-A-3196568 another structure, which is similar to the structures of JP-A-1205476 and JP-A-1228180, discussed above, is presented.

- The problem of cross-talk is approached in JP-A-1218062 by alternately forming photodiodes in recessed and protruding regions.

- In JP-A-1233777 the amount of cross-talk is reduced by diffusing mercury into detector regions thereby reducing the acceptor carrier concentration in these regions while the regions in between the detector regions are kept at a high acceptor carrier concentration.

- The method of fabrication of an imager presented in WO-A-8910007 provides for virtually all of the critical diode fabrication processes to be performed with one photomask layer in a single processing chamber. The problem of alignment of successively applied photomask layers is therefore eliminated. Furthermore, the use of one photomask layer provides for the fabrication of pn-junctions having small areas. The small pn-junction area gives a small junction capacitance and a small leakage-current. Another advantage is that repeated surface cleaning procedures are not required.

- In many applications, photodiodes operate at zero polarization. They must therefore have high dynamic resistance to ensure a high detectivity. This is achieved in US-A-4972244 by separating detector elements by trenches. Furthermore, a common electrode is provided at the bottom of the trenches thereby reducing the interconnection resistance of the photodiodes.

- A common electrode and a light shield film are combined in JP-A-2086177. The film surrounds the photodiodes of an imager.

- In JP-A-2155269 the photodiodes are created by forming n-type regions in an epitaxially grown p-type HgCdTe layer. To prevent cross-talk, the thickness of the p-type layer is smaller under the n-type regions and larger elsewhere.

- In JP-A-2213174 cross-talk between adjacent photodiodes is reduced by providing a lattice of crystal defects on the surface of a CdTe substrate before an HgCdTe detector layer is grown thereon.

- The cross-talk between adjacent photodiodes formed in an HgCdTe layer on a semi-insulating CdTe substrate, is reduced in JP-A-2248077 by providing n^+-type regions in the CdTe substrate at regions corresponding to regions between the photodiodes.

- The problem of cross-talk between adjacent photodiodes is solved in JP-A-2303160 by forming islands of HgCdTe, comprising photodiodes, on a CdTe substrate and then bonding them to a read-out device by a flip-chip bonding process.

- When detector elements of HgCdTe are provided on a first side of a supporting substrate cross-talk may occur from infrared radiation which is not absorbed in a detector element but is reflected at a side opposite to the first side. To solve this problem it is proposed in JP-A-3077373 to provide an HgCdTe layer on the side opposite to the first side of the supporting substrate. This layer will absorb radiation which is not absorbed in one of the detector elements.

- Cross-talk is reduced in JP-A-3085762 by separating photodiodes made in a first set of HgCdTe regions by a second set of HgCdTe regions which have a larger bandgap than the bandgap of the first set of HgCdTe regions.

- In JP-A-3175682 a structure is disclosed which prevents cross-talk between adjacent photodiodes. HgCdTe regions are provided which have larger energy bandgap than the energy bandgap of HgCdTe regions containing the photodiodes.

- In JP-A-3108371 cross-talk is reduced by introducing metal atoms by a laser annealing process at regions between photodiodes.

- The amount of cross-talk between adjacent photodiodes which are formed on a substrate is reduced in JP-A-3133181 by providing radiation absorbing regions in trenches formed in the substrate.

- The amount of cross-talk is reduced in JP-A-3219670 by forming photodiodes on HgCdTe islands placed in recessed regions of a GaAs substrate.

- In JP-A-3241774 individual detector elements are separated by regions having a relative high concentration of dislocations. The structure reduces the problem of cross-talk.

- The imager of EP-A-0445545 comprises a radiation absorbing layer which has a varying composition and energy bandgap. The individual detector elements are separated one from another with grooves which may be combined with a ground plane or a guard diode structure.

- When individual detector elements are separated by ion-milling grooves through an HgCdTe detector layer the HgCdTe crystal may be damaged. In JP-A-3268463 individual detector elements are formed by converting regions of a CdTe layer into HgCdTe. The detector elements are separated by the non-converted regions of the CdTe layer.

- In JP-A-3270269 individual detector elements are first formed by converting regions of a CdTe layer and then separating the detector elements by removing non-converted regions of the CdTe layer.

- The imager proposed in US-A-5030828 has parallel elongate cavities formed within a substrate, photosensitive detector elements formed within the cavities and an optical insulating layer adjacent each of the cavities to optically isolate the cavities from each other. The elongate cavities provide an increased detector element surface area which increases the sensitivity. The optical isolation reduces cross-talk among adjacent detector elements.

- An imager in which the optical energy is incident on a p-type doped region prior to being incident on a bulk n-type doped region is presented in EP-A-0485115. The short distance between the surface of the p-type region and the pn-junction causes efficient detectivity including at the short wavelength end of the spectrum of interest. The description is made for InSb detectors. HgCdTe is mentioned.

- The imager of US-A-5177580 refers to a mesa-type photodiode array having U-shaped channels formed through an n-type capping layer, a p-type radiation absorbing base layer and partially into an underlying p-type buffer layer of the type presented in WO-A-8707083. The inherently imprecise nature of the chemical etching process used to delineate the individual mesas results in the formation of ill-defined boundaries between adjacent photo-detectors. Small variations in the etched structures result in variations in the effective optically sensitive areas of the photo-detectors. Therefore, n^+-type regions are ion-implanted along the bottom of the channels, the positions of the regions being more precisely defined than the mesa geometry. These regions function as minority carrier reflectors and confine minority carriers to one side or the other of the region.

- An imager which can select a mode of either high resolution or high sensitivity is presented in EP-A-0497326.

- In JP-A-4288882 photodiodes, which are formed in an HgCdTe layer, are surrounded by regions of HgCdTe with a larger bandgap than the bandgap of the HgCdTe layer.

- The amount of cross-talk is reduced in JP-A-4313267. Individual detector elements are separated by grooves in which an absorption layer of HgCdTe is formed.

- In JP-A-4318970 the amount of cross-talk is reduced by forming a light absorbing layer and removing portions corresponding to detector elements by the use of a YAG laser.

- The imager of JP-A-4337676 comprises a multilayered structure with alternating CdTe and HgTe layers. An electron beam is used to form HgCdTe islands in which photodiodes are formed. The photodiodes are separated by the multilayered structure. Furthermore, the multilayered structure is formed on a CdTe substrate having recesses formed in its surface.

- A structure similar to the structure of JP-A-4337676, discussed above, is disclosed in JP-A-4337677. This structure comprises two multilayer structures having different composition. HgCdTe regions are formed by the use of an YAG laser beam.

- In JP-A-5055620 n-type regions are formed in a p-type layer by first providing an n-type layer in the p-type layer and then evaporating Hg from regions between pixel regions to thereby form individual n-type regions. This process provides uniform pixel characteristics because the depth of the n-type regions can be well controlled.

- In JP-A-5175476 detector element isolation is formed by evaporating mercury at isolation regions from an HgCdTe layer in which the detector elements are formed.

- A method to create high resistance regions between two adjacent photodiodes by a compensating ion-implantation is presented in JP-A-5129580.

- The amount of cross-talk between adjacent photodiodes is reduced in JP-A-5226626 by providing an inversion layer between the photodiodes.

- The amount of cross-talk is reduced in JP-A-5343727. P-type detector element regions are formed in a p^+-type layer. Photodiodes are then formed in the detector element regions.

- An imager having HgCdTe photodiodes formed in a silicon substrate is presented in JP-A-6125108.

- A structure with low amounts of crosstalk and dark current is presented in JP-A-6163969. An n-type layer and a p-type layer which form an heterojunction are provided. Individual detector elements are separated by n-type isolation regions which penetrates through the p-type layer.

- A planar detector structure is presented in JP-A-6204449. The planar structure is achieved by forming HgCdTe regions in a silicon substrate.

- The problem of separating detector elements from each other is addressed in GB-A-2284930. Several methods are disclosed. In one method regions having an high Cd doping level are formed by an electron or a proton beam.

Summary - Passivation and Leakage Current Preventing Measures

- A process for the production of photodiodes by a planar technique is presented in US-A-3988774. A thin transition layer is formed between the substrate and a dielectric masking layer necessary for local diffusion to increase the shunt resistance of the photodiodes. The shunt resistance is largely dependent on surface states of the material at the level of transition zone of n-type and p-type.

- To improve the resolution of an imager the distance between adjacent photodiodes is reduced. A problem occurs when the distance reaches the same dimensions as the diffusion length of photogenerated charge carriers. In EP-A-0024970 an insulating opaque film is introduced between the photodiodes allowing a close spacing of the photodiodes.

- The structure discussed above in EP-A-0024970 is further improved in US-A-4665609 in which an anodic oxide is applied on the surface of the detector before an opaque metal film is formed at regions surrounding the photodiodes. This anodic oxide stabilizes the pn-junctions.

- The detector elements of JP-A-61268075 are formed in mesas where the mercury to cadmium ratio is increased towards the surface of the mesas. The design provides an imager with a low level of surface leakage-currents.

- An opaque conductive shield is positioned around photodiodes of an imager presented in JP-A-62090986. An electrical potential is applied to the shield. The arrangement reduces leakage-currents in the device.

- The surface leakage-current of the imager presented in JP-A-62165973 is reduced by providing a layer of HgCdTe with a large bandgap at peripheral regions of the photodiodes of the imager.

The conversion efficiency of individual photodiode elements in an imaging array is a function of the area of the pn-junction. As the surface area of an n-type material in contact with a p-type material increases so does the conversion efficiency of infrared radiation to electrical current. However, by increasing the area of the n-type region, the amount of thermal leakage-current, or dark current, generated by the diode also increases, with a consequent reduction in the diode signal-to-noise ratio. Furthermore, the n-type regions within the diode array are typically formed by ion implantation techniques. This implantation process causes damage to the crystalline lattice thereby increasing the thermally induced diode leakage-current.

- The problems stated above are addressed in the invention of US-A-4751560. A field plate and a guard plate are formed which, when correctly biased, generate an inversion layer at a semiconductor surface surrounding each photodiode.

- The surface leakage-current of the imager presented in JP-A-62224982 is improved by forming anodic sulphide films at surface regions overlying the pn-junctions of the photodiodes of the imager.

- JP-A-62224983 also refers to forming anodic sulphide films at the surface of an HgCdTe substrate. A metal grid, covered with silicon dioxide, is formed before the film is formed.

- An imager having mesa-shaped photodiodes overlayed with a layer of passivation which contains a fixed charge is presented in WO-A-8707083. The charge creates an inversion layer within the surface of the mesa walls, thereby enlarging the pn-junction and reducing the diode leakage-current of each photodiode.

In a structure which is fabricated by etching a plurality of mesas to isolate individual photodiodes, the resulting non-planar surface is difficult to passivate. Furthermore, the pn-junctions intersect the surface of the device which gives a high leakage-current due to surface states. Another disadvantage is that the device is removed from a growth chamber in order to etch the mesas thereby increasing the susceptibility to surface contamination.

- The drawbacks of the mesa structure stated above are acknowledged in US-A-5189297. Instead, a structure having a planar surface is proposed. Isolation junctions are formed by a thermally driven process converting regions in a top layer to an opposite type of material. A method to fabricate the structure without removing it from an epitaxial reactor is also disclosed.

- The imager of EP-A-0406696 is made up of a compositionally graded p-type base layer and a compositionally graded cap layer in which pn-junctions are formed. The structure has a small amount of leakage current and a large diode impedance.

- The imager presented in GB-A-2246907 includes a transistion layer formed between an HgCdTe light absorbing layer and a CdTe substrate. The energy bandgap of the transistion layer varies through its thickness. The structure includes only one grown layer since the transistion layer is a natural product of the growth process when the light absorbing layer is grown. Moreover, the amount of surface leakage-current is low in the structure because the pn-junctions of the imager are not exposed at any surface.

- The photodiodes of the imager in JP-A-4318979 are created by first forming an n^+-type region by ion-implantation and then carrying out an annealing step. The pn-junctions of the photodiodes will be situated away from the regions which have been damaged by the ion implantation. The structure shows a high sensitivity.

Contact metals often used for photovoltaic HgCdTe sensors include gold (Au) with a nickel (Ni) overcoat for individual p-type contacts, and palladium (Pd) with a nickel overcoat for n-type common contacts. However, during high temperature storage both Au and Pd diffuse into HgCdTe, causing a high density of dislocations in the case of Pd, and shorting out the pn-junction in the case of Au, thereby degrading the performance of the device. Furthermore, the Au/Ni and Pd/Ni metal systems each have a coefficient of thermal expansion that differs from the coefficient of thermal expansion of HgCdTe.

- In US-A-5296384 a fabrication process is presented which combines passivation and contact metallization of the p and n sides of a photodiode junction in such a way that a contact metal is applied before a passivation film is formed and before a high temperture anneal is carried out. The method also includes the fabrication of a layer of dielectric overglass to eliminate the occurrence of undesirable chemical reactions during a hybridization process, and also to prevent Hg from diffusing through the passivation during high temperature storage. The fabrication process is claimed in US-A-5296384 and the corresponding device is claimed in US-A-5401986.

- A structure with low amounts of crosstalk and dark current is presented in JP-A-6163969. An n-type layer and a p-type layer which form an heterojunction are provided. Individual detector elements are separated by n-type isolation regions which penetrates through the p-type layer.

- The leakage current of the imager presented in JP-A-6237005 is reduced. N^+-type photodetector regions which have been formed in a p-type substrate are surrounded by n-type regions at the surface portion of the substrate.

Detector Arrays without Individual Detector Element Read-Out Leads

An imager having a mesa structure comprises a single n-type semiconductor block on which there are located p-type regions in the form of projections, each projection having its own connection and each projection corresponding to one detector element. An electrode common to all the detector elements is connected to the n-type semiconductor block.

US-A-3930161 (Societe Anonyme de Telecommunications, France, 30.12.75) points out that the individual connections of the detector elements are fixed in the plane of the sensitive surface which influence the effective surface of the sensitive face of the imager negatively. Therefore, it is proposed to receive the radiation through the n-type semiconductor block.

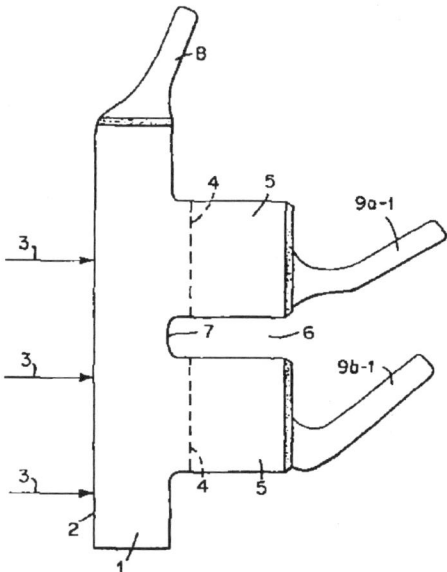

Fig. 2.2.1 (US-A-3930161 fig. 1)

An n-type region 1 has a plane face 2 which receives the radiation 3 to be detected. A junction 4 separates the n-type region from p-type regions 5. The p-type regions are separated by a squared pattern of recesses 6 which reach and form separation regions 7 in the n-type region. A common electrode of the detector elements is formed on the side of the n-type region thereby not affecting the effective surface of the sensitive face. Individual connectors 9a-1, 9b-1 are fixed to the p-type regions by soldering.

*

At the time US-A-3808435 (Texas Instruments Incorporated, USA, 30.04.74) was filed, a conventional infrared detector system had an array of infrared detectors and each detector had an amplifier for amplifying its output. The amplified output was connected to a multiplexer scan converter. Such a system required a large number of amplifiers. It is acknowledged in US-A-3808435 that an improved system would be obtained if an imager with a charge transfer device (CTD) could be used. Two problems, however, were encountered in this technique. The first problem was the fabrication of a CTD on an infrared sensitive material and the second the fact that the thermal background radiation at room temperature is many orders of magnitude greater in the infrared than in the visible (41 orders of magnitude greater at 10 µm than at 500 nm). These problems are overcome in US-A-3808435 by a.c. coupling detector signals of an infrared detector array to a CTD array. The detector array and the CTD array are formed in an HgCdTe chip and a silicon chip respectively and are connected by placing the chips on top of each other with intermediate conductor elements. This technique of connecting two chips is generally called the flip-chip technique.

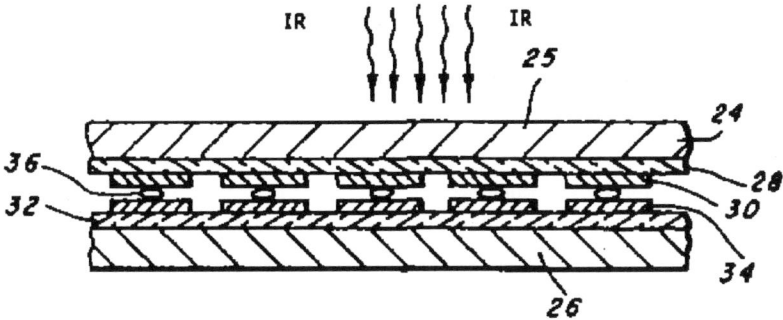

Fig. 2.2.2 (US-A-3808435 fig. 3)

The hybrid system includes an infrared detector chip 24 of HgCdTe, comprising metal-insulator-semiconductor (MIS) detectors, sandwiched with a CCD multiplexing chip 26. The thickness of the layer 24 is less than the diffusion length of charge carriers generated by the infrared radiation. A thin insulating layer 28 is formed over a surface of the semiconductor 24. An array of conductive electrodes 30 completes the detector structure. The CCD multiplexer array includes a silicon substrate 26, a silicon dioxide insulating layer 32 and an array of electrodes 34. The electrodes 34 correspond to the electrodes 30 of the detectors. The electrodes 30 and 34 are connected by ball bonds 36.

The background radiation is eliminated by utilization of a quantum differential detector (QDD). The QDD requires a chopper which is held at background temperature. The signal from the QDD is proportional to the difference between the background temperature and the scene. By removing the background, the requirements on detector uniformity and dynamic range of the associated electonics are greatly reduced.

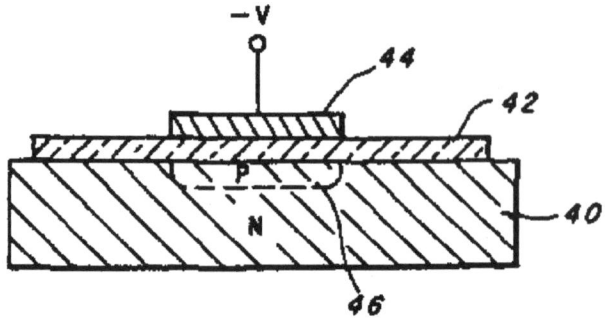

Fig. 2.2.3 (US-A-3808435 fig. 6)

Shown above is a MIS photocapacitor which can be used as a QDD. The MIS photocapacitor, when biased into inversion, is a photodiode in series with a capacitance. A substrate 40 of n-type HgCdTe is covered by an insulating layer 42. A conductor 44 is in turn formed on the insulator. An inversion region, enclosed by dashed line 46, is formed when the conductor is suitably biased with a negative voltage.

Instead of the MIS photocapacitor, it is possible to form a photodiode having a capacitor plate placed above either the p-type or the n-type region of the photodiode.

*

The invention of US-A-3902924 (Honeywell Inc., USA, 75.09.02) is concerned with low temperature growth of mercury cadmium telluride layers on insulating substrates by liquid phase epitaxy. Infrared detectors are fabricated to assist with the evaluation of the grown layers.

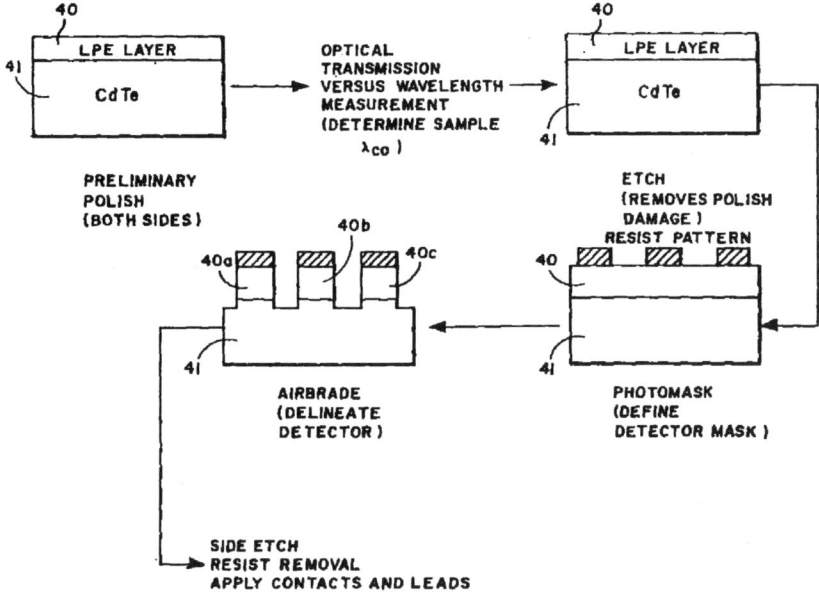

Fig. 2.2.4 (US-A-3902924 fig. 6)

The figure shows a schematic flow diagram used in fabricating detectors from HgCdTe layers grown by liquid phase epitaxy. The top surface of the HgCdTe layer 40 and the back surface of a CdTe substrate 41 first receive a preliminary polish. Optical transmission versus wavelength measurements are then made to determine the cutoff wavelength. The surface of layer 40 is etched to remove damage caused by polishing. A photo-resist pattern is deposited on layer 40 to define the pattern of the individual detectors, and the detectors are patterned by air abrasion. Finally, the sides of detectors 40a, 40b, and 40c are etched, the photo-resist is removed, and contacts and leads are applied to the individual detectors.

*

An imager with photo-conductive detectors formed in an HgCdTe body is presented in US-A-3949223 (Honeywell Inc., USA, 06.04.76). In this approach the detectors have the mechanical strength of the entire semiconductor body.

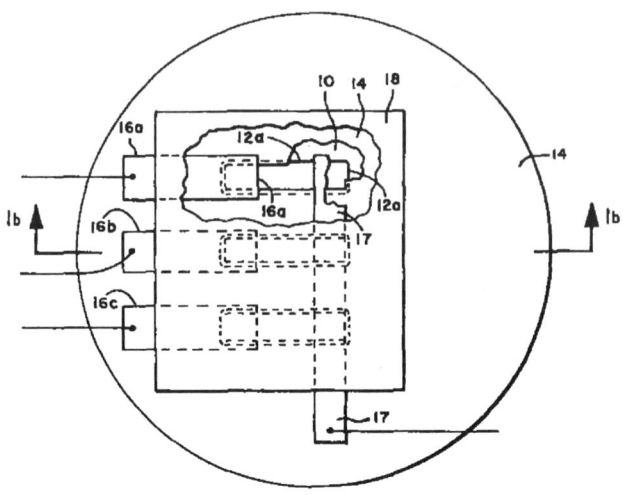

Fig. 2.2.5 (US-A-3949223 fig. 1a)

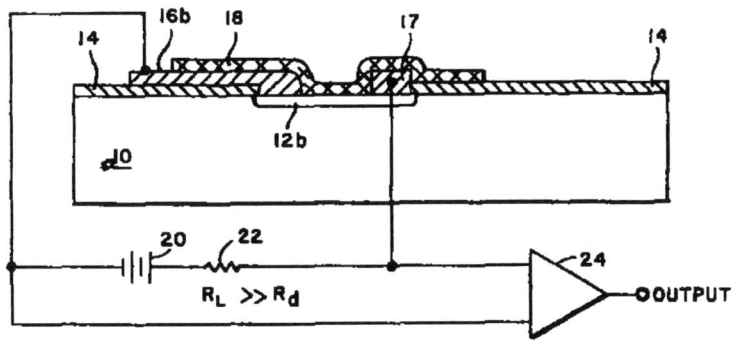

Fig. 2.2.6 (US-A-3949223 fig. 1b)

Regions 12 of one conductivity type are formed in a body of HgCdTe 10 having a second conductivity type. Different techniques to form the regions 12 are presented. The regions 12 are isolated from one another by pn-junctions. The top surface of the body is insulated by a ZnS layer 14. Electrical contacts are made to one end of the regions 12 by metal contacts 16 of indium or gold. Opposite ends of the regions 12 are electrically contacted by a common metal contact 17. A thin layer 18 of ZnS or silicon oxynitride covers the top surface and serves as a passivation layer and an anti-reflection coating.

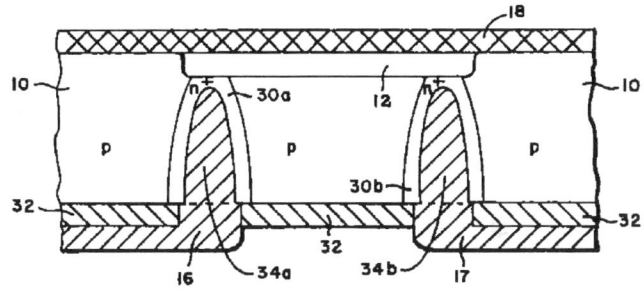

Fig. 2.2.7 (US-A-3949223 fig. 6b)

The regions 12 can also be connected by highly doped regions, formed by diffusion, particle bombardment or ion-implantation, which extend from the back surface of the substrate 10 to the backside of the regions 12. Alternatively, as shown above, cavities 34a and 34b are first formed by etching, particle bombardment or ion-milling. Then highly doped regions 30a and 30b are formed around the cavities and up to the regions 12. Electrical contacts 16 and 17 backfill the cavities and make electrical contact with the highly doped regions 30a and 30b.

The pn-junction formed between the photo-conductive detector and the substrate enhances the photo-conductive signal by essentially isolating the photogenerated minority carriers in the photo-conductive detector from the majority carriers. The minority carriers are swept across the junction while the majority carriers are allowed to flow in the photoconductive detector. This inhibits the recombination rate and extends the lifetime of the majority carriers.

*

In US-A-3980915 (Texas Instruments Incorporated, USA, 14.09.76) Schottky diodes constitute the detector elements of an imager.

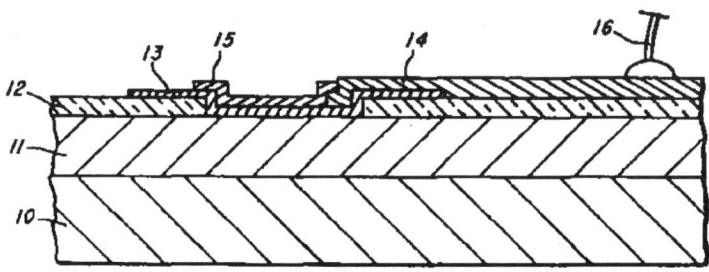

Fig. 2.2.8 (US-A-3980915 fig. 1)

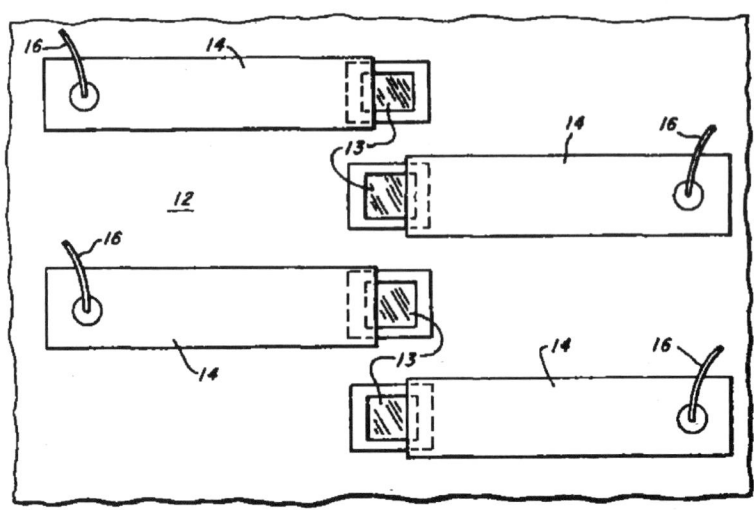

Fig. 2.2.9 (US-A-3980915 fig. 4)

A substrate 10 of HgCdTe is provided with an upper surface region 11 formed by an annealing procedure, or as an epitaxial layer or evaporated film. A layer of insulating material 12 is formed in which windows are provided. The windows are partially filled with a thin layer of metal 13 which is deposited therein to form a metal-semiconductor diode with the upper surface region. The metal layer is deposited to a thickness on the order of 10-50 nm thick and is sufficiently thin to be semi-transparent to infrared radiation. A thick layer of metal 14 is deposited to form an expanded contact and an anti-reflection coating 15 is provided. External conductors in the form of jumper wires 16 are ball bonded to the contact 14.

*

The smallest distance between two adjacent detector elements of a linear imager comprising detector elements separated by grooves etched through an HgCdTe chip is limited because the engraved walls are not completely perpendicular to the chip surface but slope obliquely toward the base of the grooves. In GB-A-1597581 (Selenia Industrie Elettroniche Associate S.p.A., Italy, 09.09.81) the distance between two detector elements is reduced by etching the chip from two opposite surfaces.

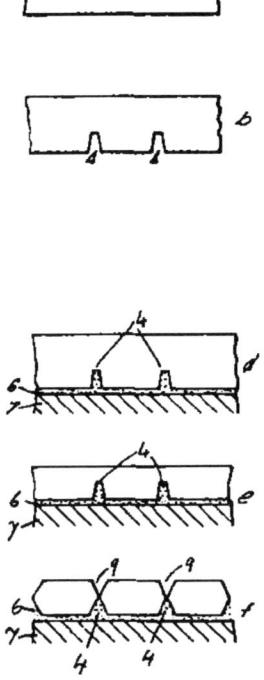

Fig. 2.2.10 (GB-A-1597581 fig. 4)

The figure shows steps used to obtain the imager viewed on a cross-section through the detector elements and parallel to the line of detectors. A first sequence of grooves 4 are formed by ionic erosion in a first surface of the chip. Simultaneously several holes or slots are made, which project deeper in the chip than the final thickness and which will be utilized as reference points for the alignment of the other face of the chip. The chip is glued by an adhesive 6 to a support base 7 with the first surface of the chip facing the support base. The chip is thinned to a desired thickness. At the end of this thinning step, the alignment holes will become visible and therefore the geometry of detectors can be defined by means of grooves 9 formed by chemical erosion on the opposite surface to the first surface of the chip. The distance between two adjacent detector elements is half the distance which would have been obtained if the separation grooves were formed from only one side.

*

A silicone rubber adhesive layer is used in US-A-4081819 (Honeywell Inc., USA, 28.03.78) to bond an HgCdTe substrate, which includes an epitaxial layer, to a second substrate. The silicone rubber adhesive reduces the risk that the substrate cracks when cooled to cryogenic temperatures.

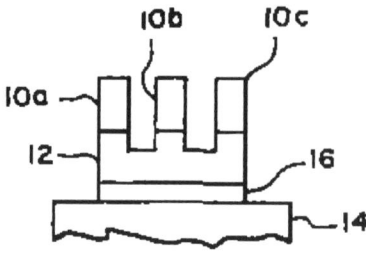

Fig. 2.2.11 (US-A-4081819 the figure)

An HgCdTe epitaxial layer is patterned to form individual devices 10a, 10b and 10c on a substrate 12 of HgCdTe. The substrate is bonded to a second substrate 14 by a silicone rubber adhesive 16.

*

Methods to form image detectors by ion-implanting Hg ions into a CdTe substrate and thereafter annealing the substrate with a high-energy laser are disclosed in US-A-4242149 (The United States of America as represented by the Secretary of the Army, USA, 30.12.80).

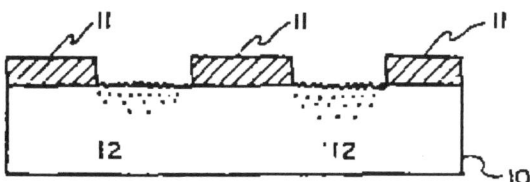

Fig. 2.2.12 (US-A-4242149 fig. 1b)

A CdTe substrate 10 is bombarded with Hg ions through an ion-opaque mask 11.

Fig. 2.2.13 (US-A-4242149 fig. 1c)

The mask is removed and a laser beam is scanned over the surface of the substrate to cause local heating. This heating melts the surface and induces both annealing of the surface and combination of the Hg with the CdTe to form the HgCdTe photo-detectors. Finally, read-out conductors are formed. In an alternative embodiment, Hg ions are implanted over the whole surface of the CdTe substrate. A laser beam is then scanned over the substrate to cause local heating. The photo-detectors are delineated by the use of a second laser beam.

*

In JP-A-57024580 (Fujitsu Ltd, Japan, 09.02.82) detector elements are formed in an HgCdTe layer which has been grown on a CdTe substrate.

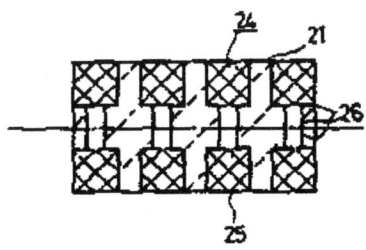

Fig. 2.2.14 (JP-A-57024580 fig. 6)

A layer of HgCdTe is grown by a liquid phase epitaxy on a CdTe substrate 21. The layer is etched by the use of a mask to form detector elements 24. Electrode regions 25 are deposited by evaporation on each side of sensitive regions 26.

*

A method to convert a region of a p-type HgCdTe body to n-type by the use of ion-etching is presented in EP-A-0062367 (Philips Electronic and Associated Industries Limited, GB, 13.10.82). A high concentration of mercury is produced from an etched-away part of the body as to act as a dopant source. The method is used to form a detector device.

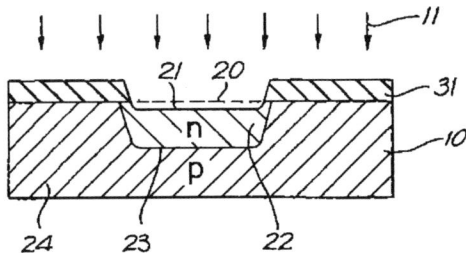

Fig. 2.2.15 (EP-A-0062367 fig. 7)

A masking layer 31 is formed on a p-type HgCdTe body 10 to restrict the etching and conductivity-type conversion. The body is bombarded with ions 11 to etch away a part 21 of the body and to produce from the etched-away part of the body an excess concentration of mercury which acts as a dopant source converting an adjacent part 22 of the body into n-type material. The conversion can be effected over a depth considerably greater than the penetration depth of the ions.

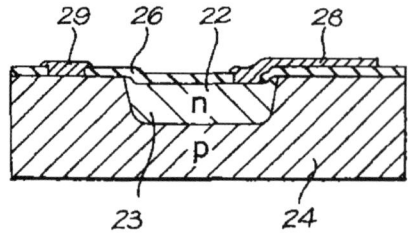

Fig. 2.2.16 (EP-A-0062367 fig. 8)

After the etching, the masking layer is removed and a passivation layer 26 and detector electrodes 28 and 29 are provided. A plurality of detectors can be formed in the body. Alternatively, the whole surface of the p-type HgCdTe body can be converted into n-type material before individual detectors are formed by mesa-etching.

*

The performace of a photodiode is affected by the resistance and the capacitance of the photodiode. A large resistance results in little thermal noise; a small capacitance allows a high operating frequency. A method to increase the resistance and to reduce the capacitance of photodiodes formed as mesas is to drill holes in the mesa as disclosed in EP-A-0064918 (Thomson-CSF, France, 17.11.82).

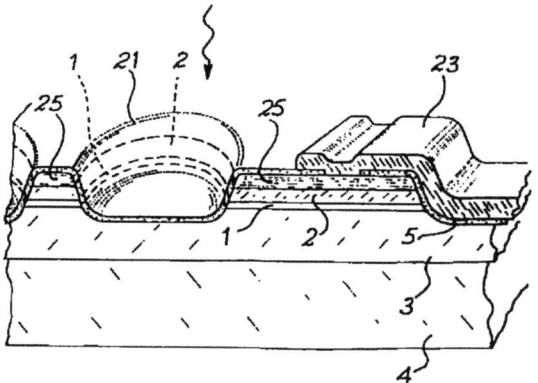

Fig. 2.2.17 (EP-A-0064918 fig. 4)

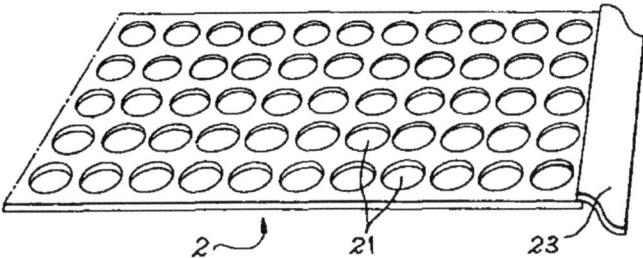

Fig. 2.2.18 (EP-A-0064918 fig. 3b)

A pn-junction 1 is formed between an n-type HgCdTe layer 2 and a p-type HgCdTe layer 3 which has been epitaxially grown on an insulating substrate 4. The detector elements are separated by the creation of mesas. An insulating layer 5 is provided and contacts are connected to the n-type regions and the p-type region. The figures refer to only one mesa and only the contact 23 to the n-type region is shown. A plurality of holes reaching the p-type layer are formed by chemical etching thereby increasing the resistance and decreasing the capacitance of the photodiodes. A thin layer of aluminium 25 may be added to form an electric connection between the electrodes 23 and the place where signal charge is generated. The holes may be circular, rectangular or honeycomb-shaped.

*

In EP-A-0068652 (The Secretary of State for Defence, GB, 05.01.83) photodiodes are formed by diffusing indium into a layer of semi-insulating CdTe which covers an HgCdTe substrate.

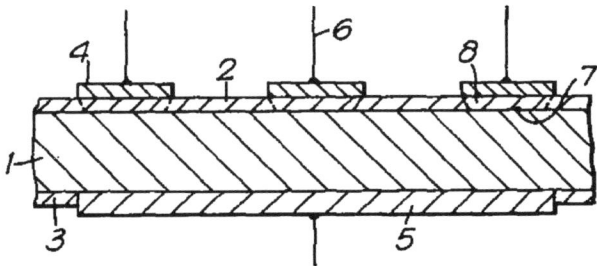

Fig. 2.2.19 (EP-A-0068652 fig. 1)

A protective passivation layer of CdTe is grown on all surfaces of a p-type HgCdTe substrate 1. Islands 4 of indium are deposited and the structure is heated to diffuse indium into the CdTe layer giving n-type heavily doped regions 8. Additionally, this heating causes inter diffusion at the HgCdTe and CdTe interface with Hg diffusing into the CdTe layer and Cd diffusing into the HgCdTe substrate thereby creating a graded heterostructure. A back electrode 5 of gold is formed on the substrate.

*

Neither a silicon multiplexer input cell nor a hybrid interconnect can be reduced in size to match the smallest size of a detector. The imager presented in US-A-4727406 (Rockwell International Corporation, USA, 23.02.88) allows several detectors to be connected to each input cell of a multiplexer. This is achieved by disposing a plurality of conducting plates around a common junction region. The conducting plates are controled, one at a time, to create an induced junction there under.

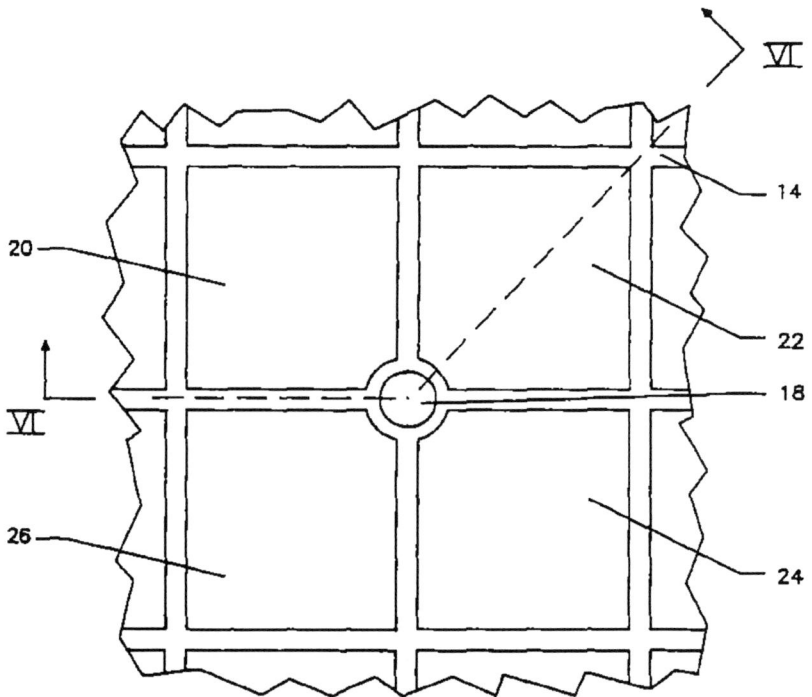

Fig. 2.2.20 (US-A-4727406 fig. 5)

A common junction region 18 with four conducting plates 20, 22, 24 and 26 are shown above. A cross-sectional side view taken along the line VI-VI is shown below.

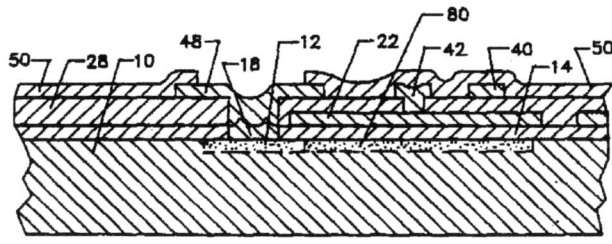

Fig. 2.2.21 (US-A-4727406 fig. 12)

N-type regions 12 are formed by ion-implantation of B or diffusion of Hg in a p-type $Hg_{0.71}Cd_{0.29}Te$ layer 10 which has been grown on a transparent substrate. Contacts 18, formed over each N-type region, and conducting plates 20, 22, 24 and 26 are formed from a first level metal. Insulation is performed by the insulating layers 14 and 28. Control lines 40, 42, 44 and 46 are formed from a second level metal and connects to the conducting plates 20, 22, 24 and 26, respectively. Interconnect bumps are added to junction contacts 48 and to control line pads 72 (see figure below). Corresponding interconnect bumps are located on a signal multiplexer chip. The detector and the multiplexer chip are assembled by using a flip-chip fabrication technique.

In operation, light is absorbed in the semiconductor layer 10. The induced junctions, such as junction 80, collect the generated charge. The charge is read out through the junction region 12.

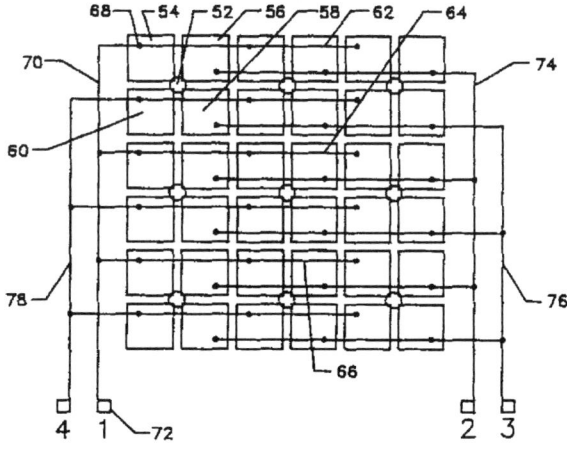

Fig. 2.2.22 (US-A-4727406 fig. 13)

The imager provides a technique for achieving a small scale pre-multiplexing of the detector array while retaining a photovoltaic structure. By cycling the potentials on the control lines to selectively create field induced junctions, the effective collection area for incoming radiation will cycle about the unit cell. Consequently, the number of resolution elements on the focal plane is multiplied.

In an alternative embodiment the conducting plates are made small and the induced junction underneath a conducting plate connects the common junction region 12 with an auxiliary junction region formed in the semiconductor layer 10. A plurality of auxiliary junction regions are disposed around a common junction region 12.

*

The photodiodes of the imagers presented in US-A-4566024 (Societe Anonyme de Telecommunications, France, 21.01.86) are formed in a p-type HgCdTe substrate by diffusing n-type impurities from two opposite faces, a front face and a rear face, of the substrate. The photodiodes can be connected to a read-out device by a flip-chip bonding process on the rear face of the substrate and still be illuminated from its front face.

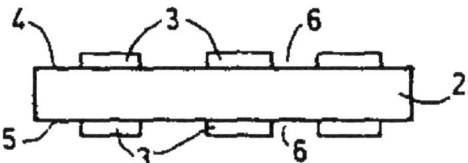

Fig. 2.2.23 (US-A-4566024 fig. 3)

A p-type substrate of HgCdTe is polished to obtain a wafer 2 with a thickness of from 20-30 µm. A masking layer of ZnS is deposited on a front face 4 and a rear face 5 of the wafer, and openings are made to form windows 6 in a mosaic configuration separated by masking bands 3. The windows of the two faces are respectively opposite one another.

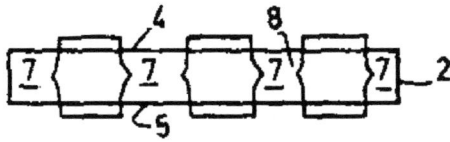

Fig. 2.2.24 (US-A-4566024 fig. 4)

An n-type diffusion is performed from the two faces 4 and 5 to obtain zones 7 extending from face 4 to face 5 of the wafer.

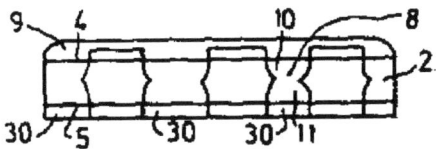

Fig. 2.2.25 (US-A-4566024 fig. 5)

The front face 4, which will be exposed to radiation, is covered with a protection layer 9. An indium layer is then deposited in the windows 6 of the rear face in order to make studs 30.

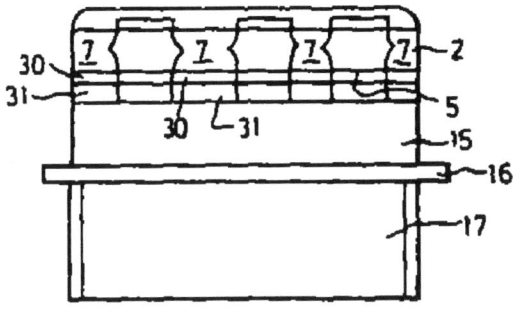

Fig. 2.2.26 (US-A-4566024 fig. 6)

The indium studs 30 are used for connection of the n-type zones by cold welding with corresponding input studs 31 of a silicon integrated circuit formed in a silicon wafer 15. The two wafers 2 and 15 are disposed on a cooled surface 16 in a cryogenic enclosure 17.

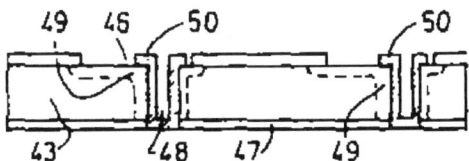

Fig. 2.2.27 (US-A-4566024 fig. 12)

In a second embodiment, a masking layer having windows 46 is formed on a thinned wafer of p-type HgCdTe. Holes 48 are etched through the wafer in the regions inside the windows by ion etching. N-type zones 49 are developed under the windows and along the walls of the holes. The n-type zones are connected to input studs of a silicon circuit via a deposited metal layer 50. The holes may be filled with a reinforced epoxy resin to perform the same function of connection.

*

An imager having detector elements supported by a resin is disclosed in JP-A-58164261 (Tokyo Shibaura Denki KK, Japan, 29.09.83).

Fig. 2.2.28 (JP-A-58164261 fig. 2b)

An n-type layer 21 is formed on a p-type substrate 22 of InSb (HgCdTe mentioned in the description). The n-type layer is polished and etched to form mesas.

Fig. 2.2.29 (JP-A-58164261 fig. 2d)

The mesas are covered by a resin layer 23. The resin layer is polished until the n-type layer is exposed.

Fig. 2.2.30 (JP-A-58164261 fig. 2e)

Indium bumps 24 are formed on the n-type regions of the mesas and the detector array is bonded to a silicon read-out chip comprising a CCD.

Fig. 2.2.31 (JP-A-58164261 fig. 2f)

Next, the HgCdTe substrate is chemically etched until the resin layer is exposed. The detector elements 27 are separated by the resin layer.

*

An imager using photocapacitive detector elements is presented in FR-A-2526227 (Thomson-CSF, France, 04.11.83). Each detector element comprises a gate with an opening and an implanted region where the amount of generated charge is measured. In this design the potential applied to the gate is not crucial and the output impedance is relatively high.

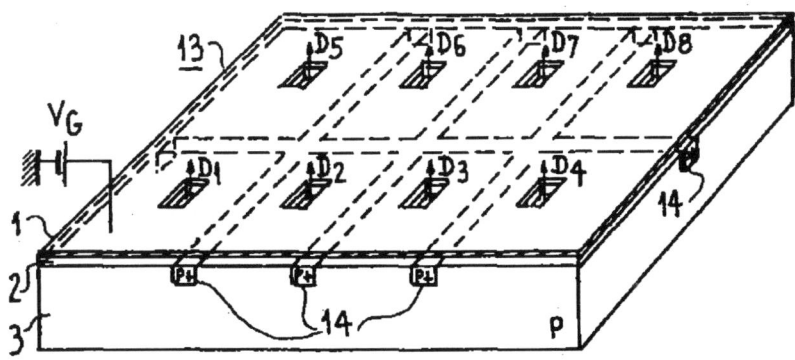

Fig. 2.2.32 (FR-A-2526227 fig. 12)

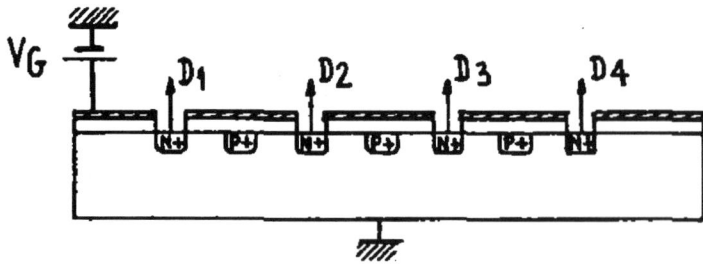

Fig. 2.2.33 (FR-A-2526227 fig. 13)

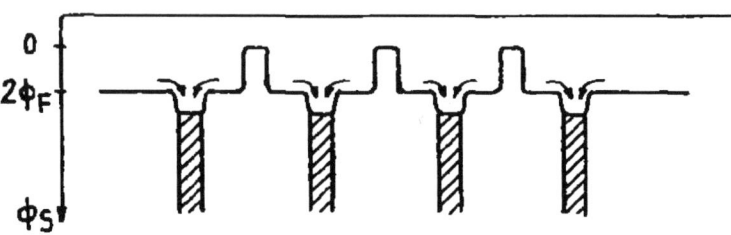

Fig. 2.2.34 (FR-A-2526227 fig. 14)

A p-type HgCdTe substrate 3 is covered with an insulating layer 2 and a detector gate 1. An opening is formed in the detector gate for each detector element D and an n-type region is formed in the substrate at a position corresponding to the opening. The n-type region is connected to a CCD formed in a separate substrate. P-type regions 14 may be formed between the detector elements. Charge generated in the substrate 3 close to a potential well under a detector gate diffuses to a region where it is read out. The p-type regions will stop charge diffusing from one detector element to an adjacent one when a detector element is exposed to a high intensity of radiation.

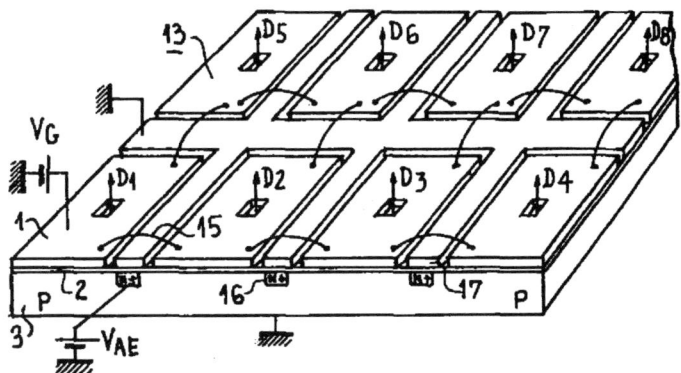

Fig. 2.2.35 (FR-A-2526227 fig. 15)

In an alternative embodiment, the detector elements are separated by n-type regions 16 and a gate 17. The detector gate 1 consists of detector gate elements connected to each other by connectors 15. When a detector element is exposed to a high intensity of radiation excess charge is drained off via the regions 16.

*

Two arrays of HgCdTe photovoltaic detectors with pn-junctions of different sizes are combined in US-A-4517464 (The United States of America as represented by the Secretary of the Air Force, USA, 14.05.85) to form a device which has an increased functionality when exposed to ionizing radiation.

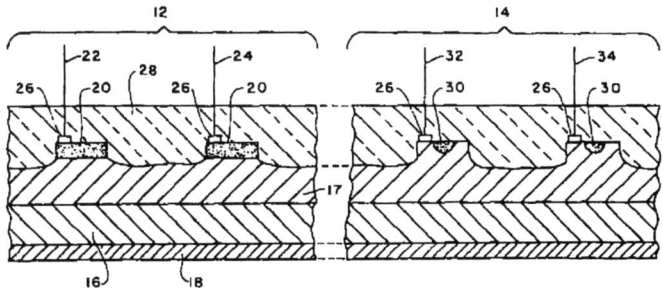

Fig. 2.2.36 (US-A-4517464 fig. 1)

A first array 12 consists of a substrate 16 composed of CdTe which has an upper liquid phase epitaxial layer 17 of p-type HgCdTe grown on it and a metallization layer 18 attached to a lower surface serving as an electrode. A plurality of mesas 20 composed of n-type HgCdTe are formed on the upper surface. Each mesa has an electrical connection 22, 24, via a metallization box 26. A passivation layer 28 is attached to the upper surface. A second array 14 is similar in nature to the first array. This array, however, is incapable of performing as a sensor system initially as each detector has a much smaller pn-junction area. As a cumulative dose of incident radiation is increased, type conversion in both arrays occurs resulting in degraded performace in the first array but an increase in performance caused by the now effective pn-junction area in the second array. A switching device, coupled with a dosimeter, selects the first or the second array depending upon the total dose of radiation received.

*

An arrangement for multi-spectral imaging of objects is presented in US-A-4596930 (Licentia Patent-Verwaltungs-GmbH, FRG, 24.06.86). Groups of detectors are disposed such that an optical system, due to its scatter circle, images each object dot on at least one detector of each group.

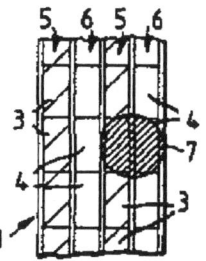

Fig. 2.2.37 (US-A-4596930 fig. 2)

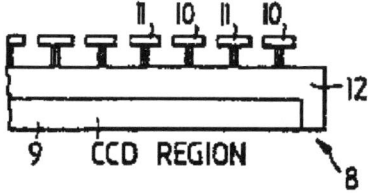

Fig. 2.2.38 (US-A-4596930 fig. 3)

A semiconductor element 1, 12, includes a CCD 9 on a first side. Series of parallel detector strips 5, 10, of HgCdTe, and 6, 11, of InSb, are bonded to the semiconductor element on a side opposite the CCD. The strips 5 and 6 are sensitive to the wavelength bands 8-12 µm and 3-5 µm, respectively. The dimensions of the strips 5 and 6 are selected such that, due to the scatter circle of an assocated optical system, the imaging of an object dot takes place on at least one individual detector 3 and 4 of each detector strip. The number of detector strips may be increased to cover further wavelength ranges.

*

A broad band detector array with homogeneous sensitivity is achieved in JP-A-60102530 (Yokokawa Hokushin Denki KK, Japan, 06.06.85) by providing a thin single crystal film of $Hg_{1-x}Cd_xTe$ having a concentration gradient of the composition ratio x.

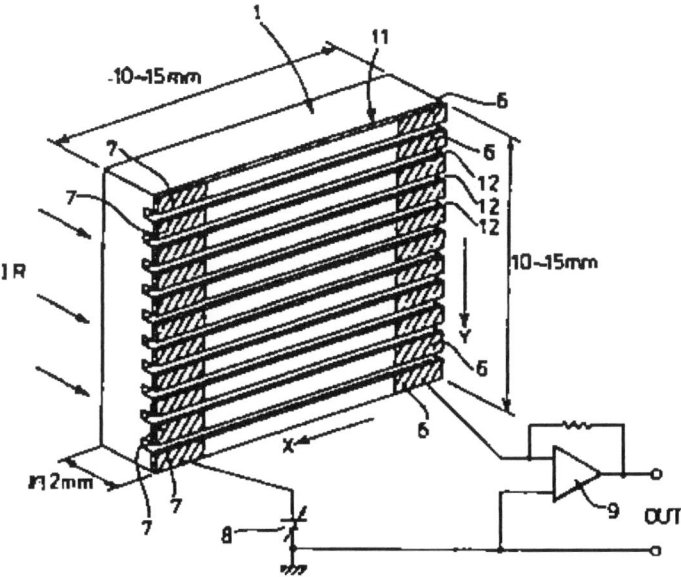

Fig. 2.2.39 (JP-A-60102530 fig. 4)

A thin film of Hg is formed on a CdTe substrate 1. A SiO_2 film is formed on the Hg film and zone annealing is performed by a heater having a specified temperature distribution thereby forming a thin film of $Hg_{1-x}Cd_xTe$ with a concentration gradient of the composition ratio x. Individual detector elements having a broad band spectral response are formed by providing separation grooves 12. The detector elements are connected by electrodes 6 and 7.

*

The invention of EP-A-0167305 (The Secretary of State for Defence, GB, 08.01.86) refers to a photo-detector which includes a photosensitive region adjacent to a minority carrier extraction region arranged when biased to depress the photosensitive region minority carrier concentration. Depression of the minority carrier concentration produces low noise and high responsivity properties as obtained by cooling, but without the need for cooling equipment. Another structure which simulates cooling is presented in WO-A-9006597 (The Secretary of

State for Defence, GB, 14.06.90). The structure, which comprises three successively disposed sections each containing three layers, a mirror and a pair of electrodes, makes up a multiple heterostructure photodetector.

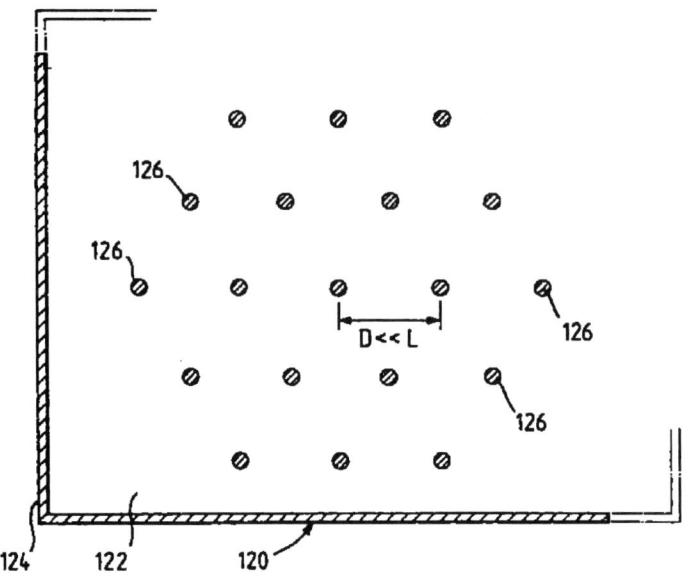

Fig. 2.2.40 (EP-A-0167305 fig. 6)

The detector 120 comprises a large p-type region 122 with a circumferential negative bias electrode 124. The p-type region is provided with a two-dimensional regular array of in-diffused n^+-type regions 126. The latter give rise to pn-junction extracting contacts and are provided with positive bias electrodes. The array is hexagonal with an n^+-type region nearest neighbour spacing $D \ll L$, where L is the minority carrier diffusion length. The parameter D here corresponds approximately to the diameter of the respective photosensitive region extending roughly D/2 from each n^+-type region. This arrangement makes it possible to extend the minority carrier extraction effect without the limitation of diffusion length. Accordingly, photosensitive regions from which minority carriers are excluded may be extended indefinitely. The photocurrent from each n^+-type region may be detected separately to provide individual pixels for a two dimensional array.

*

A method for processing a backside illuminated detector assembly, comprising the use of a temporary substrate on which a detector wafer is attached, is presented in EP-A-0171801 (Honeywell Inc., USA, 19.02.86).

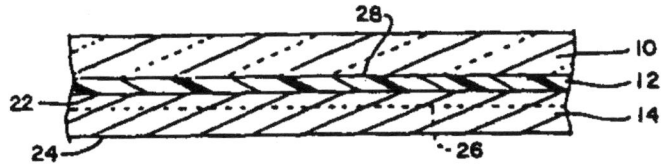

Fig. 2.2.41 (EP-A-0171801 fig. 1b)

A detector wafer 14 of HgCdTe is attached to a transmissive temporary substrate 10 of glass, quartz, alumina or sapphire by the use of a removable adhesive 12. The adhesive should not be soluble in the solvents used in the detector delineation and fabrication steps that will follow. Glycol phthalate is mentioned as an usable adhesive.

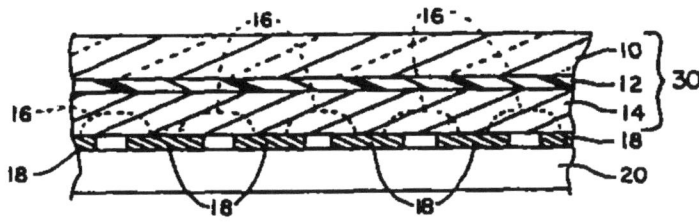

Fig. 2.2.42 (EP-A-0171801 fig. 1d)

The assembly is polished to a final thickness and individual detector elements 16 are formed.

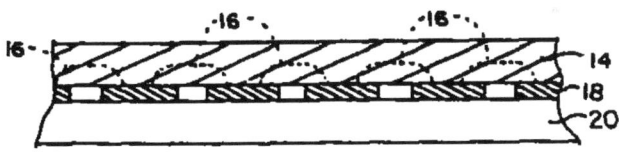

Fig. 2.2.43 (EP-A-0171801 fig. 1e)

With substrate 10 attached to detector elements 16, the complete detector assembly 30 is attached to a CCD 20 or another read-out device using bump contacts 18. Substrate 10 and adhesive 12 are then removed using an appropriate solvent such as acetone.

*

An imager having HgCdTe detectors made of different compositions to be responsive to two different infrared frequency windows, and having the detectors formed in the same focal plane is presented in US-A-4620209 (Texas Instruments Incorporated, USA, 28.10.86).

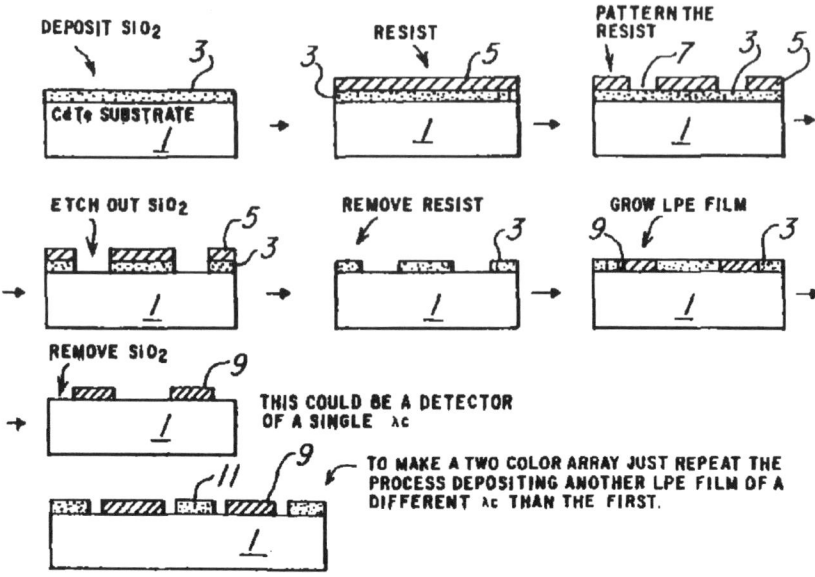

Fig. 2.2.44 (US-A-4620209 the figure)

A silicon dioxide layer 3 is formed on an insulating CdTe substrate 1. A photo-resist coating 5 is formed over the silicon dioxide layer. The photo-resist layer is patterned and the silicon layer is partly etched away. The photo-resist layer is removed and a film of HgCdTe 9 of a first mercury to cadmium ratio is deposited by liquid phase epitaxial deposition over the entire surface of the substrate. The HgCdTe film is only formed at regions where the CdTe substrate is exposed and does not adhere to the silicon dioxide. Next, the silicon dioxide layer is removed. In order to increase the window of frequency response of the detectors, the process is repeated using a second mercury to cadmium ratio different from the first ratio.

*

A shield comprising apertures corresponding to detector elements are attached to an HgCdTe detector layer in JP-A-61147118 (Fujitsu Ltd, Japan, 04.07.86) to determine the field of view of the detector elements.

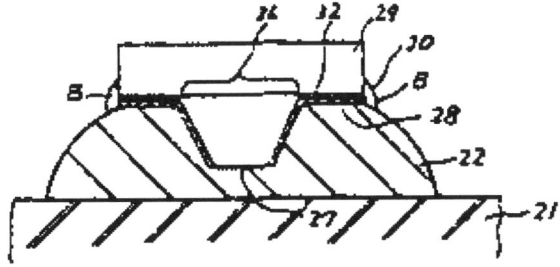

Fig. 2.2.45 (JP-A-61147118 fig. 5)

An HgCdTe substrate 22 is adhered to an insulating body 21 of sapphire. The HgCdTe substrate is thinned and a groove is formed by etching. Each detector element 27 is electrically connected by metal layers 28. A shield comprising apertures 26 positioned above the detector elements is adhered to the HgCdTe layer by an adhesive 30.

*

An imager having high resolution is presented in JP-A-61214462 (NEC Corp., Japan, 24.09.86).

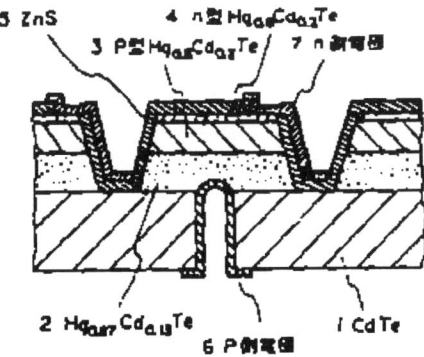

Fig. 2.2.46 (JP-A-61214462 fig. 1)

An $Hg_{0.87}Cd_{0.13}Te$ layer 2 is grown epitaxially on a CdTe substrate 1. A p-type region 3 and an n-type region 4 are comprised in an $Hg_{0.8}Cd_{0.2}Te$ layer which is grown on layer 2. A hole is formed through the CdTe substrate and the p-type region is connected, through the hole, by electrode 6. The n-type region is connected by electrode 7.

Insulation between detector elements is automatically achieved in JP-A-61222161 (Fujitsu Ltd, Japan, 02.10.86) by forming detector elements of epitaxially grown HgCdTe in recessed portions of a sapphire substrate.

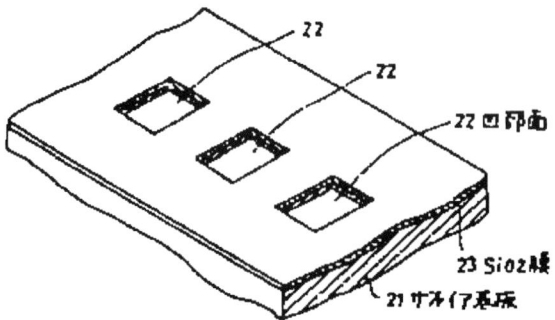

Fig. 2.2.47 (JP-A-61222161 fig. 2)

A sapphire substrate 21 which has recessed portions 22 is covered by a SiO_2 film 23 having windows corresponding to the recessed portions.

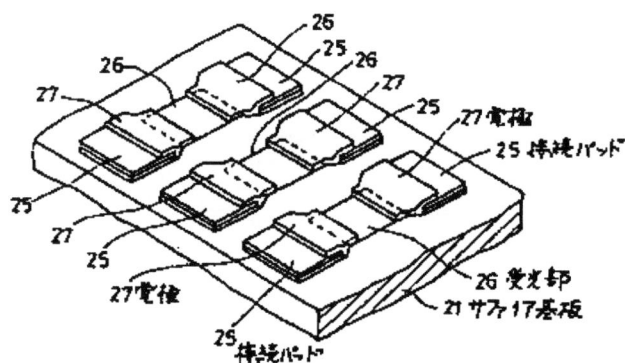

Fig. 2.2.48 (JP-A-61222161 fig. 4)

An HgCdTe layer is epitaxially grown over the surface and individual HgCdTe detector elements 26 are formed in the recessed portions when the SiO_2 layer has been removed. The surface of the HgCdTe detector elements are then polished and etched before electrodes 27 are connected to connection pads 25.

*

A comb-shaped imager having double layer metal electrodes is shown in JP-A-61251167 (Fujitsu Ltd, Japan, 08.11.86).

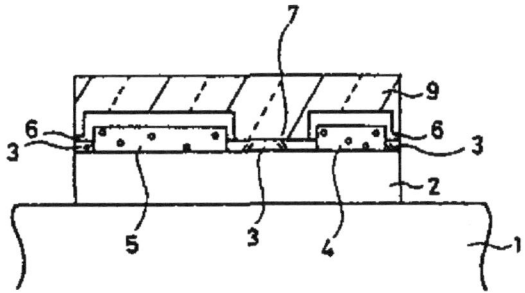

Fig. 2.2.49 (JP-A-61251167 fig. 6)

An HgCdTe layer 2 is formed on an insulating substrate 1. The surface of the HgCdTe layer is oxidized and a light-transmissive insulating film 3 having windows corresponding to electrode regions is formed thereon. A common electrode 5 of the detector elements and individual electrodes 4 are formed from a first metal layer of indium. A second metal layer 6, which overlays the electrodes 4 and 5 but is removed from detector regions 7, is formed. A photo-resist film 9 is used as a mask and element separation is performed by etching. An anodic oxide film is formed on the side surfaces of the comb-shaped HgCdTe layer 2.

*

N-type regions corresponding to photodiodes, are formed in JP-A-62011265 (Toshiba Corp., Japan, 20.01.87) by diffusing indium into a p-type HgCdTe layer. The indium is diffused from straps which also connect the formed n-type regions to a read-out device comprised in a silicon substrate. No alignment beetween the detector layer and the read-out device is needed which would be the case if the n-type regions were formed before the bonding of the detector array and the read-out device takes place.

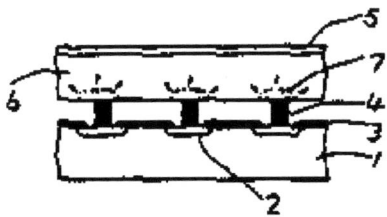

Fig. 2.2.50 (JP-A-62011265 fig. 1d)

Indium straps 4 are formed by evaporation on n-type regions formed in a p-type silicon substrate 1. A p-type HgCdTe layer 6 is formed on a CdTe substrate 5 and the layer 6 is placed to make contact with the straps 4. When the assembly is heated to 150°C in an N_2 atmosphere, indium is diffused into the p-type HgCdTe layer, thereby forming n-type regions 7. A similar method of manufacturing an imager is disclosed in JP-A-62013085 (Fujitsu Ltd, Japan, 21.01.87).

*

The conversion efficiency of individual photodiode elements in an imaging array is a function of the area of the pn-junction. As the surface area of an n-type material in contact with a p-type material increases so does the conversion efficiency of infrared radiation to electrical current. However, by increasing the area of the n-type region, the amount of thermal leakage-current, or dark current, generated by the diode also increases, with a consequent reduction in the diode signal-to-noise ratio. Furthermore, the n-type regions within the diode array are typically formed by ion implantation techniques. This implantation process causes damage to the crystalline lattice thereby increasing the thermally induced diode leakage-current.

These problems are addressed in the invention of US-A-4751560 (Santa Barbara Research Center, USA, 14.06.88). A field plate and a guard plate are formed which, when correctly biased, generate an inversion layer at a semiconductor surface surrounding each photodiode.

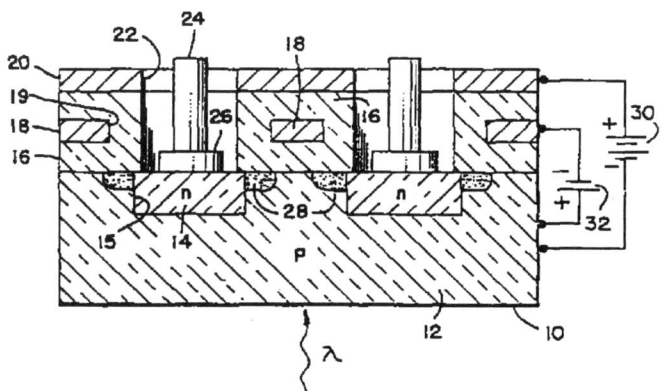

Fig. 2.2.51 (US-A-4751560 fig. 2)

N-type well regions 14 are formed within an upper surface of a p-type substrate 12 of HgCdTe creating pn-junctions which define diodes 15. An insulating layer 16 of silicon dioxide or ZnS deposited by a chemical vapor deposition process overlies the upper surface except for the regions 14. A metal layer 18 of titanium having openings 19 forms a guard plate. The openings 19 are of a larger area than the underlying regions 14. The layer 16 of insulating dielectric is further built up by a chemical vapor deposition process. A second metal layer 20 of titanium is formed on the dielectric layer. This layer has openings 22 and serves as a field plate. The areas of the openings 22 are slightly smaller than the areas of the underlying regions 14. The inequality of opening areas between the field plate and the guard plate permits the development

of an inversion layer 28 in the surface of the substrate 12 laterally adjacent to the regions 14, which surface region underlies the field plate but not the guard plate. Diode contacts 24 of indium connect the photodiodes to a second semiconductor device comprising a multiplexer.
The inversion layer operates electrically to fill up the surface energy states of the substrate 12 with electrons. This results in a reduction of the leakage, or thermal current, thereby increasing the signal-to-noise ratio. Furthermore, an n-type layer is created within the surface of substrate 12. This n-type layer effectively enlarges the area of the n-type regions 14, resulting in a larger pn-diode, hence an increased conversion efficiency of radiation to electrical current. A reduction in the amount of dark current results because the fractional area of the n-type diode junction produced by the ion implant is reduced.

*

In US-A-4868622 (Kabushiki Kaisha Toshiba, Japan, 19.09.89) isolation regions between detector elements are formed by irradiating the regions with a laser beam.

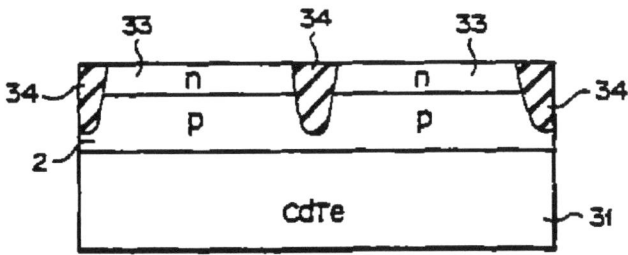

Fig. 2.2.52 (US-A-4868622 fig. 3b)

A p-type layer 32 and an n-type layer 33 each consisting of a repeating multi-layer structure of HgTe layers and CdTe layers are formed by molecular beam epitaxy on a CdTe substrate 31. The resultant wafer is fixed on an X-Y stage capable of positioning the wafer with an accuracy of 1 µm for selectively irradiating the wafer with a Nd:YAG laser beam applied with a Q switch so as to form lattice-shaped isolation regions 34. The laser beam, which has a wavelength of 1.06 µm, a pulse time of 20 ns and an output of 0.1 J/cm^2 for a single radiation, is narrowed to a diameter of 2 µm and scanned on the wafer surface for alloying the irradiated portions. The alloying permits the isolation regions to have a large bandgap. The isolation regions have a high resistance because the nominally n-type layer is undoped.

*

A SiO$_x$N$_y$ reflection preventive film is introduced on the imager presented in JP-A-63150976 (Fujitsu Ltd, Japan, 23.06.88). This film is formed on a first side of a CdTe substrate with an HgCdTe detector layer formed on an opposite side.

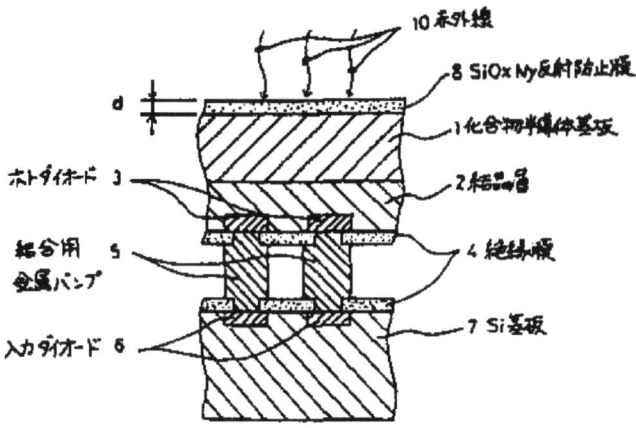

Fig. 2.2.53 (JP-A-63150976 fig. 1)

An HgCdTe layer 2, comprising photodiodes 3, is formed on a first side of a CdTe substrate 11. The detector is bonded to a silicon chip 7 by a flip-chip process. A reflection preventive film is formed on a second side, opposite to the first side, of the CdTe substrate. The film is formed by a cyclotron resonance plasma CVD method by introducing nitrogen, nitrous oxide and silane as reaction gas. The thickness of the film is selected so that the reflectivity is minimized for radiation having a wavelength to be detected by the photodiodes.

*

A process to manufacture an imager is disclosed in JP-A-63260171 (Fujitsu Ltd, Japan, 27.10.88). Photodiodes, formed in HgCdTe chips, are tested before they are mounted on a temporary substrate. The chips are further treated before they are bonded to a CCD with a flip-chip bonding process.

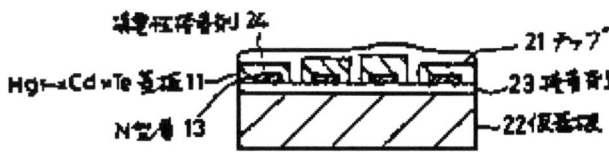

Fig. 2.2.54 (JP-A-63260171 fig. 1)

HgCdTe chips 21, comprising inspected photodiodes, are mounted on a temporary substrate 22 and the chips are electrically connected by the use of a conductive adhesive 24.

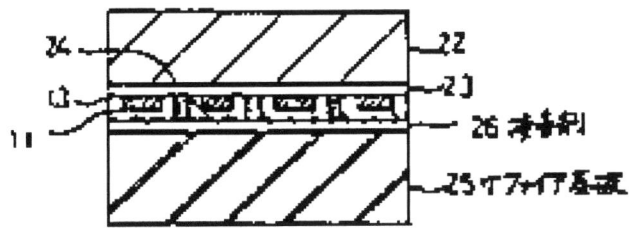

Fig. 2.2.55 (JP-A-63260171 fig. 3)

The chips 21 are treated to have a common thickness and the chips are mounted on the treated side to a transparent substrate 25. Next, the temporary substrate 22 is removed.

Fig. 2.2.56 (JP-A-63260171 fig. 4)

Finally, the exposed photodiodes are connected to a CCD by the use of connection bumps in a flip-chip process.

*

An imager having a close spacing between adjacent detector elements is disclosed in JP-A-1050557 (Fujitsu Ltd, Japan, 27.02.89).

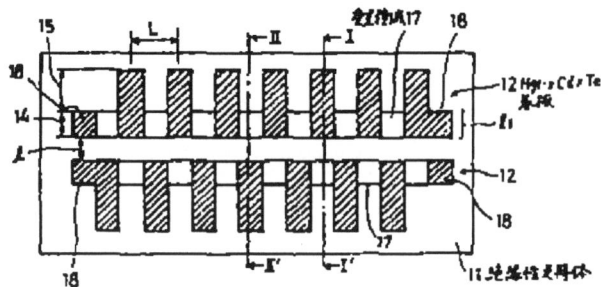

Fig. 2.2.57 (JP-A-1050557 fig. 1)

An HgCdTe layer is attached to an insulating substrate 11 by a bonding agent. The layer has projecting regions 15, which are covered with read-out electrodes 16 and sensitive regions 17. The sensitive regions form a row of detetctor elements. Common electrodes 14 and 18 are provided at opposite ends of the row of detector elements.

*

The method of fabrication of an imager presented in WO-A-8910007 (Santa Barbara Research Center, USA, 19.10.89) provides for virtually all of the critical diode fabrication processes to be performed with one photomask layer in a single processing chamber. The problem of alignment of successively applied photomask layers is therefore eliminated. Furthermore, the use of one photomask layer provides for the fabrication of pn-junctions having small areas. The small pn-junction area gives a small junction capacitance and a small leakage-current. Another advantage is that repeated surface cleaning procedures are not required.

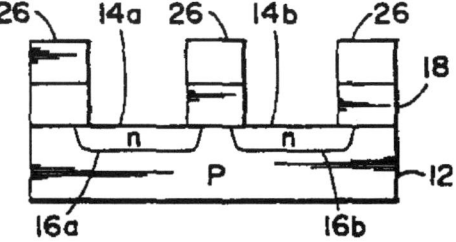

Fig. 2.2.58 (WO-A-8910007 fig. 3d)

A passivation layer 18 is disposed upon an HgCdTe layer 12. The passivation layer comprises either CdTe, HgCdTe, HgCdZnTe or $Cd_{0.96}Zn_{0.04}Te$ and has a wider bandgap than the bandgap of layer 12. A mask layer 26 having openings therethrough is deposited. Each opening defines an individual photodiode. Portions of the passivation layer are removed through the openings by ion-milling or a sputter etching technique. The lattice damage caused by the etching step also converts the p-type HgCdTe in the vicinity of the mask openings to n-type regions 14a and 14b, thereby forming pn-junctions 16a and 16b. The n-type regions extend laterally out from the opening such that edges of the pn-junctions are disposed beneath the surrounding passivation layer. Alternatively, the n-type regions may be formed by ion-implantation.

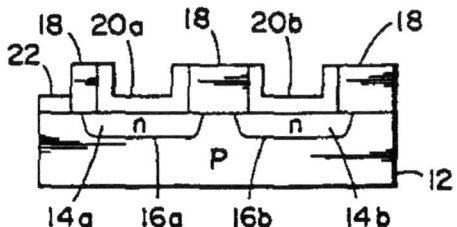

Fig. 2.2.59 (WO-A-8910007 fig. 3g)

A layer of contact metal is deposited over the surfaces of the mask layer 26 and the n-type regions 14. Thereafter, the wafer is removed from the vacuum chamber and the mask layer 26 is removed thereby removing the overlying layer of metal except where it contacts the individual n-type regions 20a and 20b. A ground contact 22 is then deposited upon the substrate 12.

*

In many applications, photodiodes operate at zero polarization. They must therefore have high dynamic resistance to ensure a high detectivity. This is achieved in US-A-4972244 (Commissariat a l'Energie Atomique, France, 20.11.90) by separating detector elements by trenches. Furthermore, a common electrode is provided at the bottom of the trenches thereby reducing the interconnection resistance of the photodiodes.

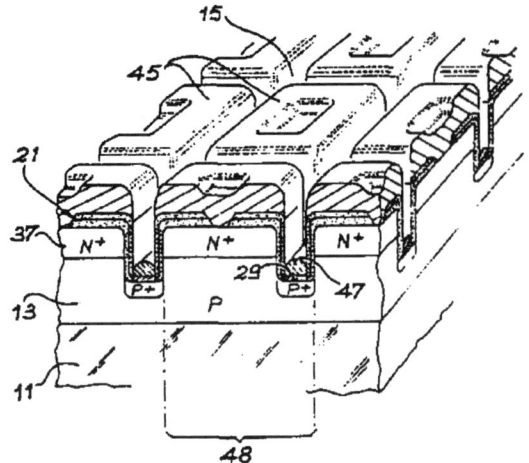

Fig. 2.2.60 (US-A-4972244 fig. 10)

A p-type HgCdTe layer 13 is formed on a CdTe insulating substrate 11. Trenches 15 are formed by an anisotropic planar etching process or by an ion-etch. P-type ions are implanted in the bottom of the trenches to form a p-type layer and n-type ions are implanted to form n-type regions 37. Metal is deposited to form a contact 47, which connects to the p-type layer, and electrical contacts 45. Contacts 45 and 47 are separated by an insulating layer 21.

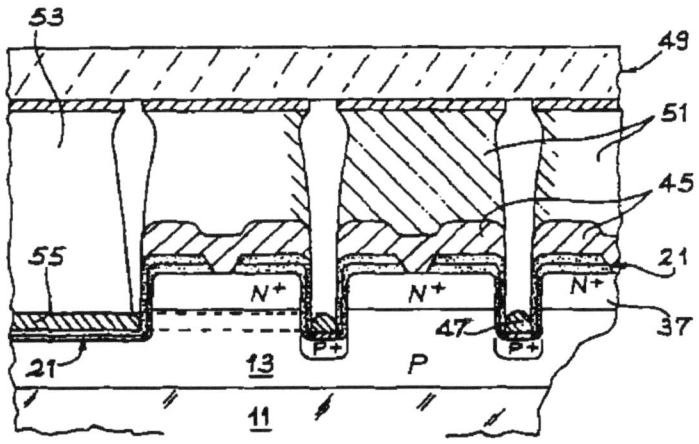

Fig. 2.2.61 (US-A-4972244 fig. 11)

The photodiodes are connected to an integrated reading and addressing circuit 49 by indium pads 51. The common electrode 47 is connected to the integrated circuit via contact 55 and indium pad 53.

*

In JP-A-2009180 (Mitsubishi Electric Corp., Japan, 12.01.90) two layers of HgCdTe having a different energy bandgap are used to form an imager which performs wavelength separation. The imager is utilized in an image tracking apparatus.

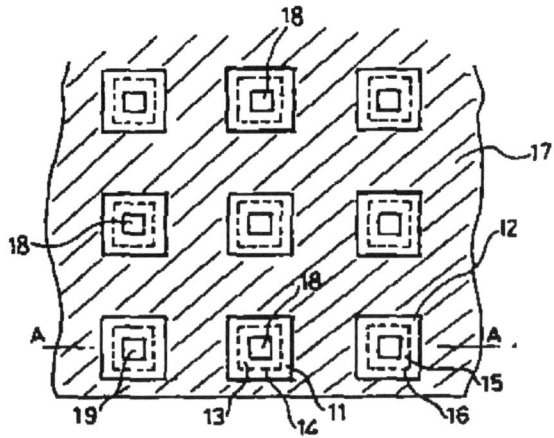

Fig. 2.2.62 (JP-A-2009180 fig. 1a)

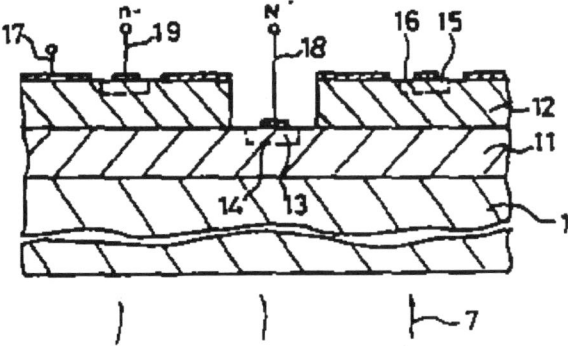

Fig. 2.2.63 (JP-A-2009180 fig. 1b)

The imager comprises a semi-insulating substrate 1 of CdTe, a p-type $Hg_{0.7}Cd_{0.3}Te$ layer 11, a p-type $Hg_{0.8}Cd_{0.2}Te$ layer 12, n-type regions 13 and 15, a common electrode 17 and individual electrodes 18 and 19. Infrared rays of 3-5 microns wavelength band is absorbed by the p-type layer 11 of wide energy bandgap, and that of 10 microns wavelength band transmits the first p-type layer 11 and is absorbed by the p-type layer 12 of narrower energy bandgap. Energy band diagrams of the structure are shown below.

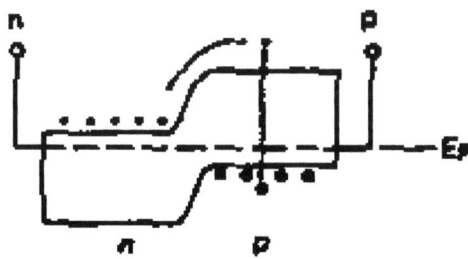

Fig. 2.2.64 (JP-A-2009180 fig. 2a)

The energy band diagram in the range of n-type region 15 and the p-type layer 12 is shown above. Carriers excited by the infrared ray of 10 microns wavelength band below the electrode 19 are detected at a region between the electrode 10 and the common electrode 17.

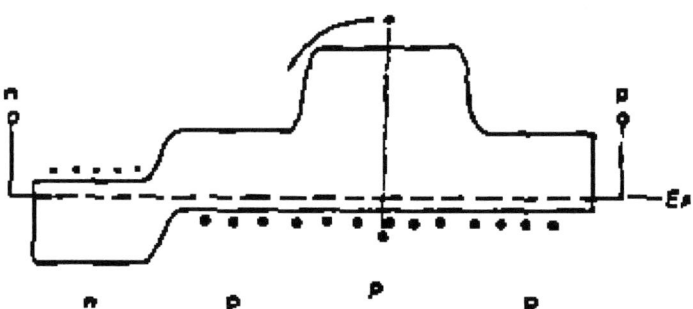

Fig. 2.2.65 (JP-A-2009180 fig. 2b)

The energy band diagram in the range of n-type region 15, p-type layer 12, p-type layer 11 and p-type layer 12 is shown above. Carriers excited by the infrared ray of 3-5 microns wavelength band absorbed below the electrode 19 are detected at between the electrode 19 and the common electrode 17.

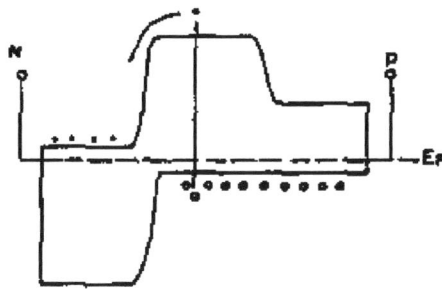

Fig. 2.2.66 (JP-A-2009180 fig. 2c)

The energy band diagram in the n-type region 13, p-type layer 11 and p-type layer 12 is shown above. Carriers are excited by the infrared radiation in the 3-5 micron wavelength band below the electrode 18, and detected in a region between the electrode 18 and the common electrode 17. The infrared ray of 10 micron wavelength below the electrode 18 is transmitted by the p-type layer 11 and is not detected.

From the above, below the electrode 19, both infrared radiation in the 10 micron and 3-5 micron wavelength bands are detected and below the electrode 18 only those in the 3-5 micron wavelength band are detected.

* *

In GB-A-2238165 (Mitsubishi Denki Kabushiki Kaisha, Japan, 22.05.91) wavelength separation is carried out by two sets of detectors having different optical path lengths.

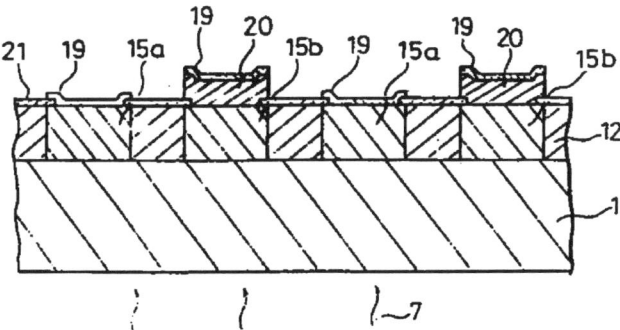

Fig. 2.2.67 (GB-A-2238165 fig. 1b)

An epitaxially grown p-type $Hg_{0.8}Cd_{0.2}Te$ layer 12 is produced on a CdTe semi-insulating substrate 1 and the p-type layer is doped by n-type impurities to produce n-type regions 15a and 15b. A protective film 21 is formed over the p-type layer but not over the n-type regions.

A light absorbing layer 20 of HgTe is formed on every second n-type region 15b and reflective metal electrodes 19 are formed on the n-type regions and the light absorbing layer. In the regions 15a incident light is reflected by the electrodes 19. Therefore, the optical path length which can absorb the light becomes twice the thickness of the n-type regions 15a. In the regions 15b incident light is absorbed in the light absorbing layer and the optical path length which can absorb the light is equal to the thickness of the regions 15b. Due to the difference in the optical path length, the characteristics of the absorption coefficient of incident light varies with the wavelength. In a second embodiment, the thickness of region 15b is reduced to increase the difference in optical path lengths.

*

A multicolour focal plane array is presented in US-A-4956555 (Rockwell International Corporation, USA, 11.09.90). A first and a second group of filters each includes a dielectric/thin metal/dielectric layer combination. The filters of the two groups differ in that the thickness of the dielectric layers are adjusted to selected wavelength bands.

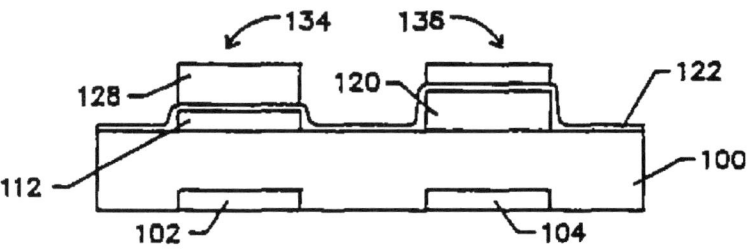

Fig. 2.2.68 (US-A-4956555 fig. 1g)

Two groups of detectors 102 and 104 are formed in a substrate 100. A first dielectric/metal/dielectric cavity layer combination 134 is formed over a first group of detectors and a second dielectric/metal/dielectric cavity layer combination 136 is formed over a second group of detectors. The absorbing metal layer 122 normally blocks all wavelengths of light. The dielectric layers, however, arranged on both sides of the metal layer produce an electric field standing wave around the metal layer and place the metal layer in a null of the electric field at one particular wavelength, allowing light to be transmitted at that wavelength. By changing the thickness of the two layers adjacent to the metal layer, a different wavelength may be selected for transmission to each of the groups of detectors within the focal plane.

*

The detector array of US-A-5293036 (Santa Barbara Research Center, USA, 08.03.94) achieves a center-surround type output from individual detectors without post-processing the detector signals externally to the array. Each detector of the detector array comprises for this purpose a central radiation absorbing area which is surrounded by a peripheral radiation absorbing area. The center-surround type output signals from the detectors are readily coupled to a neural network signal processor.

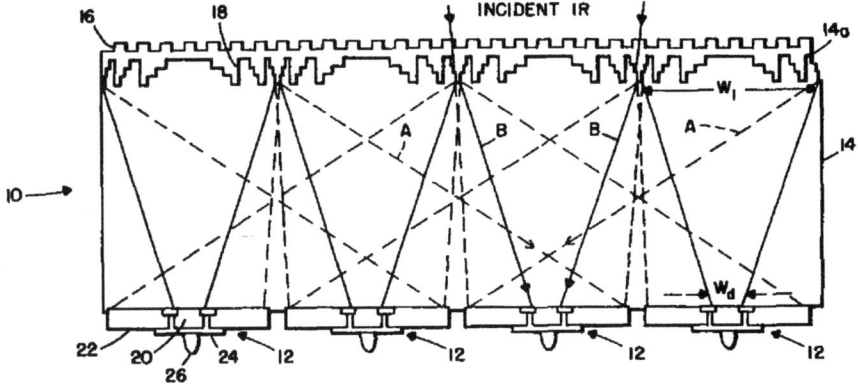

Fig. 2.2.69 (US-A-5293036 fig. 1)

The backside-illuminated detector array 10 comprises a transparent substrate 14 and center-surround detectors 12. Each of the detectors is comprised of a central detector 20 that is surrounded by a peripheral detector 22 which are coupled together by a conductor 24. Furthermore, there is provided on a radiation receiving surface such as 14a a transmission grating surface such as a diffraction grating 16 that is formed over an integral flux concentrating structure. The flux concentrating structure may be a binary optical lens element 18 that is etched directly into the surface of the substrate, a discrete lenslet formed within the surface of the substrate or a structure provided external to the array. The shape and structure of the lens 18, the diffraction grating 16 features, the substrate thickness and the detector 12 width and shape are adjusted to provide a minimal detector output signal when the illumination of the array is uniform. However, a change in the background illumination or the presence of a spot source within the field of view of the associated lens element 18 results in the generation of a detector output signal.

*

In US-A-5113076 (Santa Barbara Research Center, USA, 12.05.92) a two-terminal radiation detector which is sensitive to two spectral bands of infrared radiation is presented. Two heterojunctions are coupled in series and function electrically as two back-to-back diodes.

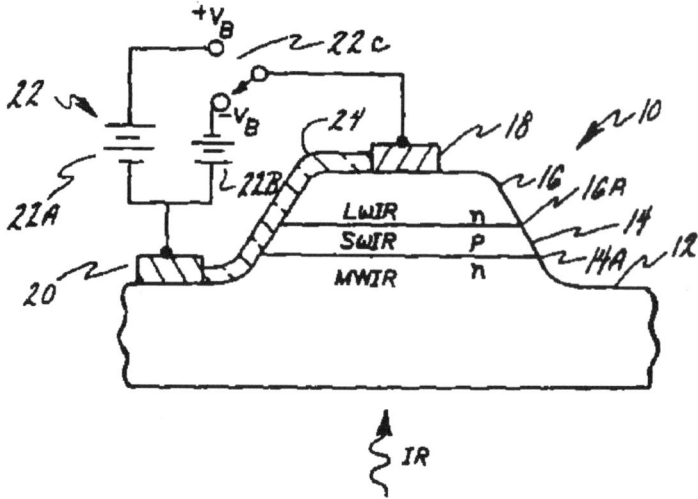

Fig. 2.2.70 (US-A-5113076 fig. 1)

Detector 10 includes an n-type $Hg_{0.7}Cd_{0.3}Te$ base layer 12 having an energy band gap responsive to mid-wavelength infrared, MWIR, radiation. Above the base layer is a heavily doped p-type $Hg_{0.6}Cd_{0.4}Te$ short-wavelength infrared, SWIR, responsive layer 14. The SWIR layer forms a heterojunction 14a with the base layer, but does not contribute significant numbers of SWIR photon-generated carriers since most SWIR radiation does not penetrate through the base layer. Above the SWIR layer is provided an n-type $Hg_{0.8}Cd_{0.2}Te$ long-wavelength infrared, LWIR, responsive layer 16. The LWIR layer is provided with a thickness great enough to absorb the LWIR radiation that has penetrated the two underlying layers. A heterojunction 16a is formed between layers 14 and 16. The detector is provided with an electrical contact in the form of a nickel pad having an indium bump 18 formed thereon and a base contact 20.

The radiation detector may be formed having a planar structure by using a diffusion or an implant/anneal process to form the heterojunctions coupled in series between two electrical teminals.

* *

A method of manufacturing a detector that is sensitive to two spectral bands of infrared radiation is shown in US-A-5149956 (Santa Barbara Research Center, USA, 22.09.92).

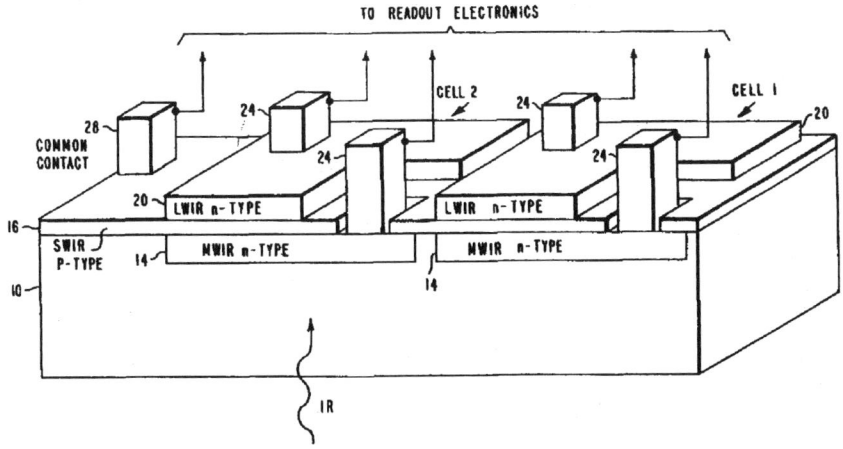

Fig. 2.2.71 (US-A-5149956 fig. 4)

An electrically insulating substrate 10 of CdZnTe is provided with wells which are etched into the substrate. Each well is formed where it is desired to fabricate an individual photodetector. The wells are filled with n-type $Hg_{0.7}Cd_{0.3}Te$ mid-wavelength infrared, MWIR, material selected for absorbing radiation of a first wavelength band. A uniform epitaxial layer 16 of p-type $Hg_{0.6}Cd_{0.4}Te$ short-wavelength infrared, SWIR, material is subsequently formed. Openings are formed in the SWIR layer in order to provide for an electrical contact to the underlying MWIR material. Next, n-type $Hg_{0.8}Cd_{0.2}Te$ long-wavelength infrared, LWIR, regions 20 are formed in registration with the underlying MWIR material. The resulting structure forms, for each photodetector element, two back-to-back photodiodes. The SWIR layer provides a common electrical contact to each of the photodetectors of the array. Electrical contact pads 22 are deposited and indium bumps 24 are formed in order to make electrical contact to external read-out electronics. An electrically insulating layer 26 of CdTe is formed over the surface of the device to reduce surface recombination effects and other surface effect noise sources. The electrical conductivity of the common SWIR layer may be enhanced by including a vertical and/or horizontal contact metal grid structure in the regions between the LWIR regions. A binary or analog lens array may also be employed on or adjacent to the illuminated surface of the substrate to improve the fill factor of the detector array.

* *

A hybrid detector responsive to visible wavelength radiation and also to at least two different wavelengths of infrared radiation is presented in US-A-5373182 (Santa Barbara Research Center, USA, 13.12.94).

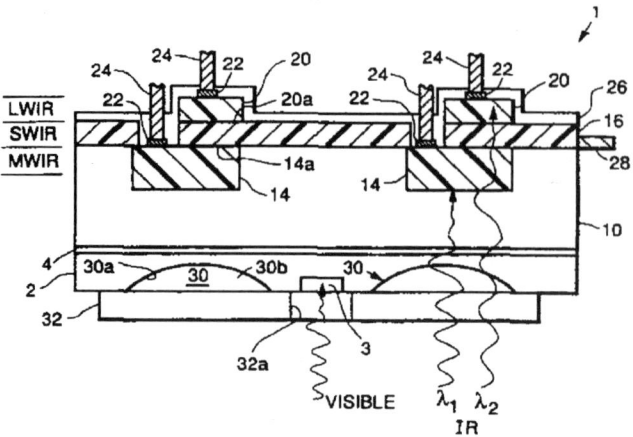

Fig. 2.2.72 (US-A-5373182 fig. 1g)

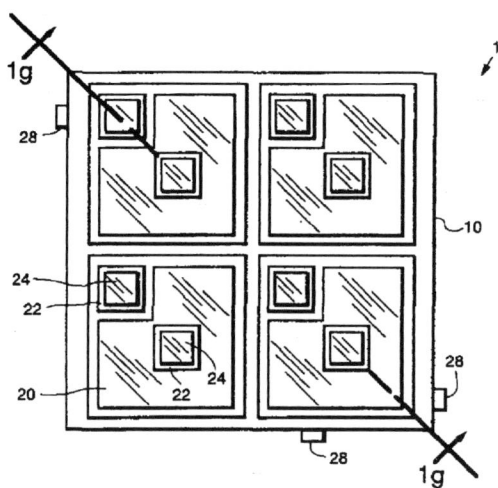

Fig. 2.2.73 (US-A-5373182 fig. 2)

A silicon substrate 2 is provided with an optional lattice-mismatch accommodation layer 4 of GaAs. Photodiodes forming visible radiation detector sites 3 are made together with read-out circuitry in the silicon substrate. An electrically insulating substrate 10 of CdZnTe is grown over the GaAs layer. A photolithographic technique is employed to define wells which are filled with a mid-wavelength infrared, MWIR, detector material 14 such as $Hg_{0.7}Cd_{0.3}Te$. The MWIR material is doped n-type. An epitaxial layer 16 comprising short-wavelength infrared, SWIR, semiconductor material such as $Hg_{0.6}Cd_{0.4}Te$ is then formed. The SWIR material is doped p-type. Openings are formed in layer 16 in order to provide for a subsequently formed electrical contact to the underlying MWIR detector material. Next, long-wavelength infrared,

LWIR, regions 20 are formed upon the SWIR layer in registration with the underlying MWIR material. The LWIR material may be formed of $Hg_{0.8}Cd_{0.2}Te$ doped n-type. The resulting structure forms two back-to-back photodiodes for each photodetector element. Indium bumps 24 are formed upon contact pads 22 and an electrically insulating layer of passivation 26 is provided over the surface of the device. An electrical connection 28 is provided at an edge of the device for electrically coupling to the common SWIR layer. Microlens elements 30 may be fabricated within the silicon substrate.

*

The invention of GB-A-2246662 (Mitsubishi Denki Kabushiki Kaisha, Japan, 05.02.92) refers to testing of an imager by the use of a test element group. Elements of the test element group are connected by indium bumps to connection pads on a read-out substrate in the same flip-chip bonding process which connects detector elements to the read-out substrate.

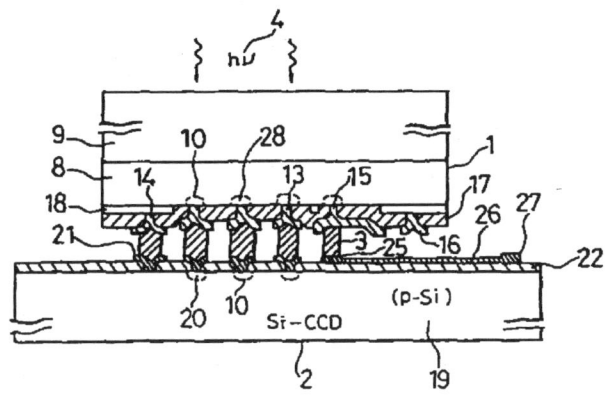

Fig. 2.2.74 (GB-A-2246662 fig. 2)

Photodiodes 10 and test group elements comprise n-type regions 28 formed in a p-type HgCdTe layer 8 disposed on a CdTe substrate 9. The photodiodes are connected to electrode pads 13 and the test group elements are connected to electrode pads 15. The test group elements are connected to connecting pads 25 via indium electrodes 3 at the same time as the photodiodes are connected to input regions 20 of a CCD 2 formed in a silicon substrate 19. The connecting pad 25 is connected by a wire 26 to a bonding pad 27. The characteristics of the test elements are measured by connecting a test probe to the bonding pad 27.

* *

From the measured I-V characteristics of the test element group, presented in GB-A-2246662 above, only the presence or absence of the pn-junction in the light responsive region is detected. Even if a photo-detector has insufficient spatial resolution caused by the excessive diffusion length of the minority charge carrier in the semiconductor layer in which the pn-junctions are formed, the insufficient space resolution is not detected from the I-V characteristics of the test element group, and this photo-detector is selected as a non-defective.

In GB-A-2274739 (Mitsubishi Denki Kabushiki Kaisha, Japan, 03.08.94) a photo-detector is presented which includes a test element group that detects the spatial resolution of the light responsive region.

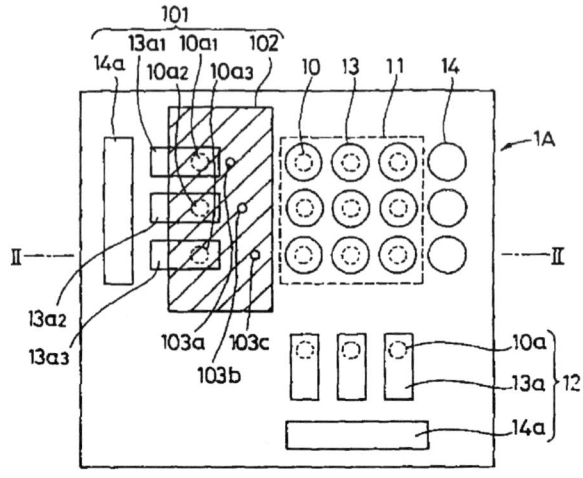

Fig. 2.2.75 (GB-A-2274739 fig. 1)

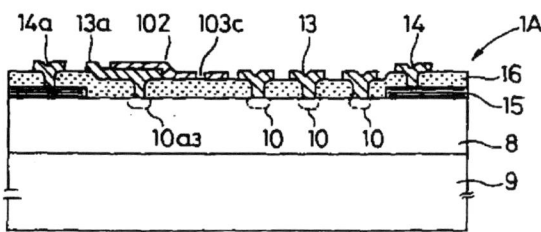

Fig. 2.2.76 (GB-A-2274739 fig. 2)

The photo-detector 1A includes a test element group 101, a part of which is covered with a mask 102 for shielding pn-junctions 10a1-10a3 from the infrared light. Infrared light transmitting windows 103a-103c, which correspond to the pn-junctions 10a1-10a3, respectively, have a diameter of 20 microns and penetrate through portions of the mask. The spacings between the pn-junction 10a1 and the window 103a, between the pn-junction 10a2 and the window 103b, and between the pn-junction 10a3 and the window 103b increase in this order.

Infrared light which is periodically interrupted by a chopper is applied to the entire surface of the detector. Only minority charge carriers, which are produced in responce to the light incident on the semiconductor layer 8 through the windows 103a-103c, reach the pn-junctions 10a1-10a3. Therefore, if signals are detected from the pn-junctions 10a1 and 10a2 when a probe is applied to the corresponding n side electrode pads 13a1 and 13a2 but no signal is detected from the pn-junction 10a3 when a probe is applied to the corresponding n side electrode pad 13a3, it is supposed that the maximum diffusion length of the minority charge carrier is less than the interval between the pn-junction 10a3 and the window 103c. The spatial resolution of the light responsive region 11 is monitored by measuring signals from the electrode pads 13a-13c corresponding to pn-junctions 10a1-10a3, respectively.

Alternative embodiments in which the transmitting window is annular with the opposite pn-junction as the center and in which only one window is provided in common for the pn-junctions 10a1-10a3 are disclosed.

*

The invention of JP-A-4133363 (Mitsubishi Electric Corp., Japan, 07.05.92) refers to an imager comprising photodiodes. The sizes of n-type regions, which are formed in a p-type layer, are made equal by making holes in the p-type layer and forming the n-type regions in the holes.

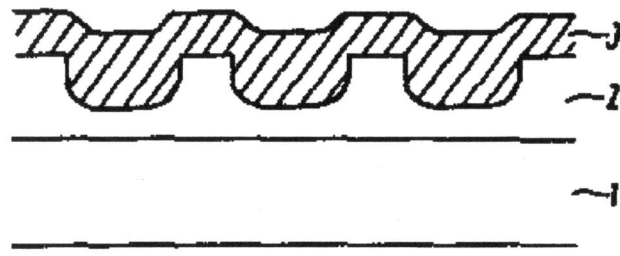

Fig. 2.2.77 (JP-A-4133363 fig. 1b)

A 15 μm thick p-type layer 2 of $Hg_{0.8}Cd_{0.2}Te$ is grown on a $Cd_{0.97}Zn_{0.03}Te$ substrate 1. Holes with a diameter of 30 μm and with a depth of 5 μm are etched in the p-type layer. Next, molecular beam epitaxy is used to form a 10 μm thick n-type layer of $Hg_{0.8}Cd_{0.2}Te$ over the structure.

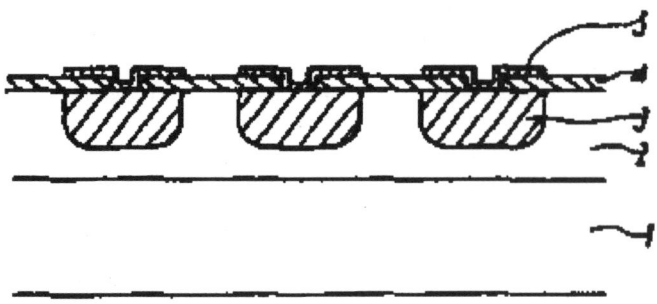

Fig. 2.2.78 (JP-A-4133363 fig. 1d)

Polishing and etching are used to reduce the n-type layer, thereby forming separated n-type islands. The surface is protected by a ZnS film 4 and the n-type regions are connected by electrodes 5.

*

The patterning accuracy of detection elements is improved in JP-A-4253344 (Fujitsu Ltd, Japan, 09.09.92) by forming the detection elements in recessed regions of a substrate.

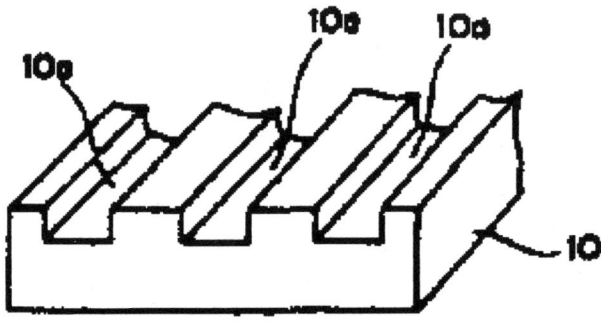

Fig. 2.2.79 (JP-A-4253344 fig. 1a)

Recessed regions 10a are formed in a substrate 10 corresponding to the shape of the detection elements.

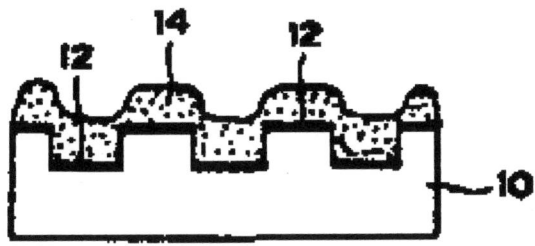

Fig. 2.2.80 (JP-A-4253344 fig. 1c)

A CdTe layer 12 is formed over the structure by the use of a vapor growth method. Thereafter, an HgCdTe layer 14 is formed over the CdTe layer.

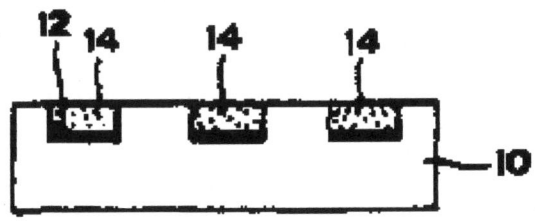

Fig. 2.2.81 (JP-A-4253344 fig. 1d)

Regions of the CdTe layer 12 and the HgCdTe layer 14 not overlaying one of the recessed regions of the substrate are removed by an abrasion method.

*

An imager which can select a mode of either high resolution or high sensitivity is presented in EP-A-0497326 (Fujitsu Limited, Japan, 05.08.92).

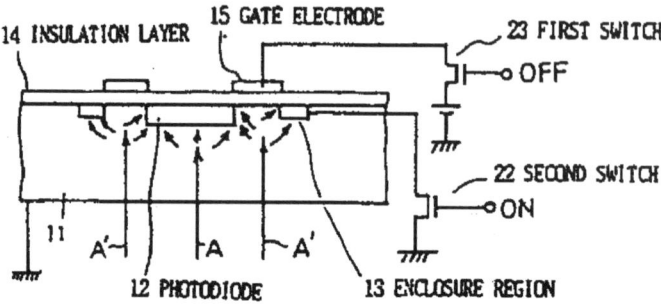

Fig. 2.2.82 (EP-A-0497326 fig. 3a)

Island-shaped surface portions of a p-type substrate 11 are doped so as to form photodiodes 12. Enclosure regions 13 surrounding but apart from the islands are also formed. Gate electrodes 15 are formed upon an insulating layer 14 so as to cover the portions in between the photodiodes and the enclosure reigons.

When a high resolution mode is required, first switches 22 are closed so as to ground the enclosure regions 13 and second switches 23 are opened so as to float gate electrodes 15. Carriers generated in the vicinity of the enclosure regions are attracted and absorbed by the junction potential of the enclosure regions. Thus, no carriers flow into adjacent photodiodes.

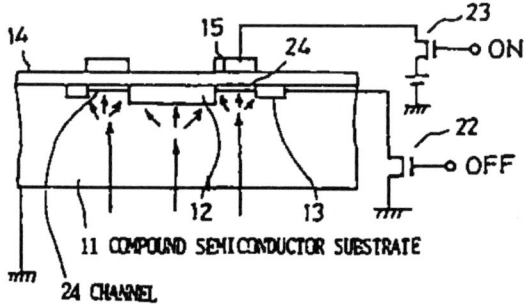

Fig. 2.2.83 (EP-A-0497326 fig. 3b)

When a high sensitivity mode is required, the first switches 22 are opened so as to float the enclosures and the second switches 23 are closed so as to apply a positive voltage to the gate electrodes 15. Due to the positive voltage of the gate electrodes, the surface of substrate 11 beneath the gate electrodes is inverted to n-type so as to form channels 24. These channels electrically connect the enclosure regions 13 to the photodiodes 12. Accordingly, all the carriers generated within the enclosure regions are gathered by the photodiodes.

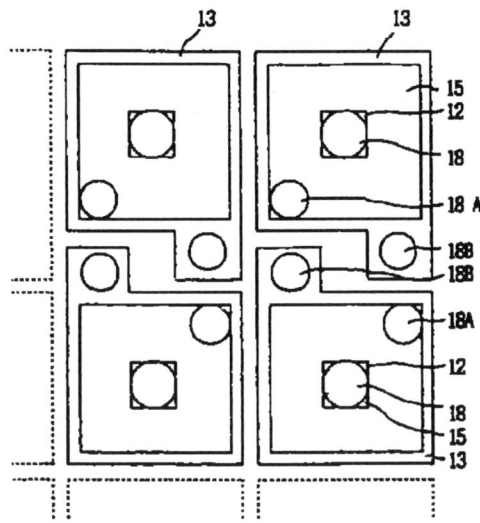

Fig. 2.2.84 (EP-A-0497326 fig. 5)

The switches 22 and 23 are MOSFETs which are integrated in a silicon substrate together with a CCD for reading out the signal charge. The two substrates are bonded together via bumps 18, 18A and 18B, which connect the photodiodes 12, the gate electrodes 15 and the enclosure regions, respectively.

*

An n-type region can be formed in a p-type layer of HgCdTe by thermally diffusing indium from an indium layer which has been deposited on the surface of the p-type layer. A diffusion mask may be used to form separate n-type regions. This technique is further developed in US-A-5198370 (Mitsubishi Denki Kabushiki Kaisha, Japan, 30.03.93) in which regions between the n-type regions are exposed during the diffusion so as to evaporate Hg. The p-type carrier concentration in these regions increases due to the lattice vacancies created. Therefore, even when indium is diffused into these regions they will remain p-type.

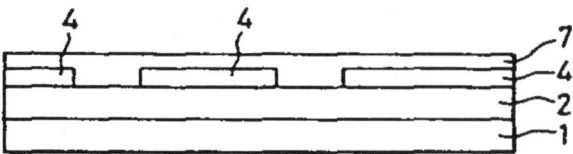

Fig. 2.2.85 (US-A-5198370 fig. 1b)

A p-type HgCdTe layer 2 is grown on a CdTe substrate 1. An insulating film 4 is formed by evaporating ZnS and openings 9 are formed in the insulating film by photolithography and etching. The openings are positioned at approximately the center of corresponding n-type regions 3 which become pixel regions in later process steps. Indium is evaporated and deposited on the entire surface to form an indium layer 7.

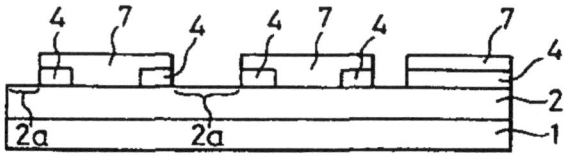

Fig. 2.2.86 (US-A-5198370 fig. 1c)

Wet etching removes the insulating film and the indium layer from regions which become non-pixel regions 2a.

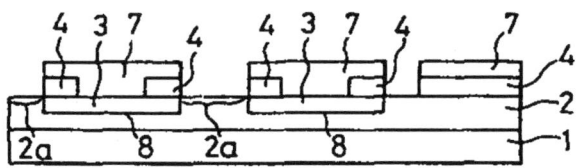

Fig. 2.2.87 (US-A-5198370 fig. 1d)

Next, indium is diffused by annealing from the indium layer 7 into the p-type HgCdTe layer 2 using the remaining insulating film 4 as a diffusion mask thereby forming n-type regions 3 and pn-junctions 8. While the solid-state diffusion advances, Hg evaporates from the regions of the p-type layer which become non-pixel regions 2a and which are exposed on the surface. Thus, lattice vacancies are produced due to the evaporation of Hg and these lattice vacancies function as acceptors so that the p-type conductivity of the layer 2 increases. As a result, even when indium is diffused into the regions in the p-type layer 2 which become non-pixel regions 2a, the doping effect of the indium is cancelled by the high concentration of p-type carriers resulting from the vacancies and these regions remain p-type without being converted to n-type.

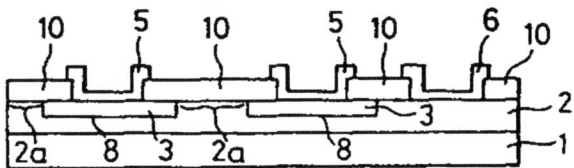

Fig. 2.2.88 (US-A-5198370 fig. 1g)

The indium layer and the insulating film are removed and an insulating film 10 of ZnS, and electrodes 5 and 6 are formed.

*

A photo-detector having photodiodes which are each formed of a main and an auxiliary region, is disclosed in EP-A-0543537 (Fujitsu Ltd, Japan, 26.05.93). The auxiliary regions reduce the blind areas of the photo-detector.

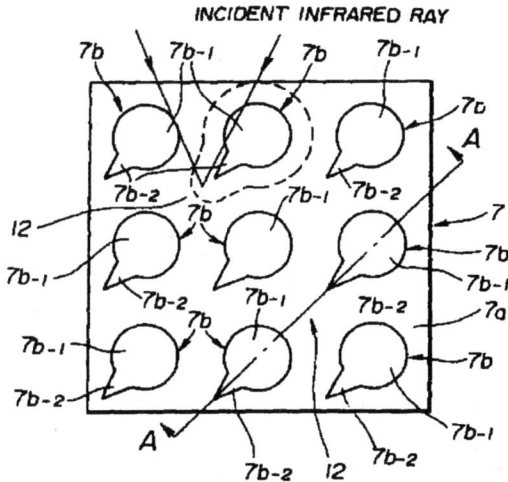

Fig. 2.2.89 (EP-A-0543537 fig. 5a)

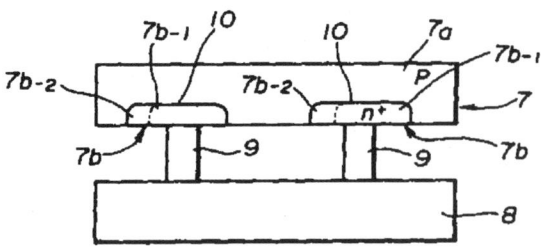

Fig. 2.2.90 (EP-A-0543537 fig. 5b)

A photo-detector part 7 includes a p-type substrate 7a of HgCdTe and n-type regions 7b, forming pn-junctions 10. A signal processing part 8 is formed by a silicon CCD which is connected to the n-type regions via electrodes 9. Each n-type region has a shape including a circular main part 7b-1 and a projecting part 7b-2. The projecting part projects towards a blind area 12 which is surrounded by four mutually adjacent n-type regions 7b. The provision of the projecting part increases the area of the effective photosensitive area compared to the effective photosensitive area which would have been formed by the main part alone. As a result, the blind area is reduced without introducing cross-talk.

When a distant object is picked up by the photo-detector, an error is introduced in the detected position of the object. However, when detecting and tracking the object, a positional error amounting to one pixel or less does not cause a problem. Furthermore, if the object comes closer to the detector, the spot of the infrared ray becomes greater than the size of one pixel, and the position of the object can be accurately detected.

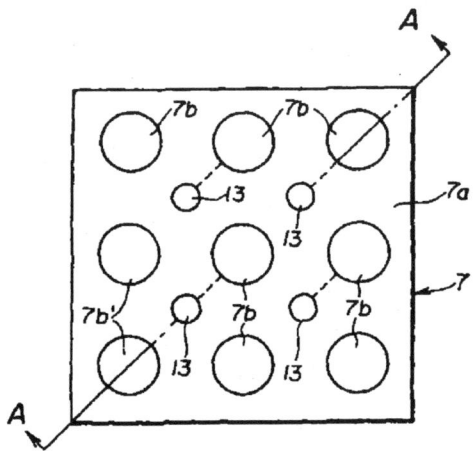

Fig. 2.2.91 (EP-A-0543537 fig. 7a)

In a second embodiment, an auxiliary n-type region 13 is provided independently to an n-type region 7b. Each auxiliary region 13 is connected to a region 7b via a connecting electrode.

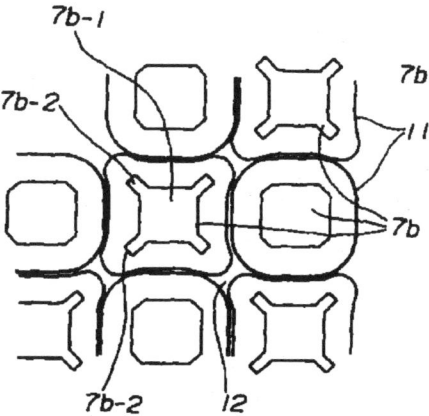

Fig. 2.2.92 (EP-A-0543537 fig. 11)

In a third embodiment, a first and a second group of diagonally arranged n-type regions 7b have mutually different shapes. However, the effective sensitive areas 11 of the detectors of the two groups are the same.

*

A method to diffuse Hg into a p-type HgCdTe substrate and thereby converting regions of the substrate to n-type is disclosed in GB-A-2261767 (Mitsubishi Denki Kabushiki Kaisha, Japan, 26.05.93). A Hg/Cd amalgam is formed on the substrate followed by a protective film. Hg vapor which is generated from the amalgam during an annealing step diffuses into the substrate while the protective film prevents the Hg vapor from diffusing into the atmosphere.

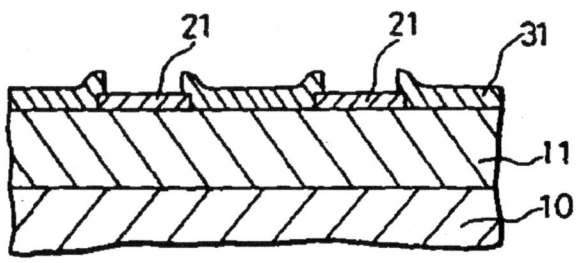

Fig. 2.2.93 (GB-A-2261767 fig. 3b)

A p-type HgCdTe layer is formed on a CdTe substrate 10. Cd patterns 21, corresponding to pixel regions, are formed on the HgCdTe layer. The thickness of the Cd pattern is chosen in accordance with the depth or a concentration of an Hg diffusion layer to be formed. Then a ZnS film is formed into pattern 31.

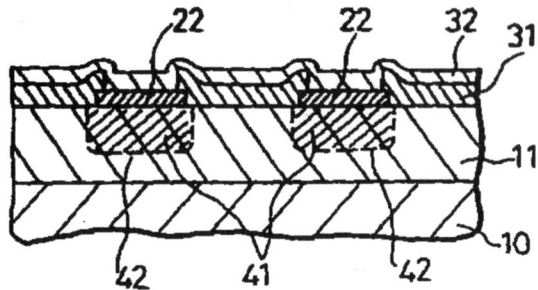

Fig. 2.2.94 (GB-A-2261767 fig. 3e)

The wafer is exposed to Hg vapor or directly brought into contact with Hg to amalgamate the Cd patterns, resulting in Cd amalgam patterns 22. At this time the ZnS film serves as a protective film so that the surface of the HgCdTe layer is not exposed to the Hg treatment. Next, a ZnS film 32 is formed on the whole surface of the wafer. The wafer is annealed at 200-250°C and Hg vapor is generated from the amalgam. The ZnS film 32 serves as a protective film against the Hg vapor and prevents the Hg vapor from diffusing into the atmosphere with the result that the Hg vapor diffuses only into the p-type HgCdTe layer and n-type regions 41 are formed.

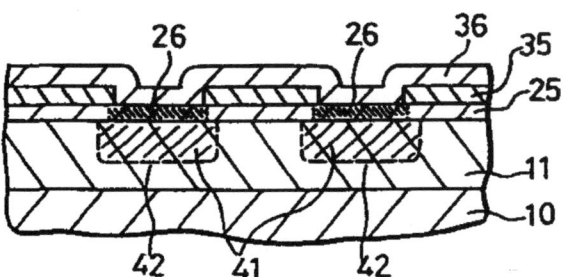

Fig. 2.2.95 (GB-A-2261767 fig. 4e)

In an alternative embodiment, a Cd layer 25 is formed over the HgCdTe layer 11. Next, a ZnS film 35 is deposited and patterned. The wafer is exposed to Hg vapor or directly brought into contact with Hg to amalgamate the Cd layer exposed on the surface, resulting in Cd amalgam regions 26. A ZnS film 36 is formed on the whole surface of the wafer and n-type regions 41 are formed in an annealing step.

*

A two colour infrared detector is described in US-A-5300777 (Texas Instruments Incorporated, USA, 05.04.94) which for each detector element comprises a heterojunction diode and a metal-insulator-semiconductor device.

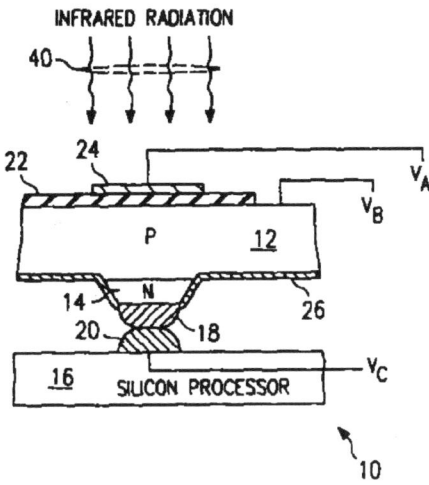

Fig. 2.2.96 (US-A-5300777 fig. 1)

Detector 10 is a heterostructure device comprising a metal insulator semiconductor, MIS, device on top of a heterojunction diode. The heterojunction diode consists of a wide-bandgap p-type HgCdTe layer 12 on top of a narrow-bandgap n-type HgCdTe layer 14. The narrow-bandgap layer is bump bonded to a silicon processor layer 16 through indium bumps 18 and 20. The MIS structure comprises an insulator layer 22 and a metal gate 24. The detector also comprises a passivation layer 26. In array form, all of the MIS devices are bused together to a common voltage source.

In operation, the diode is reverse biased with respect to the p-type layer 12 and the MIS device is pulsed to deep depletion with respect to layer 12. The wide-bandgap part of the spectrum is absorbed in the first few microns of layer 12. Electron-hole pairs are generated and the electrons diffuse to the MIS device. The narrow-bandgap radiation component of the incident radiation travels through the wide-bandgap layer to the narrow-bandgap side of the heterojunction diode where it is absorbed. Electron-hole pairs are generated and the holes diffuse to the heterojunction and out through the p-type base. This may be represented as a change in charge on a capacitor (not shown) that is connected to the n-type side of the heterojunction diode in silicon processor layer 16. After a set integration time period, the charge on the capacitor (charge from layer 14) is sensed first and the capacitor is reset. The diode is again reverse biased with respect to layer 14 and the MIS device is reset. This allows the integrated wide-bandgap photogenerated charge in layer 12 to migrate through the heterojunction diode and on to the capacitor in silicon processor layer 16. Although charge due to the two parts of the spectrum is integrated at the same time, it is sensed sequentially.

*

A method of manufacturing an imager having mesa structures is presented in JP-A-6013642 (Fujitsu Ltd, Japan, 21.01.94). The base layer of the mesas comprises a shallow n^-/n^+ structure.

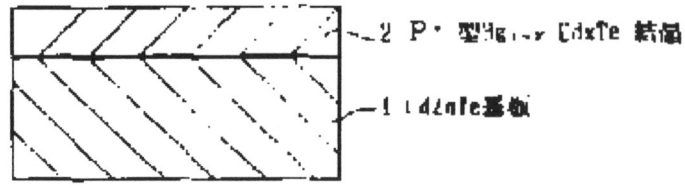

Fig. 2.2.97 (JP-A-6013642 fig. 1a)

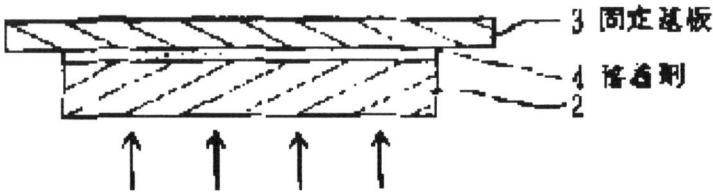

Fig. 2.2.98 (JP-A-6013642 fig. 1b)

A p^+-type HgCdTe layer 2 is first epitaxially grown on a CdZnTe substrate. The structure is bonded to a substrate 3 by an adhesive layer 4.

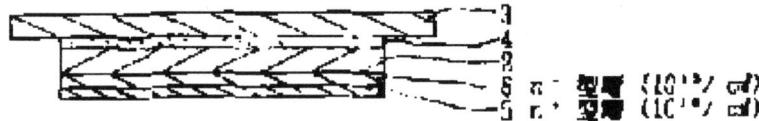

Fig. 2.2.99 (JP-A-6013642 fig. 1c)

Next, the CdZnTe substrate is removed and the p^+-type HgCdTe layer is subjected to a plasma treatment to form n^+-type and n^--type layers 5 and 6.

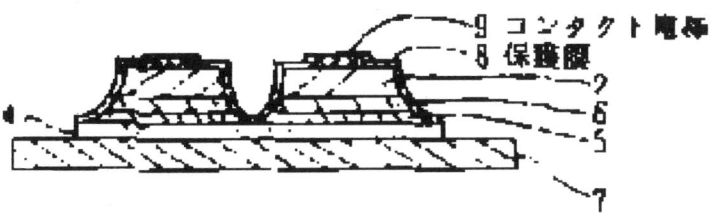

Fig. 2.2.100 (JP-A-6013642 fig. 1e)

The structure is bonded to a semiconductor substrate 7 and the substrate 3 is removed. Individual mesas are formed by etching and a protective ZnS film is applied before gold contact electrodes 9 are formed.

*

A method of fabricating photodiodes having a planar topside surface is presented in US-A-5279974 (Santa Barbara Research Center, USA, 18.01.94). The structure exhibits less retro-reflection and allows the photodiodes to have a smaller physical size than the size of the indium bumps which are connected to the photodiodes.

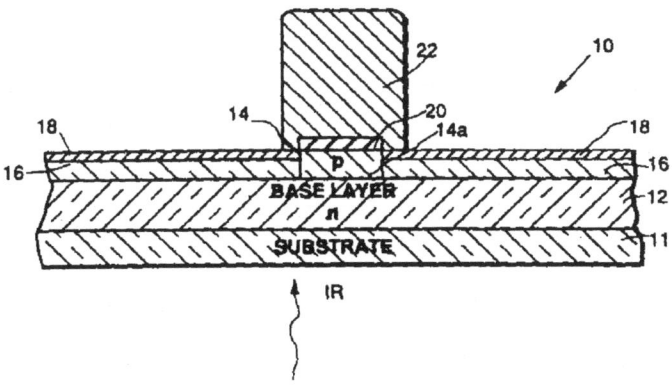

Fig. 2.2.101 (US-A-5279974 fig. 2)

An n-type radiation absorbing HgCdTe base layer 12 is grown on an electrically insulating substrate 11 of CdZnTe. The structure is thermally annealed with an Hg overpressure to remove Hg vacancies in the base layer. A CdTe layer 16 is formed to passivate the base layer. A 1500 Å thick layer 18 of SiO_2 is deposited, patterned, masked and etched to form openings having dimensions equal to the size of the photodetector junction areas. A spray etch with bromine in ethylene glycol is perfomed. This step etches completely through the exposed CdTe passivation layer 16 and is stopped at the surface of the HgCdTe base layer. A p-type HgCdTe cap layer 14 is grown. The cap layer contacts the n-type base layer forming pn heterojunctions.

14a. A protective photoresist mask is positioned over the cap layer in registration with the underlying photodiode unit cell sites and a spray etch is accomplished to remove non-masked cap layer material on top of the SiO$_2$ layer. Metal contacts 20 are formed and indium bumps 22 are provided. The indium bumps may be made significantly larger than the underlying contacts 20, with the dimensions of the indium bumps being substantially independent of the dimensions of the underlying photodetector unit cell sizes.

*

An imager comprising photo-conductors each having a charge generation region coupled to a confinement region is presented in US-A-5254850 (Texas Instruments Incorporated, USA, 19.10.93). The geometric shape of the photo-conductors increase the responsivity.

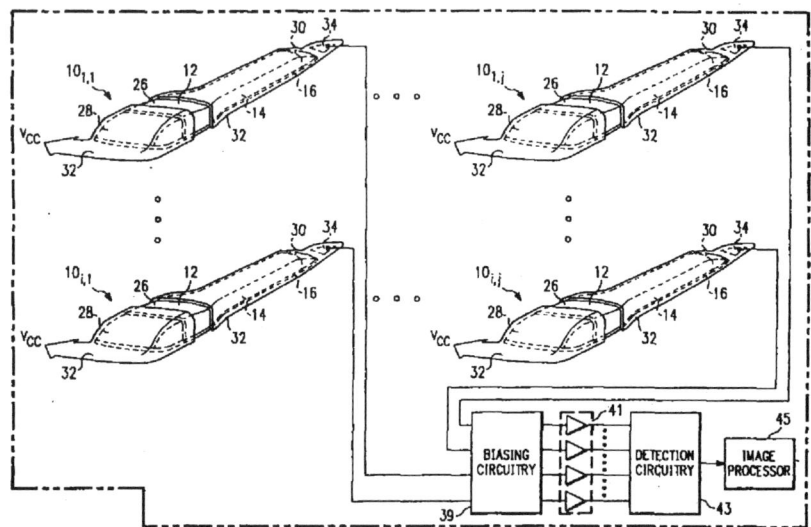

Fig. 2.2.102 (US-A-5254850 fig. 4d)

A plurality of photo-conductors 10 of HgCdTe form a focal plane array. Each photo-conductor comprises a photon collection aperture 12, a confinement region 14 and an overlap metal layer 32. The metal layer 32 forms a common electrode on one side of the aperture and an individual electrode on the other side of the aperture.

*

When detecting incident energy transmitted from a remote source, it is often desirable that the remote source is not aware that the energy it is transmitting is being detected. A problem in this regard is that focal plane arrays tend to reflect the received energy along the incident energy path. To minimize the reflection, focal plane arrays are provided with a slot shield which operates on the theory that most of the energy impinging upon the detector is travelling at an acute or obtuse angle relative thereto. A slot shield which is fabricated as part of the focal plane array fabrication procedure is disclosed in US-A-5298733 (Texas Instruments Incorporated, USA, 29.03.94).

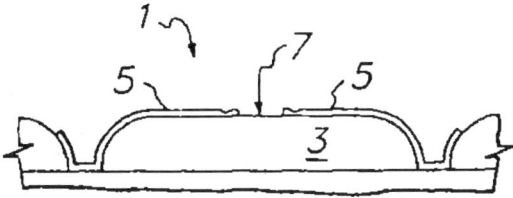

Fig. 2.2.103 (US-A-5298733 fig. 1a)

A cross-section of a detector element of a focal plane array is shown. The detector element includes a mercury cadmium telluride substrate 3 with a reflective metal layer 5 thereover which can be used as an interconnect. A portion of the reflective layer is removed over the optically sensitive portion of the substrate in the form of windows 7 so that the only energy reaching the substrate passes through the windows.

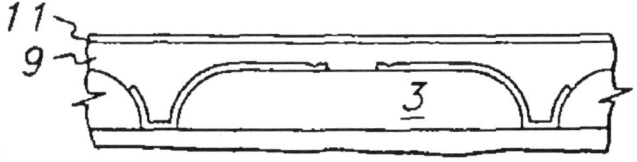

Fig. 2.2.104 (US-A-5298733 fig. 1b)

An epoxy is spun to planarise the surface. A reflective metal layer 11 of tungsten is then formed over the epoxy layer 9. The reflective layer is then patterned with a resist and the portion disposed over the windows 7 and a small additional amount thereof is etched away. The portion of the epoxy layer directly below the portion of the reflective layer which has been etched away is also etched away anisotropically down to the reflective layer 5.

*

An imager having HgCdTe mesa structures is presented in JP-A-6209096 (Fujitsu Ltd, Japan, 26.07.94). The mesas are formed by dry etching using a mask. The etching ratio of the mask to HgCdTe is such that an uniform layer of type converted HgCdTe is formed at the surface of the mesas.

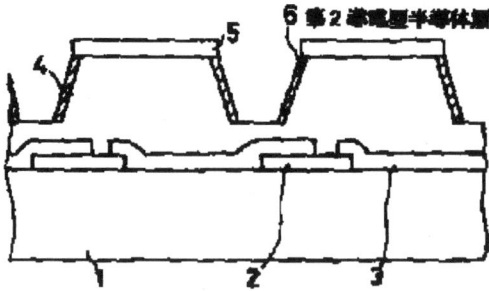

Fig. 2.2.105 (JP-A-6209096 fig. 1B)

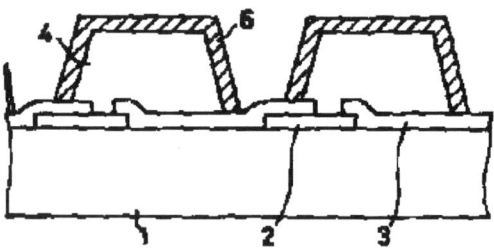

Fig. 2.2.106 (JP-A-6209096 fig. 1C)

Electrodes 2, an insulating layer 3 and an HgCdTe layer 4 are formed on a substrate 1. A mask 5 is provided and the structure is dry etched. The etching ratio of the mask to HgCdTe is selected so as to form an uniform layer 6 of type converted HgCdTe.

*

Radiation incident to a detector array, which is not absorbed but reflected back into space by the detector array, is referred to as the "light signature", LS. In order to minimize the light signature of a detector array having trench walls or mesa structures, these wall-sides must be reduced to be much smaller than an optical blur diameter. The optical blur diameter is given by 1.22 times the wavelength, divided by the numerical aperture. In US-A-5414294 (Santa Barbara Research Center, USA, 09.05.95) the width of the mesa and trench walls are reduced to a fraction of the optical blur diameter. Each mesa comprises a photodiode and the effective area of a pixel is made larger than the area of one mesa region by coupling photodiodes in parallel.

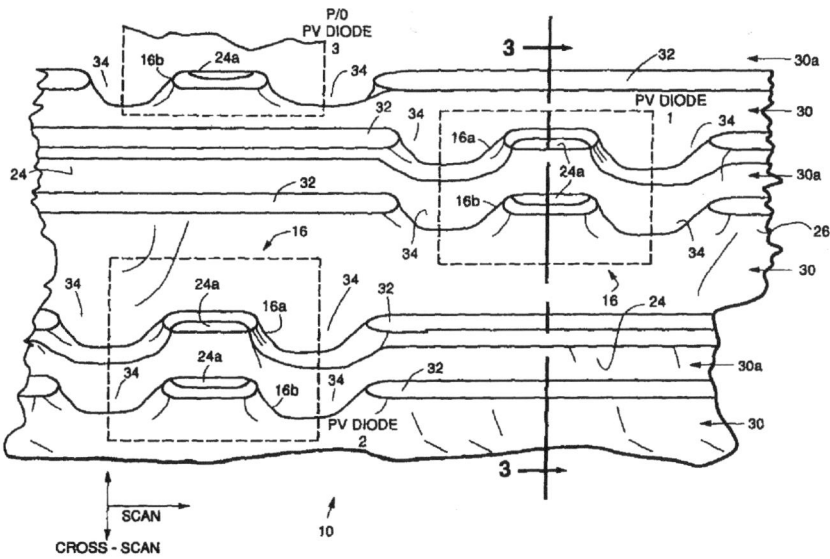

Fig. 2.2.107 (US-A-5414294 fig. 2)

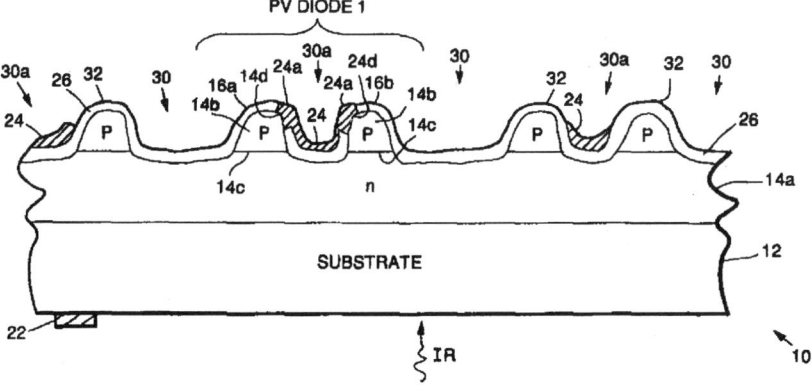

Fig. 2.2.108 (US-A-5414294 fig. 3)

Infrared radiation is incident on and passes through a substrate 12. The radiation is absorbed within an n-type absorbing layer 14a. The n-type layer is overlied by a p-type layer. The n-type layer and the p-type layer are divided into mesa structures 16, which have sub-mesa structures 16a and 16b each containing a portion of the p-type layer as a p-type cap layer 14b. Trenches 30 are etched or milled to a depth that extends completely through the p-type cap layer and partially into the n-type layer. The trench width is approximately 10 microns. Trench walls 32 are etched or milled to form regions 34 resulting in the formation of sub-mesas 16a and 16b. A passivation layer of CdTe is formed over the cap layer 14b, the mesa sidewalls, the surface of the trenches 30 and 30a, and over the surface of the regions 34. A subsequent metallization step deposits contact metal as linear metallic traces 24 within the trenches 30a, and also deposits contact metal up and over the sub-mesa sidewalls to contact the exposed portions 14d of the cap layer. The metal tab 24a has a small area so the contact metallization reflects little light radiation.

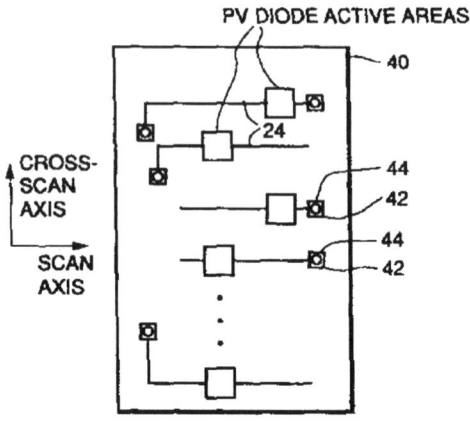

Fig. 2.2.109 (US-A-5414294 fig. 5)

A top view of a focal plane array is shown above. Metallic traces 24 are aligned parallel to the focal plane array scan axis. Each of the metallic traces is terminated at a peripheral portion of the array with a contact pad 42 having an indium bump interconnect 44 disposed thereon. The placement of indium bumps separately from the active areas of the focal plane array permits these to be shielded from the scanned radiation. Unwanted reflections from the contact pads and the indium bump interconnects are therefore reduced or eliminated.

*

An imager having a mesa structure is presented in JP-A-7079008 (Fujitsu Ltd, Japan, 20.03.95). The mesas are provided with an n⁺-type surface layer.

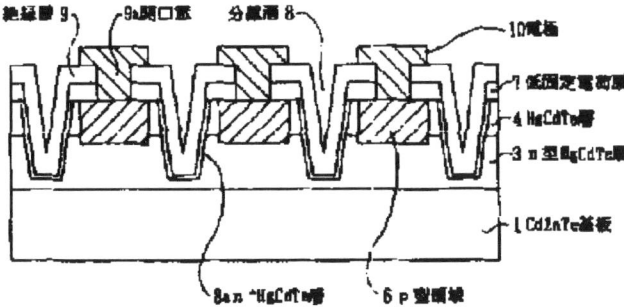

Fig. 2.2.110 (JP-A-7079008 fig. 1)

An n-type HgCdTe layer 3 is formed on a CdZnTe substrate 1. An HgCdTe layer 4 having a larger band gap than the n-type HgCdTe layer 3 is formed on the layer 3. P-type regions 6 and an electrically charged layer 7 are provided. Grooves 8 and n⁺-type surface layers 8a are formed.

*

In JP-A-7094693 (Toshiba KK, Japan, 07.04.95) an RC-network is formed between each detector element and a corresponding indium bump which connects the element to a read-out circuit.

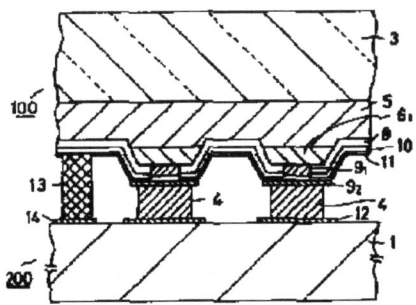

Fig. 2.2.111 (JP-A-7094693 fig. 1)

An n-type HgCdTe layer 4 is formed on a CdZnTe substrate 3. P-type mesas 61 are formed on the n-type layer. A CdTe layer 8 is formed on the surface of the detector substrate. A capacitor is formed between electrodes 91 and 92 using a layer 10 of CdTe as a dielectric. A resistor is formed from the electrode 91 by a resistor film 10. The resistor is connected to ground potential.

*

Cross-Talk Preventing Measures

To improve the resolution of an imager the distance between adjacent photodiodes is reduced. A problem occurs when the distance reaches the same dimensions as the diffusion length of photogenerated charge carriers. In EP-A-0024970 (Thomson-CSF, France, 11.03.81) an insulating opaque film is introduced between the photodiodes allowing a close spacing of the photodiodes.

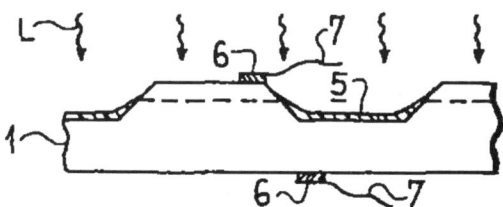

Fig. 2.2.112 (EP-A-0024970 fig. 5)

Mesa shaped photodiodes are formed in an HgCdTe substrate 1. A thin film 5 of an opaque insulating material is formed at regions surrounding the photodiodes. Electrical connections 7 are made to connection pads 6 which are connected to each photodiode and to the substrate.

* *

The structure shown above in EP-A-0024970 is further improved in US-A-4665609 (Thomson-CSF, France, 19.05.87) in which an anodic oxide is applied on the surface of the detector before an opaque metal film is formed at regions surrounding the photodiodes. This anodic oxide stabilizes the pn-junctions.

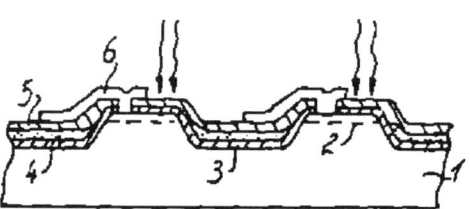

Fig. 2.2.113 (US-A-4665609 fig. 1e)

PN-junctions 2 are formed in a substrate 1. An oxide is grown by anodic oxidation over the whole surface of the substrate. A metallic layer 4 is deposited on the anodic oxide. This layer may consist of a first layer which has a good adhesion to the oxide and a second layer which is opaque to infrared radiation. The metallic layer is shaped by photolithography in order to cover only zones surrounding the detector regions. A dielectric layer 5 is deposited covering the whole surface and the photodiodes are connected by metal connections 6. An embodiment with a planar structure is also shown.

*

An imager using photocapacitive detector elements is presented in FR-A-2526227 (Thomson-CSF, France, 04.11.83). Each detector element comprises a gate with an opening and an implanted region where the amount of generated charge is measured. In this design the potential applied to the gate is not crucial and the output impedance is relatively high.

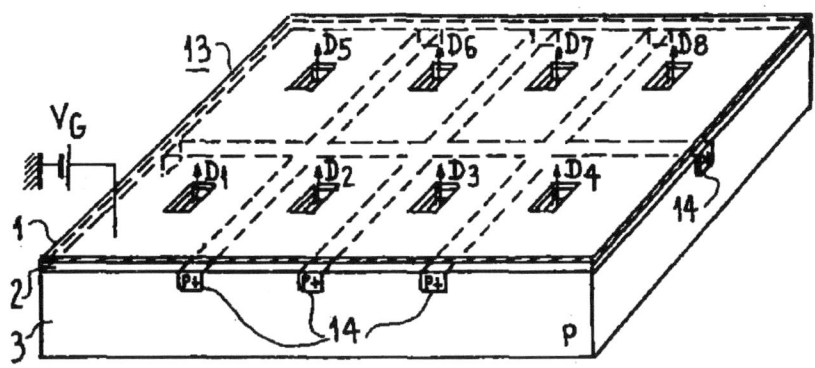

Fig. 2.2.114 (FR-A-2526227 fig. 12)

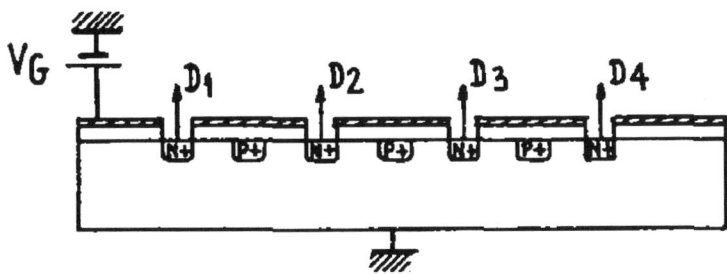

Fig. 2.2.115 (FR-A-2526227 fig. 13)

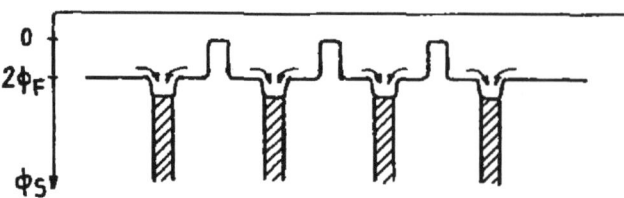

Fig. 2.2.116 (FR-A-2526227 fig. 14)

A p-type HgCdTe substrate 3 is covered with an insulating layer 2 and a detector gate 1. An opening is formed in the detector gate for each detector element D and an n-type region is formed in the substrate at a position corresponding to the opening. The n-type region is connected to a CCD formed in a separate substrate. P-type regions 14 may be formed between the detector elements. Charge generated in the substrate 3 close to a potential well under a detector gate diffuses to a region where it is read out. The p-type regions will stop charge diffusing from one detector element to an adjacent one when a detector element is exposed to a high intensity of radiation.

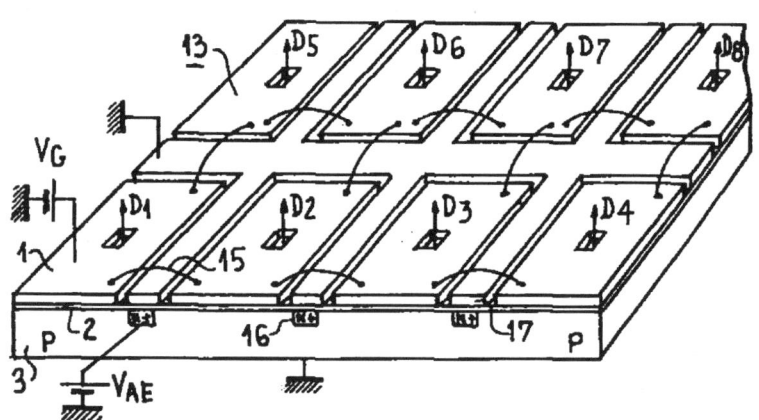

Fig. 2.2.117 (FR-A-2526227 fig. 15)

In an alternative embodiment, the detector elements are separated by n-type regions 16 and a gate 17. The detector gate 1 consists of detector gate elements connected to each other by connectors 15. When a detector element is exposed to a high intensity of radiation excess charge is drained off via the regions 16.

*

The detector elements of GB-A-2132017 (The Secretary of State for Defence, GB, 27.06.84) are separated by polycrystalline regions of HgCdTe.

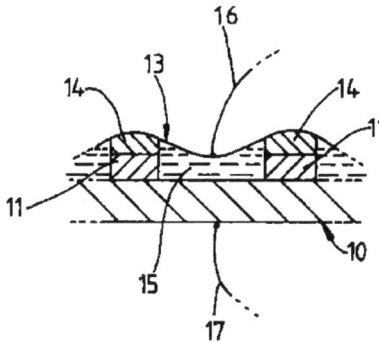

Fig. 2.2.118 (GB-A-2132017 fig. 2)

A ZnS layer 11 is formed on an (001) face InSb single-crystal substrate. The ZnS layer, which is polycrystalline, is removed from islands where the detector elements will be formed. Molecular beam epitaxy is then used to form an HgCdTe layer on the structure, which will produce single crystal and polycrystalline layers on detector element regions 15 and the ZnS layer 14 respectively. The detector elements are separated by the polycrystalline layer, which is electrically insulating.

*

The problem of cross-talk between two adjacent detector elements is approached in JP-A-61198787 (Fujitsu Ltd, Japan, 03.09.86) in which a p^+-type substrate is used and each photodiode is formed in a p-type region.

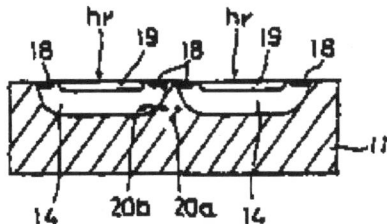

Fig. 2.2.119 (JP-A-61198787 fig. 1d)

An HgCdTe substrate is heat-treated in a mercury containing atmosphere to form a p^+-type substrate 11. An insulating layer is used as a mask and the device is again heat-treated in a mercury containing atmosphere to form p-type regions 14. Next, intense light is used to form

p^+-type layers 18. The photodiodes are created by forming n^+-type layers 19 in the p-type regions. The charge carriers 20a photogenerated between photodiodes have a short diffusion length. The problem of cross-talk is therefore reduced.

*

In US-A-4646120 (The United States of America as represented by the Secretary of the Army, USA, 24.02.87) an ohmic contact layer is introduced which allows individual detector elements to be completely separated thereby reducing the problem of cross-talk.

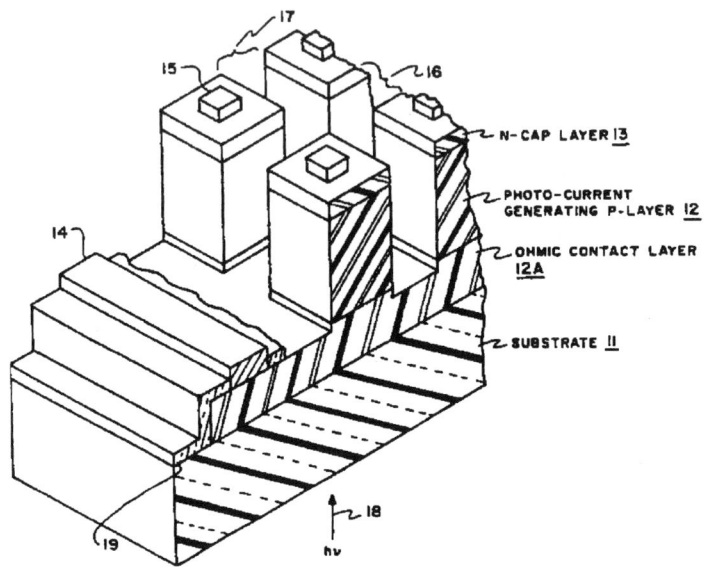

Fig. 2.2.120 (US-A-4646120 fig. 2)

The imager is built on a cadmium telluride or cadmium zinc telluride substrate 11. One surface of the substrate is covered with an ohmic contact layer 12A of $Hg_{0.7}Cd_{0.3}Te$ which is doped to a high conductivity and has a bandgap too large to absorb radiation in a selected operating band. A p-type photocurrent layer 12 of $Hg_{0.8}Cd_{0.2}Te$ is formed to a thickness of 10-15 microns. An n-type junction cap layer 13 of $Hg_{0.7}Cd_{0.3}Te$ 13 completes the photodiodes. The junction cap layer is 1-2 microns thick. Two series of parallel grooves 16 and 17 are formed entirely through the photocurrent layer and the junction cap layer. This completely isolates the photocurrents induced in layer 12 as well as the storage process taking place in the junction cap layer. Finally, metal contacts 14 and 15 and a passivation layer 19 are formed.

*

Insulation between detector elements is automatically achieved in JP-A-61222161 (Fujitsu Ltd, Japan, 02.10.86) by forming detector elements of epitaxially grown HgCdTe in recessed portions of a sapphire substrate.

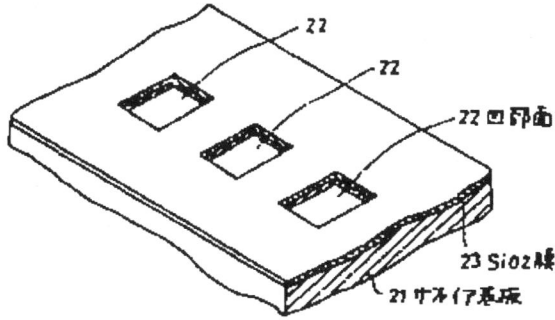

Fig. 2.2.121 (JP-A-61222161 fig. 2)

A sapphire substrate 21 which has recessed portions 22 is covered by a SiO_2 film 23 having windows corresponding to the recessed portions.

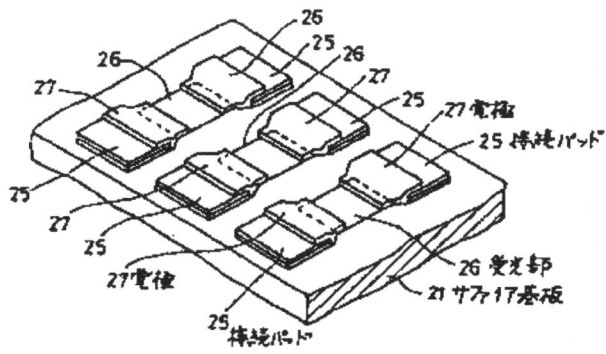

Fig. 2.2.122 (JP-A-61222161 fig. 4)

An HgCdTe layer is epitaxially grown over the surface and individual HgCdTe detector elements 26 are formed in the recessed portions when the SiO_2 layer has been removed. The surface of the HgCdTe detector elements are then polished and etched before electrodes 27 are connected to connection pads 25.

*

In JP-A-62036858 (Fujitsu Ltd, Japan, 17.02.87) reflectors are provided in V-grooves which separate individual detector elements. The amount of cross-talk between adjacent detector elements is thereby reduced.

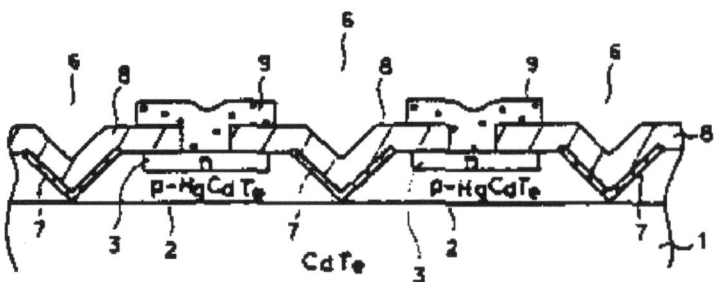

Fig. 2.2.123 (JP-A-62036858 fig. 1)

A p-type HgCdTe layer 2 is formed on a CdTe substrate 1. Individual detector element regions are formed by ion beam milling V-grooves 6. Light reflecting layers 7 are formed by depositing gold. N-type regions 3 are formed in the detector element regions and indium electrodes 9 are provided. Radiation incident in the device region will penetrate the CdTe substrate. Radiation incident in the V-groove regions will be reflected by the reflecting layers towards a photodiode.

*

Light shields are provided between a CdTe substrate and an HgCdTe film in JP-A-62104163 (Fujitsu Ltd, Japan, 14.05.87) to reduce the problem of cross-talk.

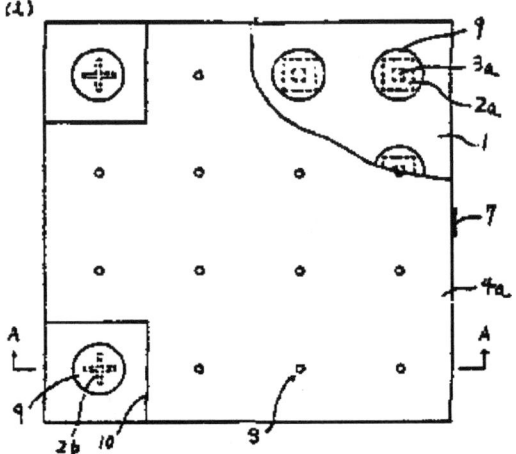

Fig. 2.2.124 (JP-A-62104163 fig. 1a)

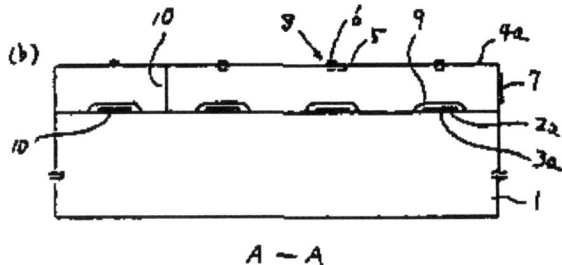

Fig. 2.2.125 (JP-A-62104163 fig. 1b)

Square metallic light shields 2a having openings 3a are formed on a surface of a CdTe substrate 1. Alignment marks 2b are formed at the same time. The light shields and the alignment marks are covered by regions 9 of SiO_2. An p-type HgCdTe layer 4a is epitaxially grown on the light shields and alignment marks. Regions of the HgCdTe layer over the alignment marks are removed. N-type regions 5 are provided in the HgCdTe layer at a position directly over the openings in the light shields. The HgCdTe layer and the n-type regions are connected to electrodes 7 and 6, respectively.

*

Absorption regions of HgCdTe are formed at regions between detector elements in the invention of JP-A-63043366 (Fujitsu Ltd, Japan, 24.02.88). The design reduces the problem of cross-talk.

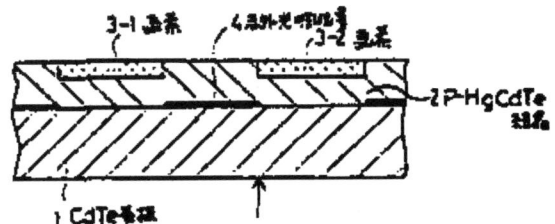

Fig. 2.2.126 (JP-A-63043366 fig. 1)

Absorption regions 4 of $Hg_{0.85}Cd_{0.15}Te$ are formed on a CdTe substrate. A p-type HgCdTe layer is formed on the absorption regions and on the CdTe substrate. Detector elements are made by forming n-type regions 3-1, 3-2 in the p-type HgCdTe layer. The absorption regions are positioned between the detector elements and they absorb radiation, which has passed through the substrate and which falls between the detector elements.

* *

The structure of JP-A-63043366, shown above, is further improved in GB-A-2229036 (Mitsubishi Denki Kabushiki Kaisha, Japan, 12.09.90). An absorption layer is combined with n-type photodiode regions which penetrate through a p-type layer.

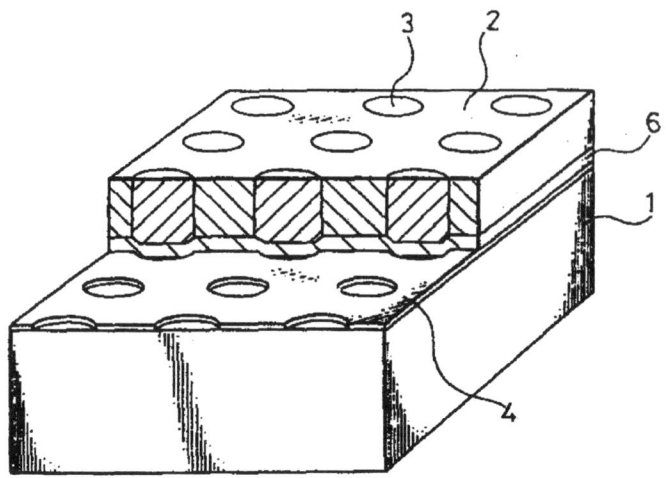

Fig. 2.2.127 (GB-A-2229036 fig. 2)

A CdTe insulating layer 6 is provided between a p-type HgCdTe layer 2 and a CdTe substrate 1 covering an HgTe light absorption layer 4. N-type HgCdTe portions 3 are provided so as to reach the insulating layer 6. Cross-talk is reduced because light falling between the portions 3 is absorbed in the light absorption layer. If an insulating material such as polysilicon is used for the light absorption layer, the insulating layer 6 is not needed.

*

A structure of an imager having a low amount of cross-talk between adjacent photodiodes is disclosed in JP-A-63133580 (Fujitsu Ltd, Japan, 06.06.88). The structure comprises islands of p-type material formed in a p^+-type HgCdTe substrate. A photodiode is provided in each island by forming an n^+-type region in each p-type island.

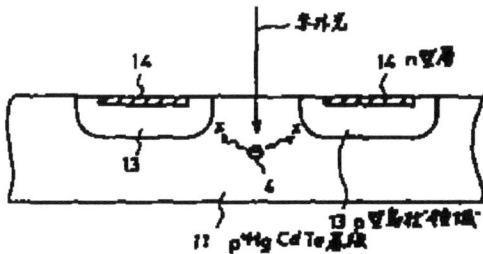

Fig. 2.2.128 (JP-A-63133580 fig. 1)

Hg is diffused selectively into a p^+-type substrate 11 by the use of a SiO_2 mask to from p-type islands 13. Next, boron ions are injected to form n^+-type regions in the p-type islands.

*

Cross-talk may occur in an imager comprising an HgCdTe layer formed on a first surface of a transparent substrate when photons which are not absorbed in a particular detector element in the HgCdTe layer are reflected at a second surface of the substrate, opposite to the first surface, and thereafter absorbed in an adjacent detector element. In JP-A-63170960 (Fujitsu Ltd, Japan, 14.07.88) this kind of cross-talk is prevented by the use of a substrate which absorbes infrared radiation.

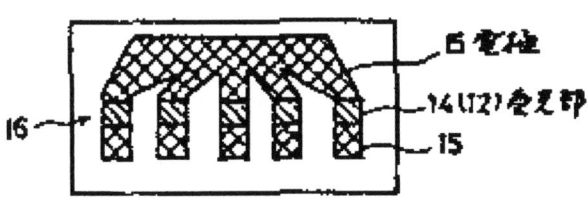

Fig. 2.2.129 (JP-A-63170960 fig. 4b)

An HgCdTe layer is epitaxially grown on a first substrate. An adhesive is used to attach the HgCdTe layer to a sapphire substrate, which absorbes infrared radiation. The first substrate is removed and the epitaxially grown HgCdTe layer is shaped by etching to form detector regions 14, which are connected by electrodes 15. Photons which are not absorbed by the detector elements will be absorbed in the sapphire substrate.

*

The problem of cross-talk, presented in JP-A-63170960 above, is solved in JP-A-63229751 (Fujitsu Ltd, Japan, 26.09.88) where an infrared light absorbing layer is provided on a substrate, opposite the surface holding an HgCdTe detector layer.

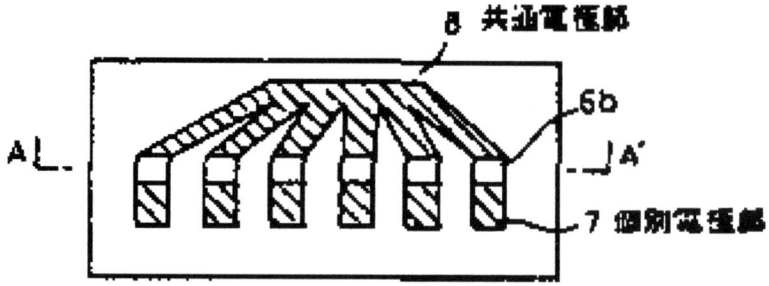

Fig. 2.2.130 (JP-A-63229751 fig. 3)

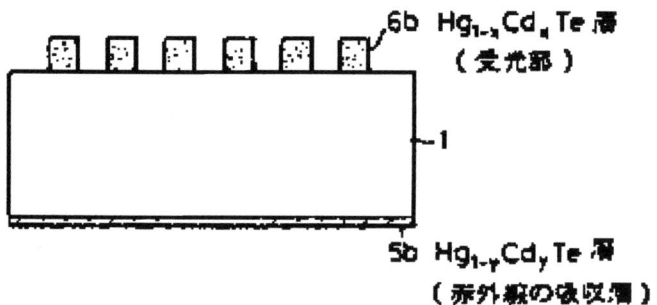

Fig. 2.2.131 (JP-A-63229751 fig. 2)

A vapor-phase diffused infrared light absorbing layer 5b of $Hg_{1-y}Cd_yTe$ is formed on a first side of a CdTe substrate 1. An $Hg_{1-x}Cd_xTe$ detector layer 6b is then formed on a second side of the substrate, opposite to the first side, by liquid-phase epitaxy. The detector layer is shaped to form detector elements which are connected by electrodes 7 and 8. The absorbing layer 5b will absorb photons which have not been absorbed by the detector elements.

*

A backside illuminated imager is presented in JP-A-63273365 (Mitsubishi Electric Corp., Japan, 10.11.88). The amount of cross-talk between adjacent photodiodes, which have been separated by grooves, is reduced by forming an infrared blocking film in the groove regions.

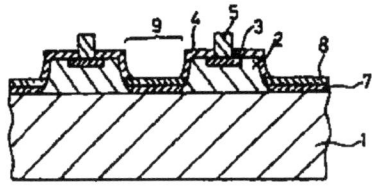

Fig. 2.2.132 (JP-A-63273365 fig. 1)

A p-type HgCdTe layer 2 is formed by liquid phase epitaxy on a CdTe substrate 1. Photodiodes are formed by forming n-type regions 3 in the HgCdTe layer by ion-implantation. The photodiodes are separated by etching grooves 9 through the HgCdTe layer to expose portions of the CdTe surface. An infrared blocking film 7 made of SiO_2 is deposited by an electron beam depositing method on these surface portions. Next, bumps 5 are connected to the n-type regions and a film 4 is formed to stabilize the surface. The opaque film is covered with a metal layer 8 to electrically connect the p-type HgCdTe islands.

*

The problem of cross-talk between adjacent photodiodes and the problem of different thermal expansion coefficients between a silicon substrate and an HgCdTe substrate are approached in JP-A-63281460 (Fujitsu Ltd, Japan, 17.11.88). A detector is made up of HgCdTe wells, comprising photodiodes, formed in a CdTe substrate. The detector is bonded to a silicon substrate by a flip-chip process, and finally the CdTe substrate is etched away.

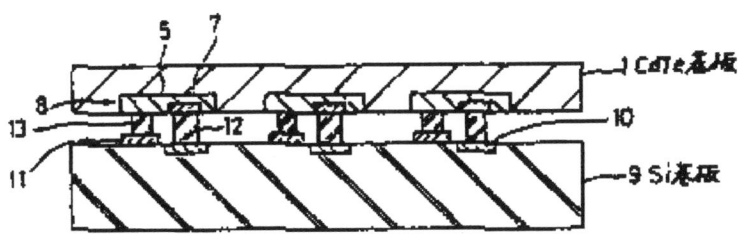

Fig. 2.2.133 (JP-A-63281460 fig. 6)

HgCdTe wells 5, comprising photodiodes 7, are formed in a CdTe substrate 1. The anodes and cathodes of the photodiodes are connected by indium columns 12 and 13 to a silicon substrate 9 by a flip-chip bonding process.

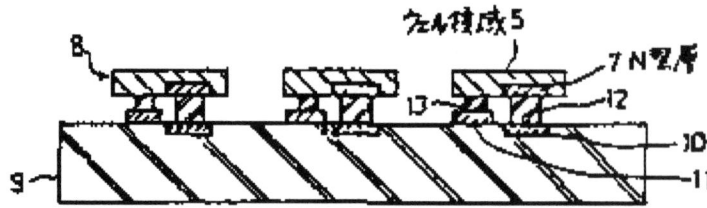

Fig. 2.2.134 (JP-A-63281460 fig. 1)

Next, the CdTe substrate is removed by dipping the device in an etching liquid comprising hydrofluoric acid (HF), nitric acid (HNO$_3$), acetic acid (CH$_3$COOH) and water (H$_2$O) with a weight ratio of 2-5:3-5:6:6.

*

A similar structure to the structure shown in JP-A-63281460 above is disclosed in JP-A-63296272 (Fujitsu Ltd, Japan, 02.12.88). In the latter, the HgCdTe wells are not connected by indium bumps but by Au films supported by an epoxy resin.

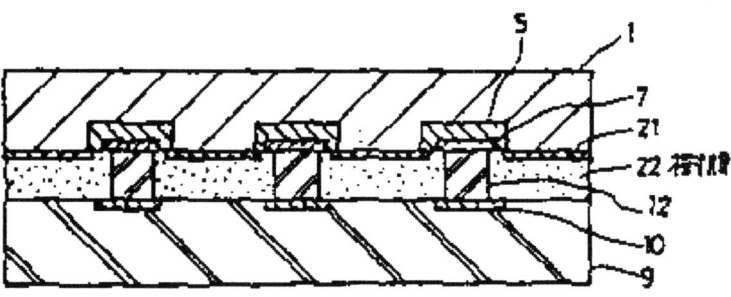

Fig. 2.2.135 (JP-A-63296272 fig. 3)

P-type HgCdTe well regions 5 are formed in a CdTe substrate 1. Photodiodes are created by forming n-type regions in the p-type wells. The p-type well regions 5 are connected to each other by an Au film 21 forming a common electrode. The n-type regions are connected to input diodes 10 prepared in a silicon substrate 9 by indium columns 12 in a flip-chip process. The space between the CdTe substrate and the silicon substrate is filled with an epoxy resin 22.

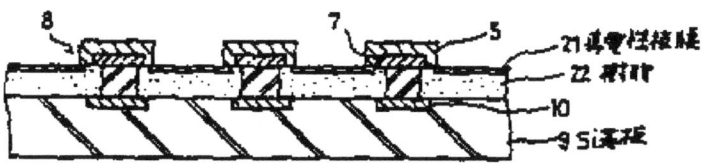

Fig. 2.2.136 (JP-A-63296272 fig. 4)

Next, the CdTe substrate is etched away by using the same etching liquid as is used in JP-A-63281460 above.

* *

The problem of difference in thermal expansion coefficients between an HgCdTe substrate and a silicon read-out substrate is addressed in JP-A-2214159 (Mitsubishi Electric Corp., Japan, 27.08.90) where a design similar to the design of JP-A-63296272 is presented.

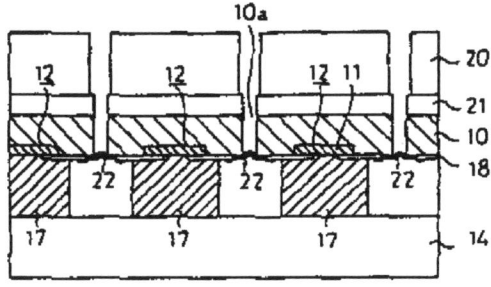

Fig. 2.2.137 (JP-A-2214159 fig. 1)

N-type regions 11 are formed in a p-type substrate 10 of HgCdTe thereby forming detector elements 12 which are separated by grooves 10a except for metal electrodes 22 which bridge the grooves. The metal electrodes connect the p-type regions of all detector elements. The detector elements are connected to a silicon read-out substrate 14 via electrodes 17.

* *

Cross-talk is reduced in US-A-5376558 (Fujitsu Limited, Japan, 27.12.94) by forming a three layered light absorption structure in the areas between the photodiodes of the array. The second layer has a refractive index which is greater than that of the first layer and the third layer is a metal layer.

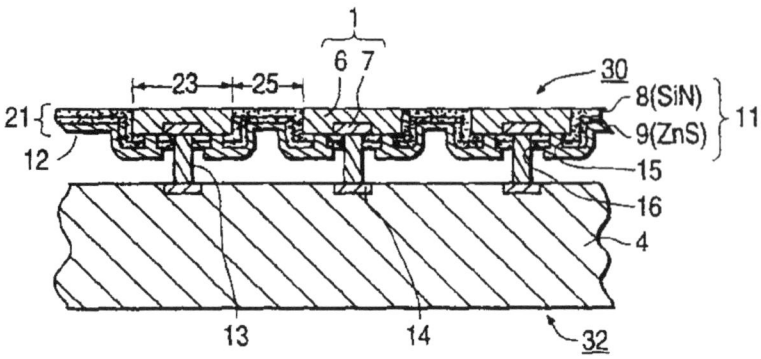

Fig. 2.2.138 (US-A-5376558 fig. 2)

Detector elements 1 comprising p-type HgCdTe regions 6 and n-type regions 7 are insulated from neighboring detector elements by a double insulation layer 11 composed of a silicon nitride layer 8 (refraction index approx. 2.0) and a ZnS layer 9 (refraction index approx. 2.2). A metal layer 12 of aluminium covers the insulation layer 11. The metal layer and the insulation layer form a light shield layer 21 which covers an outer surface of the p-type regions 7 except at light input ports 23. Holes 15 and 16 are formed through the insulation layer, and the metal layer is in contact with the p-type regions through holes 15, and a bump is formed through holes 16. The photo-sensitive array 30 is assembled with a signal processing substrate 32 using bumps 13.

Infrared rays which are incident to regions 25 penetrate into the silicon nitride layer and ZnS layer are thereby partly attenuated and reflected, and are finally reflected from the metal layer. The incident rays and the reflected rays cause a complex interference with each other, and an apparent overall reflectance of the light shield layer is reduced if proper thickness and refraction index are chosen for each layer of the insulation layer. Preferable, the refractive index of the first layer 8 is less than the refractive index of the second layer 9.

A method of manufacturing the structure is also disclosed. In short, islets of HgCdTe are formed from an HgCdTe layer which has been formed on a CdTe substrate. Next, the double insulation layer and the metal layer are formed. The bumps are provided before the CdTe substrate is removed by etching.

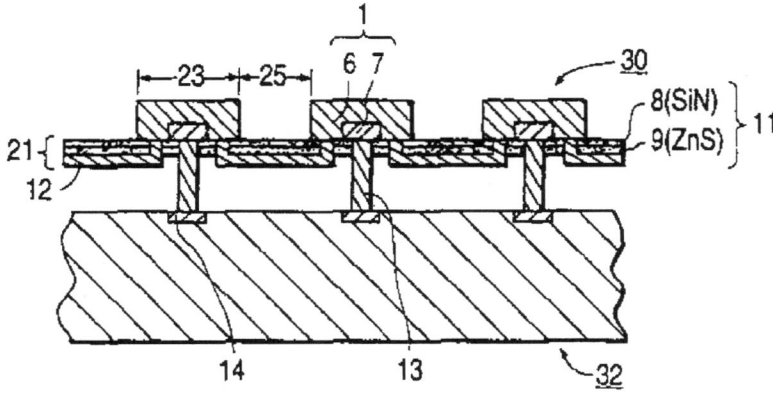

Fig. 2.2.139 (US-A-5376558 fig. 3)

In a second embodiment, the detector elements 1 are protruding from a flat light shield layer 21. Bumps 13 are used to connect the array to a signal processing substrate 32. The light shield layer comprises the same layers as in the first embodiment.

A method of manufacturing the structure of the second embodiment is also disclosed. In short, buried HgCdTe regions are formed from the surface of a CdTe substrate. Next, the double insulation layer and the metal layer are formed. The bumps are provided before the CdTe substrate is removed leaving the buried HgCdTe regions protruding from the from the light shield layer.

*

The problem of difference in thermal expansion coefficients between a silicon read-out substrate and an HgCdTe detector substrate is approached in US-A-4783594 (Santa Barbara Research Center, USA, 08.11.88) by filling the space between the two substrates with a resilient electrically insulating polymeric material and thereafter separating the detector elements from each other by removing a layer of the HgCdTe detector substrate.

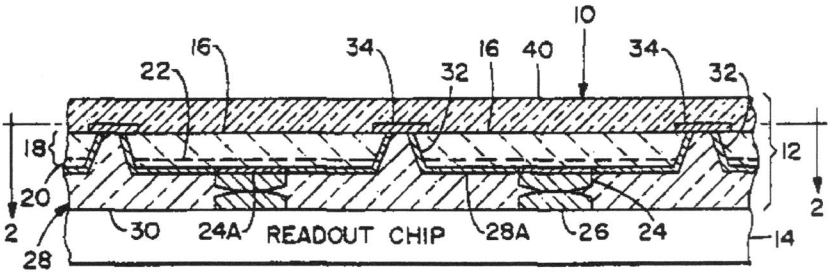

Fig. 2.2.140 (US-A-4783594 fig. 1)

A detector array 12, which is composed of detectors 16 arranged in rows and columns, is disposed on a silicon read-out chip 14. Each detector includes two HgCdTe layers 18 and 20, which form a pn-junction. A set of upper contacts 24 and a set of lower contacts 26 of indium are provided for connecting the detectors to the read-out chip. A resilient support 28 of a resilient polymer, such as silicone elastomer, envelops and holds the detectors. An insulating layer 28A of silicon dioxide separates the detectors from the support and passivates the surfaces of the HgCdTe layers to prevent development of electric charge by interaction with polymer material of the support. Electrical connection through the insulating layer 28A is made by contact metal 24A. An electrically conductive grid 34 is disposed on top of the array to form an electrically conductive ohmic contact with each detector followed by a radiation-transmissive coating 40 of ZnS. The process for constructing the assembly 10 is shown below.

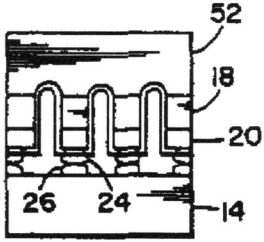

Fig. 2.2.141 (US-A-4783594 fig. 5h)

A p-type layer 18 and an n-type layer 20 of HgCdTe are grown epitaxially on a substrate 52 of cadmium zinc tellurium. A set of through holes are formed which extend through the layers 18 and 20 up to the surface of the substrate. A silicon dioxide layer is applied and at each detector a window is etched. A contact metal 24A is applied within each window. Contacts 24 and 26 are constructed on the detectors and on the read-out chip 14. The chip and the substrate 52 are pushed together to compress the contacts 24 and 26 against each other which cold welds them to each other.

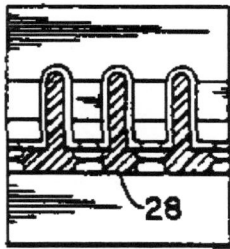

Fig. 2.2.142 (US-A-4783594 fig. 5i)

The spaces between the welded sets of contacts are filled with a polymer material in liquid form. The polymer material is then allowed ot cure and solidify to provide the support 28.

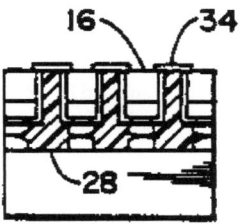

Fig. 2.2.143 (US-A-4783594 fig. 5k)

The substrate 52 is removed and a grid 34 is deposited followed by an anti-reflection coating. Due to the increased resilience of the support relative to the chip, any differential expansion or contraction of the support relative to the chip due to changes in temperature can occur without the development of excessive stress and strain in the contacts. Furthermore, the structure eliminates a significant amount of cross-talk between signals of neighboring detectors.

*

An infrared light absorbing layer is formed at regions between detector elements in the imager presented in JP-A-1201971 (Fujitsu Ltd, Japan, 14.08.89).

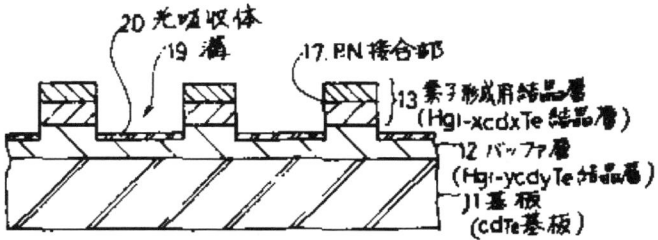

Fig. 2.2.144 (JP-A-1201971 fig. 1)

A buffer layer 12 of HgCdTe is formed on a CdTe substrate 11. Detector elements 13 are formed by providing pn-junctions 17 between two HgCdTe layers. Cross-talk is reduced by providing an infrared light absorbing layer 20 at regions between the detector elements.

*

Photo-detector regions are formed in JP-A-1205476 (Fujitsu Ltd, Japan, 17.08.89) by laser annealing a multi-layer structure of HgTe and CdTe. Radiation which has generated charge carriers between two detector elements will be absorbed in the multi-layer structure thereby reducing cross-talk.

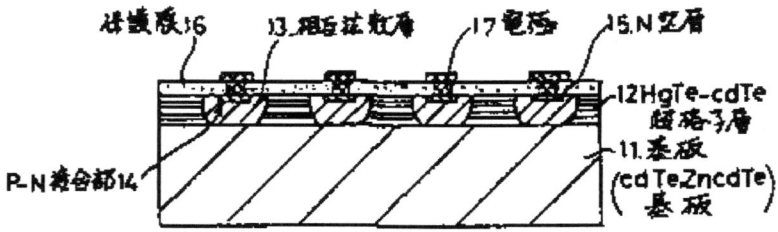

Fig. 2.2.145 (JP-A-1205476 fig. 1)

A multi-layer structure 12 of HgTe and CdTe is formed on a substrate 11 of CdTe or CdZnTe. A metal mask is applied and laser annealing is performed to form island-shaped regions 13 of $Hg_{0.7}Cd_{0.3}Te$. N-type regions 15 are ion-implanted in the island-shaped regions thereby forming photodiodes. A protective film 16 is formed and the n-type regions are connected to electrodes 17.

* *

A structure which reduces the amount of cross-talk and which is similar to the structure proposed in JP-A-1205476, shown above, is presented in JP-A-1228180 (Fujitsu Ltd, Japan, 12.09.89).

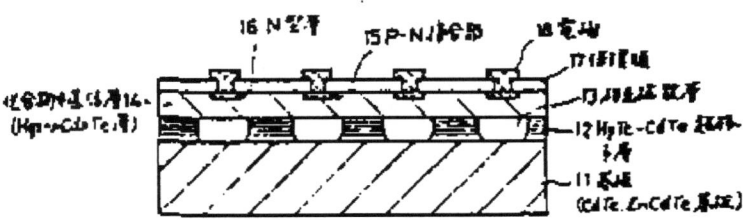

Fig. 2.2.146 (JP-A-1228180 fig. 1)

A multi-layer structure 12 of HgTe and CdTe is formed on a substrate 11. Islands 13 are subjected to a heat treatment to form HgCdTe regions. Next, a p-type HgCdTe layer 14 is formed in which n-type regions 16 are formed. PN-junctions 15 are formed between the layer 14 and the n-type regions. The n-type regions are connected to electrodes 18. Radiation, which has passed through the substrate and which has generated charge carriers between two detector elements, will be absorbed in the multi-layer structure thereby reducing cross-talk.

* *

In JP-A-3196568 (Mitsubishi Electric Corp., Japan, 28.08.91) another structure, which is similar to the structures of JP-A-1205476 and JP-A-1228180, shown above, is presented.

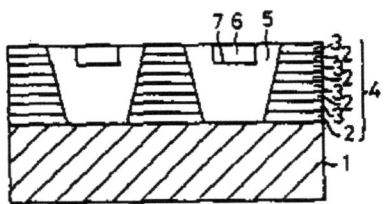

Fig. 2.2.147 (JP-A-3196568 fig. 2)

A multi-layer absorption structure 4 of CdTe 2 and HgTe 3 layers are formed on a substrate 1. The CdTe and HgTe layers are diffused together thereby forming HgCdTe regions 5. PN-junctions 7 are formed in the HgCdTe regions. The structure has a reduced amount of cross-talk.

*

The problem of cross-talk is approached in JP-A-1218062 (Fujitsu Ltd, Japan, 31.08.89) by alternately forming photodiodes in recessed and protruding regions.

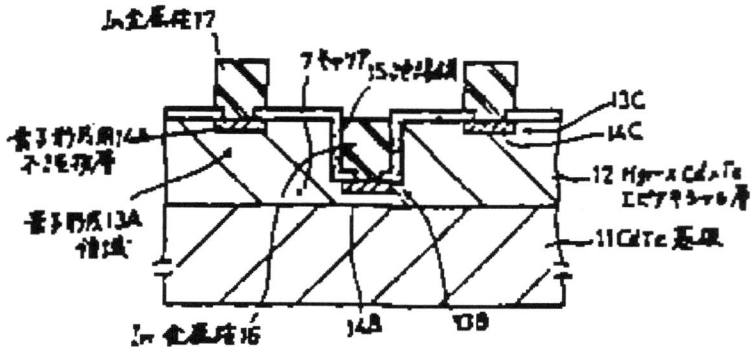

Fig. 2.2.148 (JP-A-1218062 fig. 1)

A p-type HgCdTe layer 12 is formed epitaxially on a substrate 11 of CdTe. Recessed regions 13B are formed between protruding regions 13A and 13B. N-type regions 14A-14C are formed to which connections are made by indium electrodes 16 and 17 after an insulating film 15 of ZnS has been formed.

*

In JP-A-1233777 (Fujitsu Ltd, Japan, 19.09.89) the amount of cross-talk is reduced by diffusing mercury into detector regions thereby reducing the acceptor carrier concentration in these regions while the regions in between the detector regions are kept at a high acceptor carrier concentration.

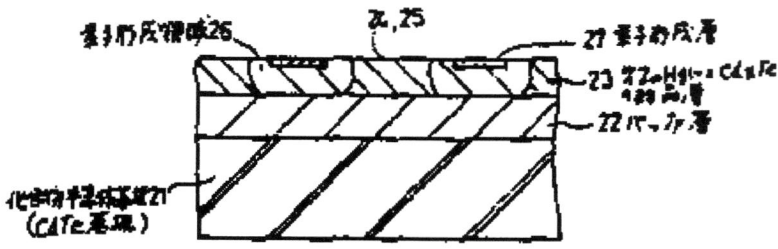

Fig. 2.2.149 (JP-A-1233777 fig. 1)

An $Hg_{0.7}Cd_{0.3}Te$ buffer layer 22 is formed on a CdTe substrate 21. An $Hg_{0.8}Cd_{0.2}Te$ layer 23 having an acceptor carrier concentration of 10^{17} cm^{-3} is formed on the buffer layer. A mask is used to selectively diffuse mercury into regions 26 to fill vacancies and thereby reduce the acceptor carrier concentration in these regions to 10^{16} cm^{-3}. N-type regions 27 are formed in the regions 26, which correspond to detector regions. Minority carriers, which are generated between the detector elements, will recombine in the layer 23.

*

The method of fabrication of an imager presented in WO-A-8910007 (Santa Barbara Research Center, USA, 19.10.89) provides for virtually all of the critical diode fabrication processes to be performed with one photomask layer in a single processing chamber. The problem of alignment of successively applied photomask layers is therefore eliminated. Furthermore, the use of one photomask layer provides for the fabrication of pn-junctions having small areas. The small pn-junction area gives a small junction capacitance and a small leakage-current. Another advantage is that repeated surface cleaning procedures are not required.

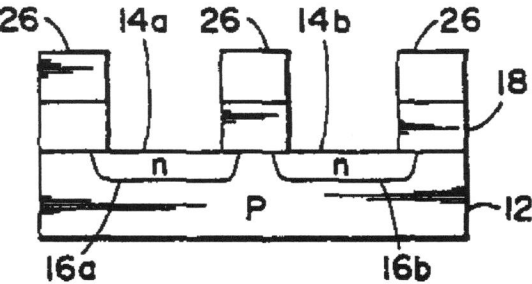

Fig. 2.2.150 (WO-A-8910007 fig. 3d)

A passivation layer 18 is disposed upon an HgCdTe layer 12. The passivation layer comprises either CdTe, HgCdTe, HgCdZnTe or $Cd_{0.96}Zn_{0.04}Te$ and has a wider bandgap than the bandgap of layer 12. A mask layer 26 having openings therethrough is deposited. Each opening defines an individual photodiode. Portions of the passivation layer are removed through the openings by ion-milling or a sputter etching technique. The lattice damage caused by the etching step also converts the p-type HgCdTe in the vicinity of the mask openings to n-type regions 14a and 14b, thereby forming pn-junctions 16a and 16b. The n-type regions extend laterally out from the opening such that edges of the pn-junctions are disposed beneath the surrounding passivation layer. Alternatively, the n-type regions may be formed by ion-implantation.

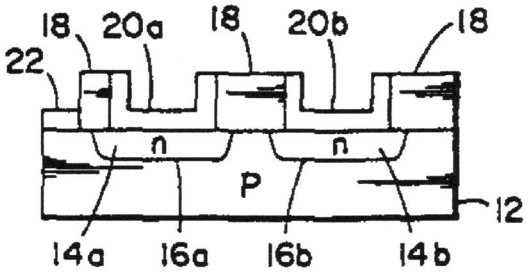

Fig. 2.2.151 (WO-A-8910007 fig. 3g)

A layer of contact metal is deposited over the surfaces of the mask layer 26 and the n-type regions 14. Thereafter, the wafer is removed from the vacuum chamber and the mask layer 26 is removed thereby removing the overlying layer of metal except where it contacts the individual n-type regions 20a and 20b. A ground contact 22 is then deposited upon the substrate 12.

*

In many applications, photodiodes operate at zero polarization. They must therefore have high dynamic resistance to ensure a high detectivity. This is achieved in US-A-4972244 (Commissariat a l'Energie Atomique, France, 20.11.90) by separating detector elements by trenches. Furthermore, a common electrode is provided at the bottom of the trenches thereby reducing the interconnection resistance of the photodiodes.

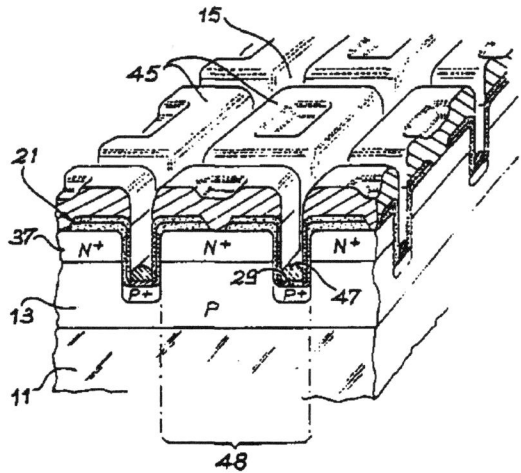

Fig. 2.2.152 (US-A-4972244 fig. 10)

A p-type HgCdTe layer 13 is formed on a CdTe insulating substrate 11. Trenches 15 are formed by an anisotropic planar etching process or by an ion-etch. P-type ions are implanted in the bottom of the trenches to form a p-type layer and n-type ions are implanted to form n-type regions 37. Metal is deposited to form a contact 47, which connects to the p-type layer, and electrical contacts 45. Contacts 45 and 47 are separated by an insulating layer 21.

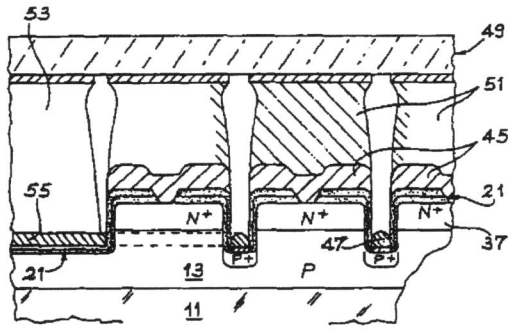

Fig. 2.2.153 (US-A-4972244 fig. 11)

The photodiodes are connected to an integrated reading and addressing circuit 49 by indium pads 51. The common electrode 47 is connected to the integrated cicuit via contact 55 and indium pad 53.

*

A common electrode and a light shield film are combined in JP-A-2086177 (Fujitsu Ltd, Japan, 27.03.90). The film surrounds the photodiodes of an imager.

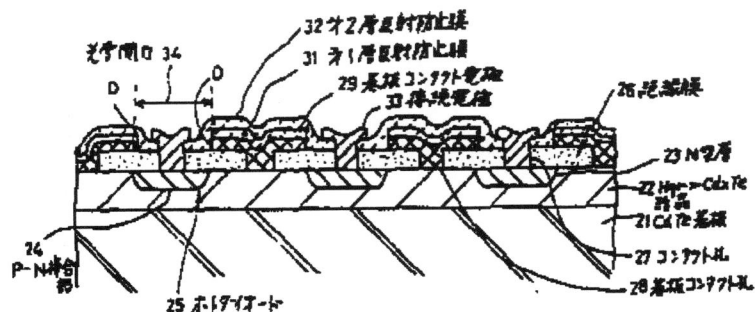

Fig. 2.2.154 (JP-A-2086177 fig. 2)

A p-type HgCdTe layer 22 is formed on a CdTe substrate 21. Photodiodes are created by forming n-type regions 23 in the p-type layer. An insulating film 26 having openings 28 at regions surrounding the n-type regions is formed. The substrate is connected to a common electrode 29 by an ohmic contact which is formed in the openings 28. The common electrode also acts as a light shield. After reflection preventing films 31 and 32 are formed, openings 27 are made and the n-type regions are connected to indium electrodes 33.

*

In JP-A-2155269 (Fujitsu Ltd, Japan, 16.06.90) the photodiodes are created by forming n-type regions in an epitaxially grown p-type HgCdTe layer. To prevent cross-talk, the thickness of the p-type layer is smaller under the n-type regions and larger elsewhere.

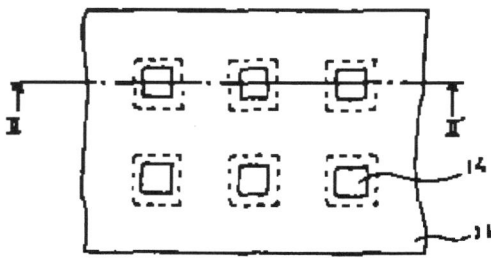

Fig. 2.2.155 (JP-A-2155269 fig. 2a)

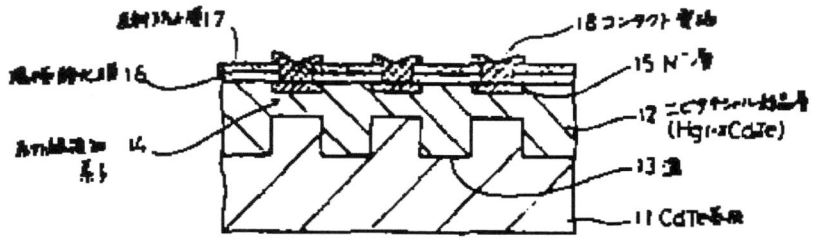

Fig. 2.2.156 (JP-A-2155269 fig. 2b)

An p-type HgCdTe layer 12 is epitaxially grown on a CdTe substrate 11 in which grooves 13 are formed by ion-etching. N-type regions 15 are formed in the p-type layer over protruding parts of the substrate. An anodic film 16 and a reflection preventing film 17 are formed. The n-type regions are connected by electrodes 18.

*

In JP-A-2213174 (Mitsubishi Electric Corp., Japan, 24.08.90) cross-talk between adjacent photodiodes is reduced by providing a lattice of crystal defects on the surface of a CdTe substrate before an HgCdTe detector layer is grown thereon.

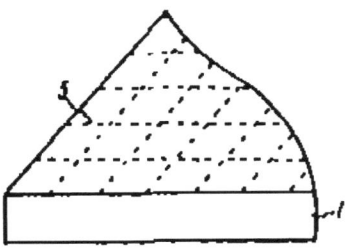

Fig. 2.2.157 (JP-A-2213174 fig. 2)

A lattice of crystal defects 5 are formed on the surface of a CdTe substrate 1 by producing flaws in the surface mechanically by means of a needle or by the use of an electron beam.

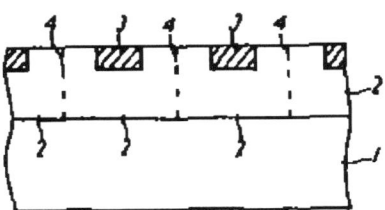

Fig. 2.2.158 (JP-A-2213174 fig. 1)

An HgCdTe layer 2 is grown on the CdTe substrate 1 and photodiodes are created by forming a region 3 of a type opposite to the type of the substrate in each section of the lattice.

*

The cross-talk between adjacent photodiodes formed in an HgCdTe layer on a semi-insulating CdTe substrate, is reduced in JP-A-2248077 (NEC Corp., Japan, 03.10.90) by providing n^+-type regions in the CdTe substrate at regions corresponding to regions between the photodiodes.

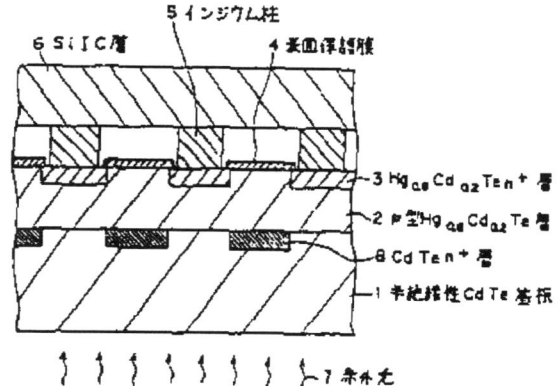

Fig. 2.2.159 (JP-A-2248077 fig. 1)

N^+-type regions 8 are formed in a semi-insulating substrate 1 of CdTe by diffusion of indium. A p-type layer 2 of HgCdTe is epitaxially grown on the CdTe substrate, and photodiodes are created by forming n-type regions 3. The photodiodes are connected via electrodes 5 to a read-out device formed in a silicon substrate 6. Incident infrared radiation at regions between the photodiodes is absorbed in the n^+-type regions 8.

*

The problem of cross-talk between adjacent photodiodes is solved in JP-A-2303160 (Fujitsu Ltd, Japan, 17.12.90) by forming islands of HgCdTe, comprising photodiodes, on a CdTe substrate and then bonding them to a read-out device by a flip-chip bonding process.

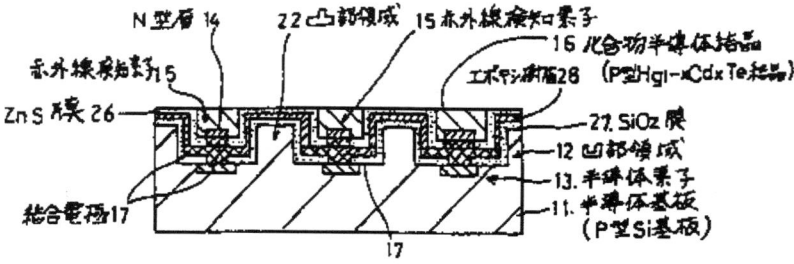

Fig. 2.2.160 (JP-A-2303160 fig. 2)

A p-type layer 15 of HgCdTe having n-type regions 14 formed therein corresponding to photodiodes is formed on a CdTe substrate. The photodiodes are separated by forming grooves in the p-type layer. A second substrate of silicon is prepared with input diodes at recessed regions 12. Indium electrodes 17 are provided and the photodiodes are connected to the input diodes by pressure-bonding. Finally, the CdTe substrate is removed.

*

When detector elements of HgCdTe are provided on a first side of a supporting substrate cross-talk may occur from infrared radiation which is not absorbed in a detector element but is reflected at a side opposite to the first side. To solve this problem it is proposed in JP-A-3077373 (Fujitsu Ltd, Japan, 02.04.91) to provide an HgCdTe layer on the side opposite to the first side of the supporting substrate. This layer will absorb radiation which is not absorbed in one of the detector elements.

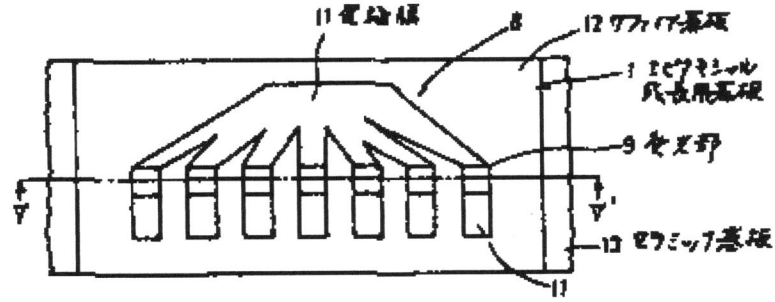

Fig. 2.2.161 (JP-A-3077373 fig. 5a)

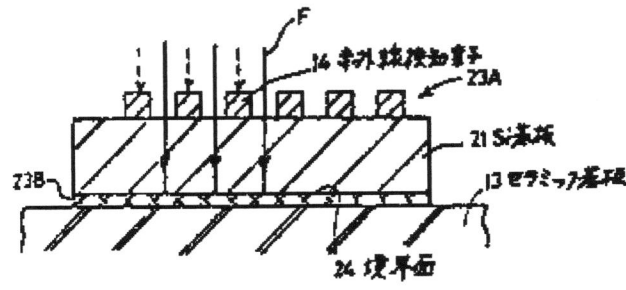

Fig. 2.2.162 (JP-A-3077373 fig. 2c)

Detector elements 14 of HgCdTe are formed on a first side of a silicon substrate 21. An HgCdTe layer 23b is formed on a second side of the silicon substrate, opposite to the first side. Infrared radiation not absorbed in one of the detector elements will not be reflected at the second side of the substrate because the refractive index of silicon is 3.4 and the refractive index of HgCdTe is 3.55. Instead, this radiation will be absorbed in the HgCdTe layer 23b.

*

Cross-talk is reduced in JP-A-3085762 (Fujitsu Ltd, Japan, 10.04.91) by separating photodiodes made in a first set of HgCdTe regions by a second set of HgCdTe regions which have a larger bandgap than the bandgap of the first set of HgCdTe regions.

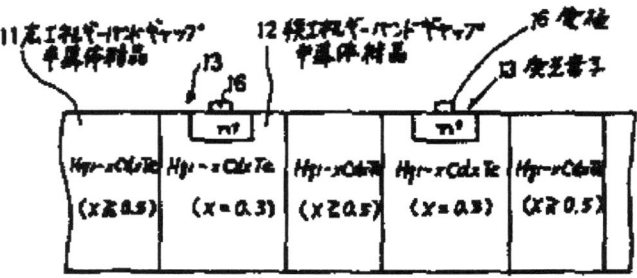

Fig. 2.2.163 (JP-A-3085762 fig. 3)

A first set of p-type $Hg_{0.7}Cd_{0.3}Te$ regions 12 and a second set of $Hg_{0.5}Cd_{0.5}Te$ regions 11 are formed alternately. The bandgap of the $Hg_{0.7}Cd_{0.3}Te$ is narrower than the bandgap of the $Hg_{0.5}Cd_{0.5}Te$ regions. Photodiodes 13 are made in the regions 12 by forming n-type regions. The n-type regions are connected by indium electrodes 16.

* *

In JP-A-3175682 (Mitsubishi Electric Corp., Japan, 30.07.91) a structure is disclosed which prevents cross-talk between adjacent photodiodes. HgCdTe regions are provided which have larger energy bandgap than the energy bandgap of HgCdTe regions containing the photodiodes.

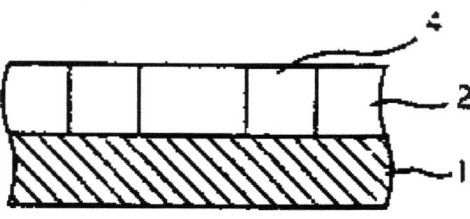

Fig. 2.2.164 (JP-A-3175682 fig. 2c)

A p-type HgCdTe layer 2 is epitaxially grown on a CdTe or CdZnTe substrate 1 and trenches 5 are formed in the p-type layer. The trenches are filled with p⁺-type HgCdTe material 4 having a larger energy bandgap than the energy bandgap of the p-type layer.

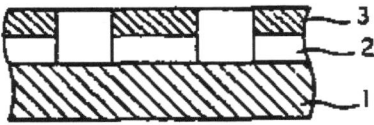

Fig. 2.2.165 (JP-A-3175682 fig. 2d)

Next, photodiodes are created by forming n-type regions 3 at the top surface of the p-type layer 2.

*

In JP-A-3108371 (Fujitsu Ltd, Japan, 08.05.91) cross-talk is reduced by introducing metal atoms by a laser annealing process at regions between photodiodes.

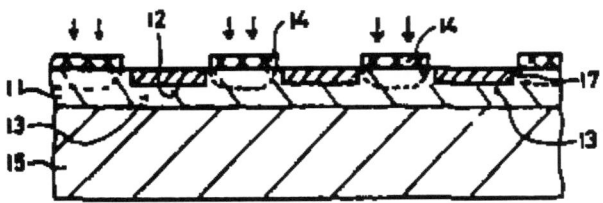

Fig. 2.2.166 (JP-A-3108371 fig. 1c)

A p-type HgCdTe layer 11 is epitaxially grown on a CdZnTe substrate 15. N-type regions 17 are formed in the HgCdTe layer, thereby creating photodiodes 13 with pn-junctions 12. A metallic film 14 is formed on and in contact with the p-type layer. Metal atoms of the film are introduced in the p-type HgCdTe layer by a laser annealing process. The increase of the impurity concentration in these regions reduces the mean lifetime of minority charge carriers, thereby reducing cross-talk.

*

The amount of cross-talk between adjacent photodiodes which are formed on a substrate is reduced in JP-A-3133181 (Fujitsu Ltd, Japan, 06.06.91) by providing radiation absorbing regions in trenches formed in the substrate.

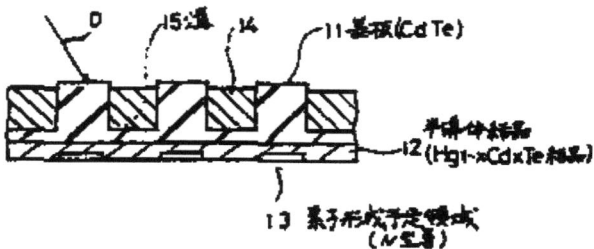

Fig. 2.2.167 (JP-A-3133181 fig. 1b)

N-type regions 13 are formed by ion-implantation in a p-type $Hg_{1-x}Cd_xTe$ layer 12 which has been formed on a CdTe substrate 11. Trenches 15 are formed in the substrate by ion-milling at regions which do not correspond to the n-type regions. Next, radiation absorbing regions 14 of $Hg_{1-y}Cd_yTe$, where $y < x$, are formed in the trenches.

*

The amount of cross-talk is reduced in JP-A-3219670 (Fujitsu Ltd, Japan, 27.09.91) by forming photodiodes on HgCdTe islands placed in recessed regions of a GaAs substrate.

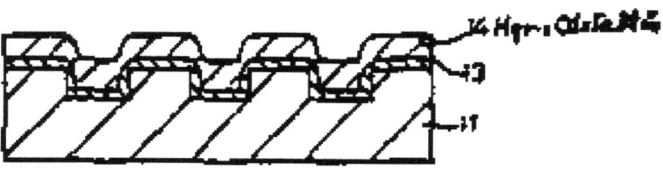

Fig. 2.2.168 (JP-A-3219670 fig. 1d)

Recessed regions are formed in a GaAs substrate 11. A CdTe layer 13 and an HgCdTe layer 14 are formed over the surface of the substrate.

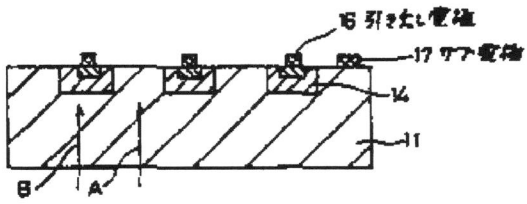

Fig. 2.2.169 (JP-A-3219670 fig. 1g)

The device is subjected to a heat treatment and the HgCdTe layer outside the recessed regions is removed. Thus, p-type HgCdTe islands are formed in the recessed regions. Next, n-type regions are formed in the HgCdTe islands and individual electrodes 16 and a common electrode 17 are provided. Incident infrared radiation B, which has penetrated the GaAs substrate, will be absorbed in an HgCdTe island, while radiation A falling between two islands will not be absorbed.

*

In JP-A-3241774 (Fujitsu Ltd, Japan, 28.10.91) individual detector elements are separated by regions having a relative high concentration of dislocations. The structure reduces the problem of cross-talk.

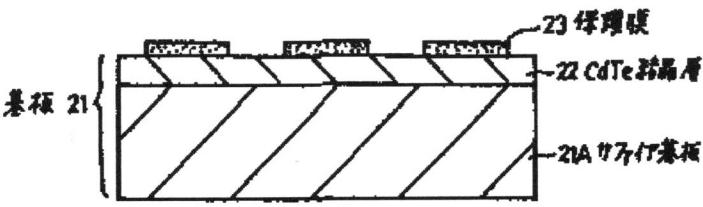

Fig. 2.2.170 (JP-A-3241774 fig. 1a)

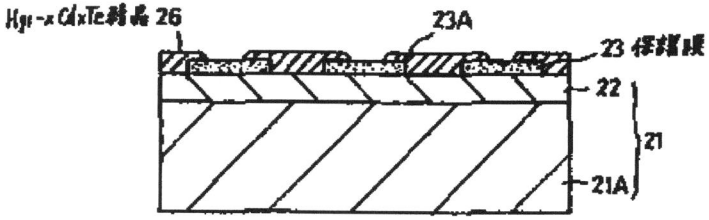

Fig. 2.2.171 (JP-A-3241774 fig. 1b)

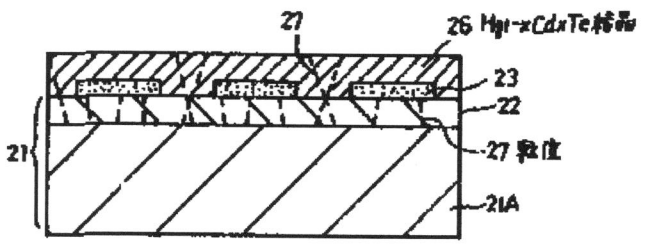

Fig. 2.2.172 (JP-A-3241774 fig. 1c)

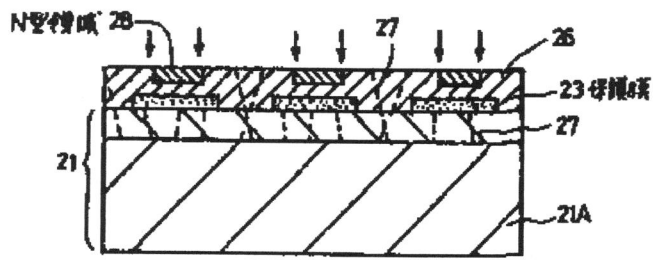

Fig. 2.2.173 (JP-A-3241774 fig. 1d)

The imager comprises a substrate 21A, a CdTe layer 22, a protective film 23, a p-type HgCdTe layer 26 and n-type regions 28 formed in the p-type HgCdTe layer. The n-type regions are positioned above the protective film. The protective film assures the HgCdTe film formed thereon to have a low concentration of dislocations. On the other hand, regions 27 have a high concentration of dislocations thereby capturing minority charge carriers generated in these regions.

*

The imager of EP-A-0445545 (Santa Barbara Research Center, USA, 11.09.91) comprises a radiation absorbing layer which has a varying composition and energy bandgap. The individual detector elements are separated one from another with grooves which may be combined with a ground plane or a guard diode structure.

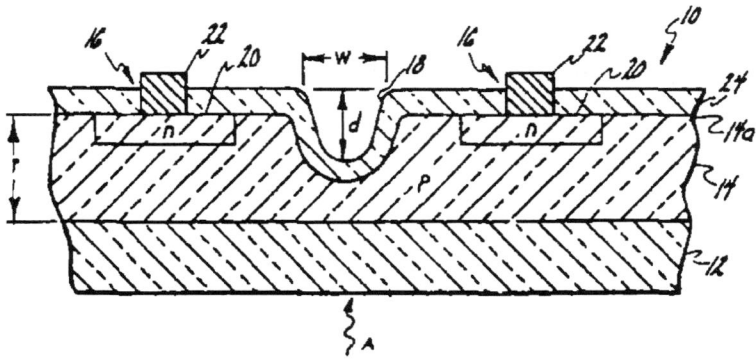

Fig. 2.2.174 (EP-A-0445545 fig. 1b)

A radiation detecting layer 14 overlays a substrate 12. The detecting layer is comprised of $Hg_xCd_{1-x}Te$ where x varies from 0.9 near the substrate to 0.2-0.3 at the surface 14a. Adjacent photodetecting active regions 16 are separated by a groove 18 which extends only partially through the detecting layer. The groove serves to physically prevent a charge carrier created near one active region from diffusing to an adjacent active region. The function of the groove is assisted by constructing the detecting layer as a compositionally graded layer, which causes a reduction of the energy bandgap at the surface as compared to the energy bandgap at the substrate interface. This lowering of the energy bandgap causes the charge carriers to be constrained to exist within the upper surface 14a region of the detecting layer. Each of the active regions includes an n-type region 20 having an associated electrical terminal 22. A passivation layer 24 is applied to the upper surface 14a and the surface of the groove 18 to reduce surface recombination losses of photocarriers.

*

When individual detector elements are separated by ion-milling grooves through an HgCdTe detector layer the HgCdTe crystal may be damaged. In JP-A-3268463 (Fujitsu Ltd, Japan, 29.11.91) individual detector elements are formed by converting regions of a CdTe layer into HgCdTe. The detector elements are separated by the non-converted regions of the CdTe layer.

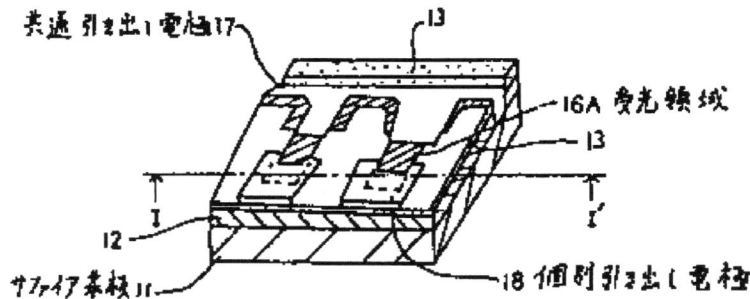

Fig. 2.2.175 (JP-A-3268463 fig. 1d)

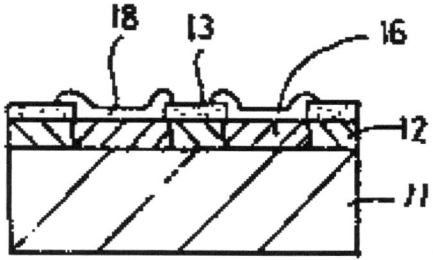

Fig. 2.2.176 (JP-A-3268463 fig. 1e)

A CdTe layer 12 is formed on a sapphire substrate 11. An insulating film 13 is formed at regions not corresponding to the detector elements. An isothermal vapor phase growth method is used to convert regions 16 corresponding to the detector elements.

*

In JP-A-3270269 (Fujitsu Ltd, Japan, 02.12.91) individual detector elements are first formed by converting regions of a CdTe layer and then separating the detector elements by removing non-converted regions of the CdTe layer.

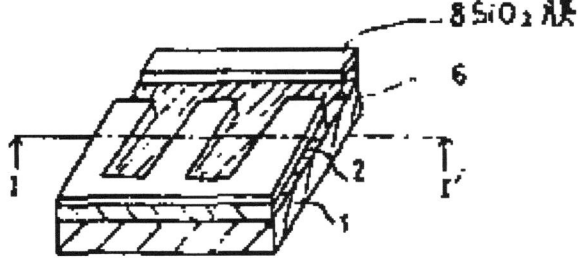

Fig. 2.2.177 (JP-A-3270269 fig. 1c)

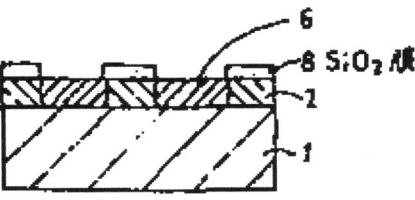

Fig. 2.2.178 (JP-A-3270269 fig. 1d)

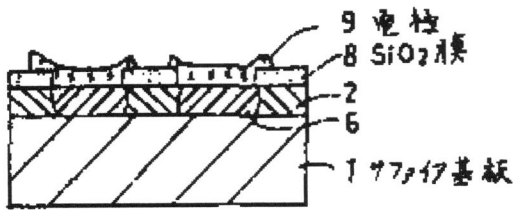

Fig. 2.2.179 (JP-A-3270269 fig. 1e)

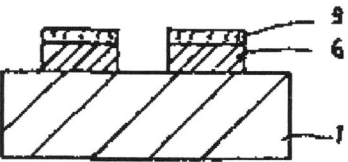

Fig. 2.2.180 (JP-A-3270269 fig. 1f)

A CdTe layer 2 is deposited on a sapphire substrate 1. An insulating film 8 is formed at regions not corresponding to the detector elements. An isothermal vapor phase growth method is used to convert regions 6 to HgCdTe. Electrodes 9 are formed and the insulating film 8 and the non-converted CdTe are removed thereby separating the detector elements.

*

The imager proposed in US-A-5030828 (Grumman Aerospace Corporation, USA, 09.07.91) has parallel elongate cavities formed within a substrate, photosensitive detector elements formed within the cavities and an optical insulating layer adjacent each of the cavities to optically isolate the cavities from each other. The elongate cavities provide an increased detector element surface area which increases the sensitivity. The optical isolation reduces cross-talk among adjacent detector elements.

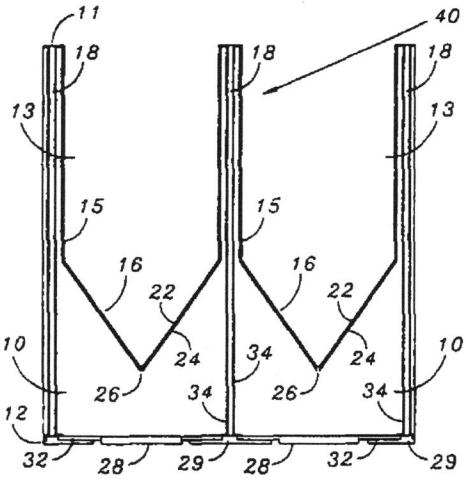

Fig. 2.2.181 (US-A-5030828 fig. 2)

Parallel elongate cavities 13 and a layer of optical isolating material 18, surrounding each of the cavities, are formed within a substrate 10 of CdTe. The cavities each have a vertical wall 15 and a pyramidal floor 16. The wall and floor of each cavity has a body of detector material, HgCdTe, formed as a layer thereon. The body of detector material is comprised of a layer of a first type 22 and a layer of a second type 24. Individual electrical contacts 28 and a common electrical contact 29 are provided. An insulating layer 32 insulates the common contact from the substrate. A diffusion layer 34 of semiconductor material of the second type provides electrical communication between the common contact and the material of the second type formed in the cavity. The cavities may have cylindrical walls and a round floor.

*

An imager in which the optical energy is incident on a p-type doped region prior to being incident on a bulk n-type doped region is presented in EP-A-0485115 (Cincinnati Electronics Corporation, USA, 13.05.92). The short distance between the surface of the p-type region and the pn-junction causes efficient detectivity including at the short wavelength end of the spectrum of interest. The description is made for InSb detectors. HgCdTe is mentioned.

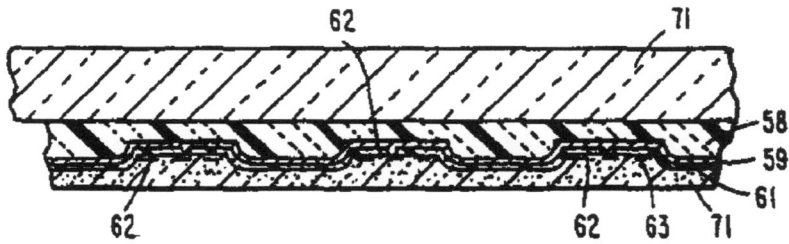

Fig. 2.2.182 (EP-A-0485115 fig. 6)

P-type regions 62 are formed in an n-type substrate 71. A metal grid 59, which connects the p-type regions, is formed over an insulating layer 61. The substrate is bonded to a dielectric plate 57 by an epoxy glue layer 58, which overlays the metal grid. Next, the thickness of the substrate is reduced.

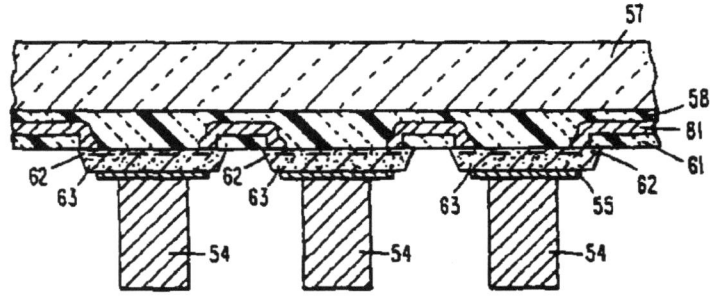

Fig. 2.2.183 (EP-A-0485115 fig. 8)

Trenches are formed in the n-type substrate, thereby separating the photodiodes on n-type islands 63.

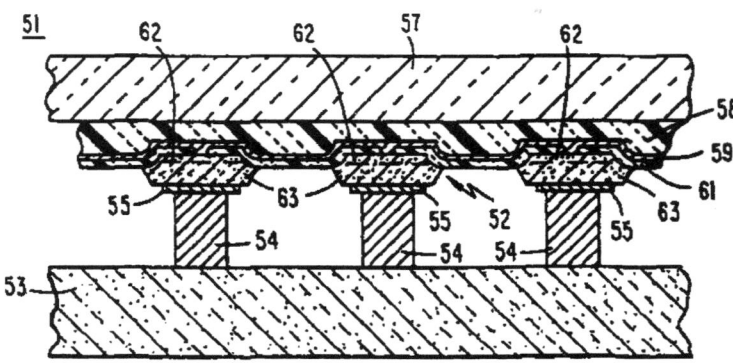

Fig. 2.2.184 (EP-A-0485115 fig. 3)

Metal ohmic contact pads 55 are formed on the n-type islands and the detector is bonded to a multiplexer integrated circuit substrate 53 via indium columns 54.

*

The imager of US-A-5177580 (Santa Barbara Research Center, USA, 05.01.93) refers to a mesa-type photodiode array having U-shaped channels formed through an n-type capping layer, a p-type radiation absorbing base layer and partially into an underlying p-type buffer layer of the type presented in WO-A-8707083. The inherently imprecise nature of the chemical etching process used to delineate the individual mesas results in the formation of ill-defined boundaries between adjacent photo-detectors. Small variations in the etched structures result in variations in the effective optically sensitive areas of the photo-detectors. Therefore, n^+-type regions are ion-implanted along the bottom of the channels, the positions of the regions being more precisely defined than the mesa geometry. These regions function as minority carrier reflectors and confine minority carriers to one side or the other of the region.

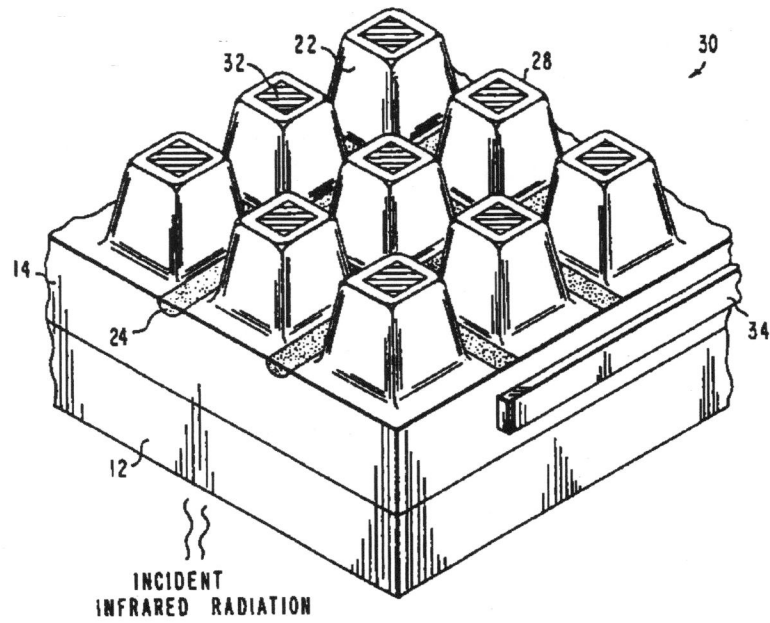

Fig. 2.2.185 (US-A-5177580 fig. 4)

A radiation absorbing layer 14 of n-type HgCdTe overlies a CdZnTe substrate 12. A p-type HgCdTe layer is formed on the radiation absorbing layer. Channels 22 are formed by the use of a wet chemical etching technique. The bottom of the channels is located within the radiation absorbing layer. The same mask used during the etching is also used during an ion-implanting step which produces n^+-type regions 24 along and within only the bottom portion of the channels. Contact metallization pads 32 are formed on the mesas 28 and a common contact metallization 34 is also provided. The n^+-type regions not only reduce cross-talk but also reduce the impedance of the radiation absorbing layer 14.

*

An imager which can select a mode of either high resolution or high sensitivity is presented in EP-A-0497326 (Fujitsu Limited, Japan, 05.08.92).

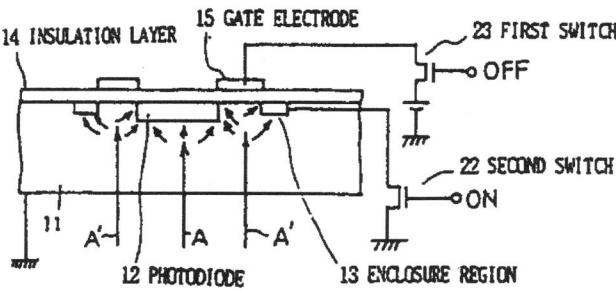

Fig. 2.2.186 (EP-A-0497326 fig. 3a)

Island-shaped surface portions of a p-type substrate 11 are doped so as to form photodiodes 12. Enclosure regions 13 surrounding but apart from the islands are also formed. Gate electrodes 15 are formed upon an insulating layer 14 so as to cover the portions in between the photodiodes and the enclosure reigons.

When a high resolution mode is required, first switches 22 are closed so as to ground the enclosure regions 13 and second switches 23 are opened so as to float gate electrodes 15. Carriers generated in the vicinity of the enclosure regions are attracted and absorbed by the junction potential of the enclosure regions. Thus, no carriers flow into adjacent photodiodes.

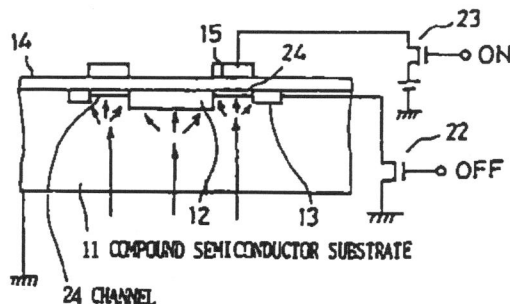

Fig. 2.2.187 (EP-A-0497326 fig. 3b)

When a high sensitivity mode is required, the first switches 22 are opened so as to float the enclosures and the second switches 23 are closed so as to apply a positive voltage to the gate electrodes 15. Due to the positive voltage of the gate electrodes, the surface of substrate 11 beneath the gate electrodes is inverted to n-type so as to form channels 24. These channels electrically connect the enclosure regions 13 to the photodiodes 12. Accordingly, all the carriers generated within the enclosure regions are gathered by the photodiodes.

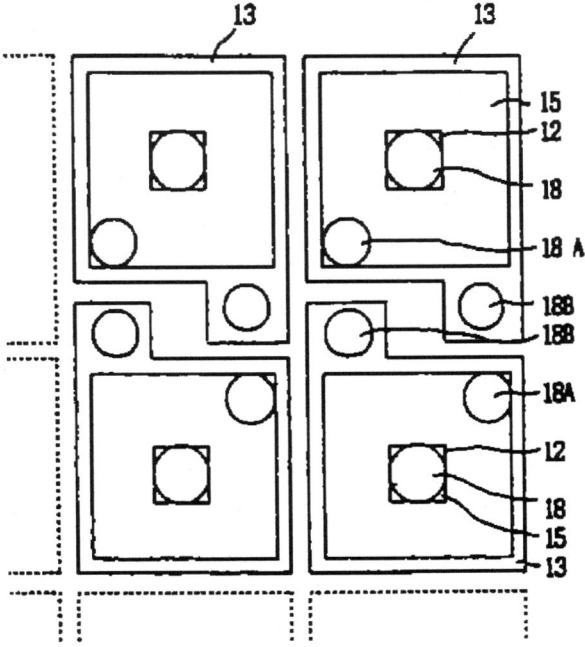

Fig. 2.2.188 (EP-A-0497326 fig. 5)

The switches 22 and 23 are MOSFETs which are integrated in a silicon substrate together with a CCD for reading out the signal charge. The two substrates are bonded together via bumps 18, 18A and 18B, which connect the photodiodes 12, the gate electrodes 15 and the enclosure regions, respectively.

*

In JP-A-4288882 (Mitsubishi Electric Corp., Japan, 13.10.92) photodiodes, which are formed in an HgCdTe layer, are surrounded by regions of HgCdTe with a larger bandgap than the bandgap of the HgCdTe layer.

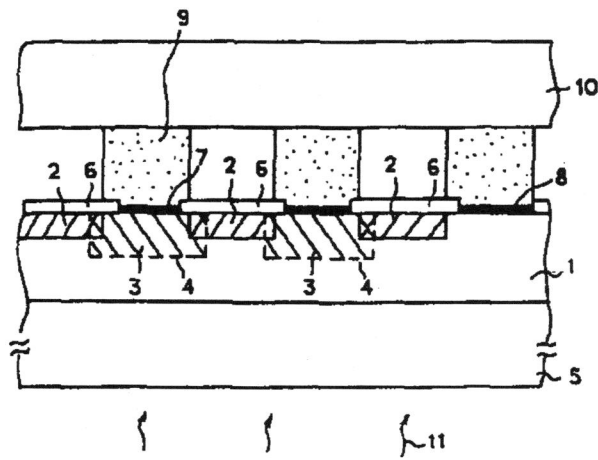

Fig. 2.2.189 (JP-A-4288882 fig. 1)

Protruding regions of $Hg_{0.8}Cd_{0.2}Te$ are formed by a MOCVD method at detector element regions on an $Hg_{0.8}Cd_{0.2}Te$ layer 1 which has been formed on a substrate 5. Regions 2 of $Hg_{0.7}Cd_{0.3}Te$, having a larger bandgap than the $Hg_{0.8}Cd_{0.2}Te$ regions, are formed at the regions surrounding the protruding regions. Next, pn-junctions 4, an insulating layer 6 and electrodes 7 and 8 are formed. The detector array is bonded via bumps 9 to a chip 10 comprising a CCD.

*

The amount of cross-talk is reduced in JP-A-4313267 (Mitsubishi Electric Corp., Japan, 05.11.92). Individual detector elements are separated by grooves in which an absorption layer of HgCdTe is formed.

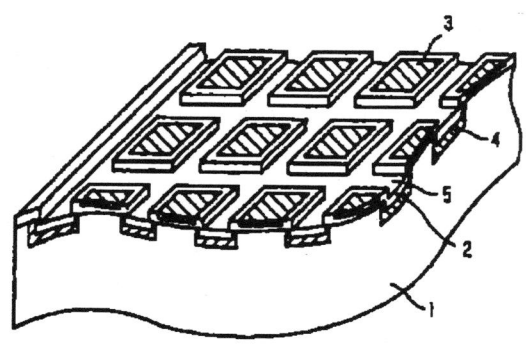

Fig. 2.2.190 (JP-A-4313267 fig. 1)

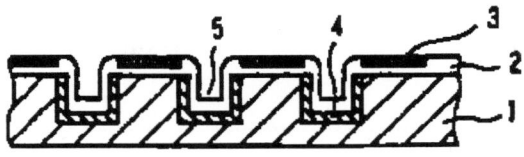

Fig. 2.2.191 (JP-A-4313267 fig. 2d)

Grooves 5 are formed in a substrate 1 of either cadmium telluride, cadmium zinc telluride, gallium arsenic, silicon or sapphire. A layer 4 of $Hg_{1-y}Cd_yTe$ is formed at the bottom and at the sides of the grooves. Next, the whole structure is covered by a p-type $Hg_{1-x}Cd_xTe$ layer (y < x), in which n-type regions 3 are formed.

*

In JP-A-4318970 (Mitsubishi Electric Corp., Japan, 10.11.92) the amount of cross-talk is reduced by forming a light absorbing layer and removing portions corresponding to detector elements by the use of a YAG laser.

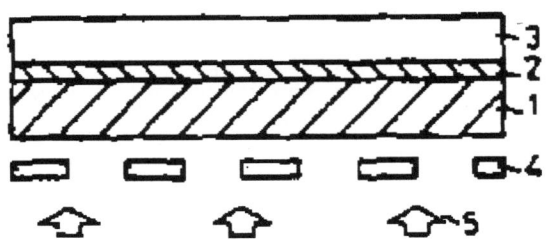

Fig. 2.2.192 (JP-A-4318970 fig. 1c)

An HgTe light absorbing layer 2 is grown on a CdTe substrate 1, followed by a p-type HgCdTe layer 3. The structure is irradiated 5 by a YAG laser at detector element regions. The irradiated HgTe regions are converted to HgCdTe.

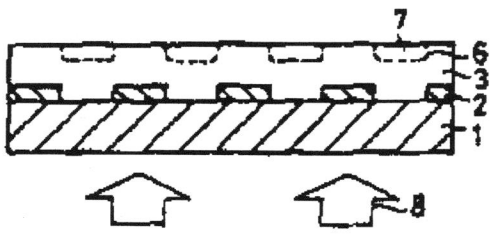

Fig. 2.2.193 (JP-A-4318970 fig. 1e)

Detector elements are created by forming N-type HgCdTe regions 7 in the p-type HgCdTe layer 3.

*

The imager of JP-A-4337676 (Mitsubishi Electric Corp., Japan, 25.11.92) comprises a multilayered structure with alternating CdTe and HgTe layers. An electron beam is used to form HgCdTe islands in which photodiodes are formed. The photodiodes are separated by the multilayered structure. Furthermore, the multilayered structure is formed on a CdTe substrate having recesses formed in its surface.

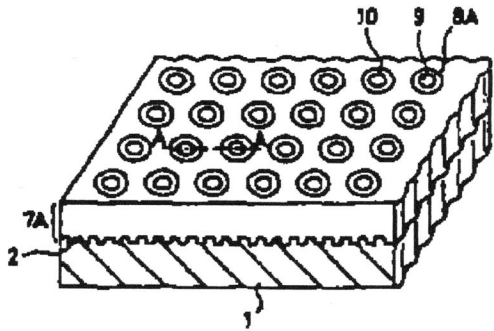

Fig. 2.2.194 (JP-A-4337676 fig. 1)

A multilayer structure 7A of alternating CdTe and HgTe layers is formed on a CdTe substrate 1 which has been prepared with recesses 2. P-type HgCdTe regions 8A are formed by the use of an electron beam and a mask. Photodiodes are created by forming an n-type region in each region 8A.

*

A structure similar to the structure of JP-A-4337676, shown above, is disclosed in JP-A-4337677 (Mitsubishi Electric Corp., Japan, 25.11.92). This structure comprises two multilayer structures having different composition. HgCdTe regions are formed by the use of an YAG laser beam.

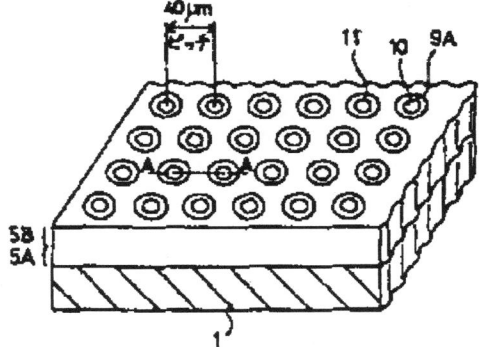

Fig. 2.2.195 (JP-A-4337677 fig. 1)

First and second multilayer structures 5A and 5B of different compositions are formed on a CdTe substrate 1. P-type HgCdTe regions 9A are formed by the use of an YAG laser beam and a mask.

*

In JP-A-5055620 (Mitsubishi Electric Corp., Japan, 05.03.93) n-type regions are formed in a p-type layer by first providing an n-type layer in the p-type layer and then evaporating Hg from regions between pixel regions to thereby form individual n-type regions. This process provides uniform pixel characteristics because the depth of the n-type regions can be well controlled.

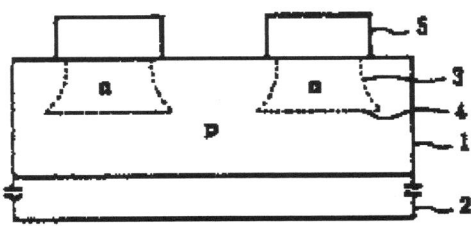

Fig. 2.2.196 (JP-A-5055620 fig. 1d)

An n-type layer 3 is formed by diffusing Hg into a p-type HgCdTe layer 1 which has been formed on a CdTe substrate 2. A cap layer 5 is applied on regions corresponding to pixel regions. The regions between the pixel regions are returned to p-type by heating the structure. The cap layer is removed and the individual n-type regions are connected.

* *

In JP-A-5175476 (Fujitsu Ltd, Japan, 13.07.93) detector element isolation is formed by evaporating mercury at isolation regions from an HgCdTe layer in which the detector elements are formed.

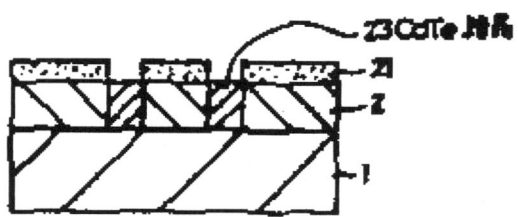

Fig. 2.2.197 (JP-A-5175476 fig. 1c)

A protective film 21 having openings corresponding to isolation regions are formed on an HgCdTe layer 2 which has been formed on a substrate 1. The structure is heat treated and the mercury at the isolation regions evaporates, thereby forming isolation regions of CdTe.

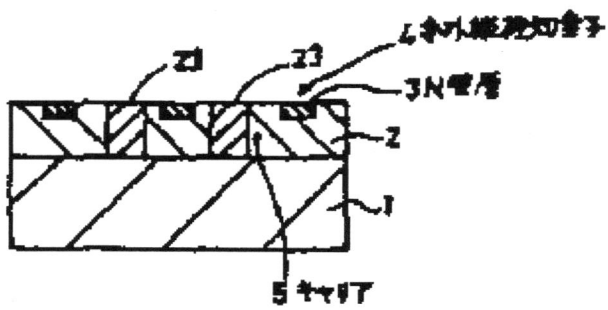

Fig. 2.2.198 (JP-A-5175476 fig. 1d)

The protective film is removed and detector elements are created by forming n-type regions 3 in the HgCdTe layer 2.

*

A method to create high resistance regions between two adjacent photodiodes by a compensating ion-implantation is presented in JP-A-5129580 (Fujitsu Ltd, Japan, 25.05.93).

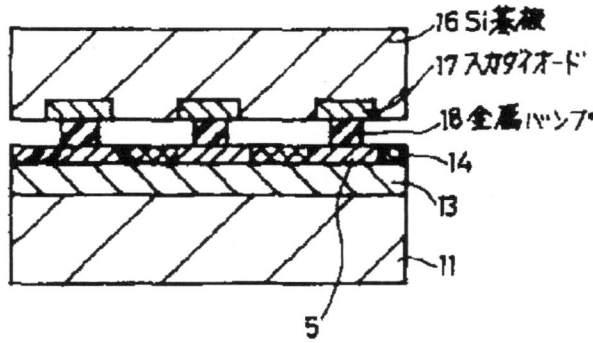

Fig. 2.2.199 (JP-A-5129580 fig. 2d)

A first epitaxial layer of HgCdTe having a first conductivity type is grown on a CdTe substrate 11, followed by a second epitaxial layer of HgCdTe having an opposite conductivity type. Detector element regions 5 are covered by a mask and a compensating ion-implantation is carried out in the second layer to form high resistive regions 14 between the detector elements. A read-out circuit formed in a silicon substrate is connected to the detector elements via connection electrodes 18.

*

The amount of cross-talk between adjacent photodiodes is reduced in JP-A-5226626 (Fujitsu Ltd, Japan, 03.09.93) by providing an inversion layer between the photodiodes.

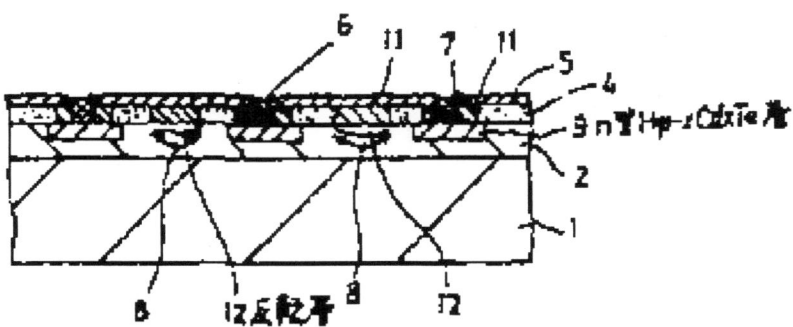

Fig. 2.2.200 (JP-A-5226626 fig. 1b)

An anodic oxide film 11 is selectively formed on a p-type HgCdTe layer 2 which comprises n-type regions 9. Contact holes 6 are provided to the n-type regions. Inversion layers 12 are formed between the n-type regions.

*

The amount of cross-talk is reduced in JP-A-5343727 (NEC Corp., Japan, 24.12.93) where p-type detector element regions are formed in a p^+-type layer. Photodiodes are then formed in the detector element regions.

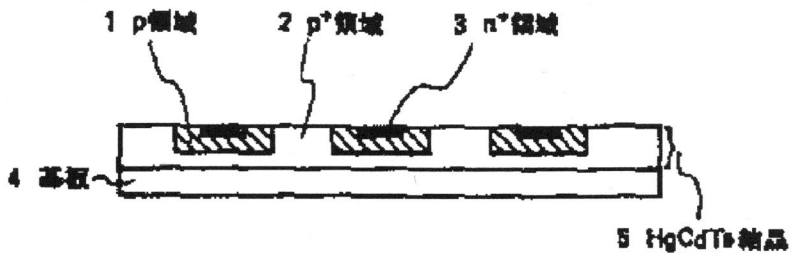

Fig. 2.2.201 (JP-A-5343727 fig. 1)

The imager comprises a p^+-type HgCdTe layer 5 formed on a substrate 4, p-type detector element regions 1 having lower impurity concentration compared to the impurity concentration of the p^+-type layer 5, and n^+-type regions 3 formed in the detector element regions.

*

An imager having HgCdTe photodiodes formed in a silicon substrate is presented in JP-A-6125108 (Fujitsu Ltd, Japan, 06.05.94).

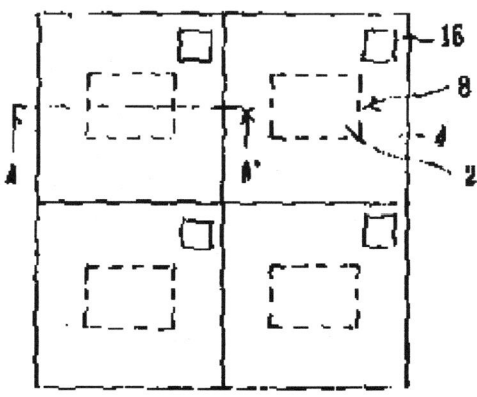

Fig. 2.2.202 (JP-A-6125108 fig. 1a)

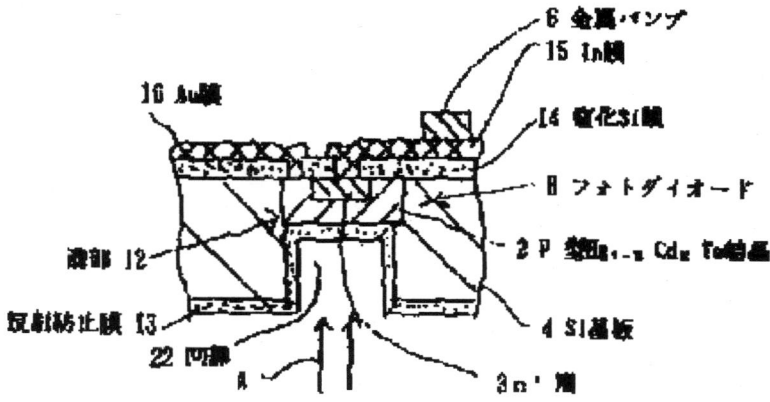

Fig. 2.2.203 (JP-A-6125108 fig. 1b)

Wells 12 are formed in a silicon substrate 4. The wells are filled with HgCdTe material in which photodiodes 8 are formed. Recesses 22 are formed in the silicon substrate down to the HgCdTe material. Metal connections 15, bumps 6 and reflection preventing film 13 are finally provided.

*

A structure with low amounts of crosstalk and dark current is presented in JP-A-6163969 (Fujitsu Ltd, Japan, 10.06.94).

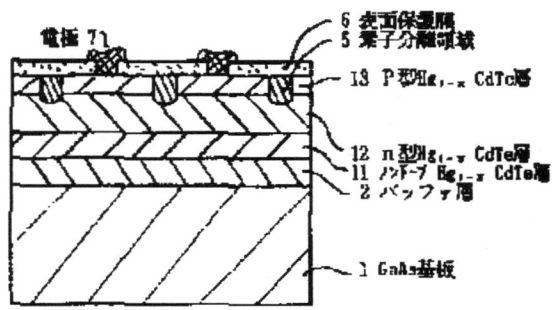

Fig. 2.2.204 (JP-A-6163969 fig. 1a)

A buffer layer 2 of CdTe is grown formed on a GaAs substrate 1. An HgCdTe layer 11, an n-type HgCdTe layer 12 and a p-type HgCdTe layer 13 are formed on the CdTe layer. The n-type layer and the p-type layer form an heterojunction. The individual detector elements are separated by n-type insulation regions 5 which penetrate through the p-type layer. Electrodes 7 connect the p-type regions of the detector elements.

*

A planar detector structure is presented in JP-A-6204449 (Fujitsu Ltd, Japan, 22.07.94). The planar structure is achieved by forming HgCdTe regions in a silicon substrate.

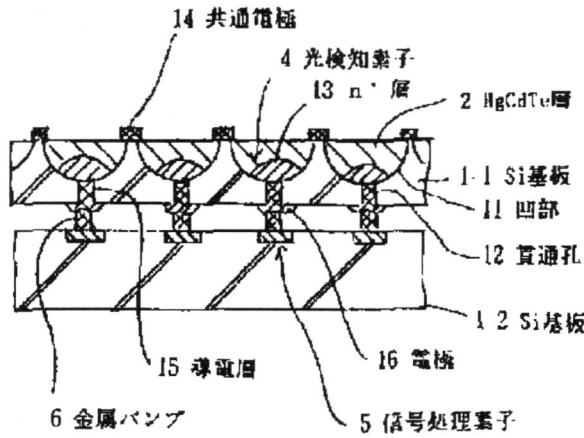

Fig. 2.2.205 (JP-A-6204449 fig. 1)

HgCdTe regions 2 are formed in recessed regions 11 of a silicon substrate 1-1. Holes 12 which reach the HgCdTe regions are formed through the silicon substrate. Impurity atoms are introduced through the holes to form pn-regions in the HgCdTe regions. A common electrode 14 is formed on the array. The holes are filled with a conducting material 15 which connects the HgCdTe regions and outer electrodes 16. Bumps 6 are used to connect the detector array to a silicon read-out chip 1-2.

*

The problem of separating detector elements from each other is addressed in GB-A-2284930 (Mitsubishi Denki Kabushiki Kaisha, Japan, 21.06.95). Several methods are disclosed. In one method regions having an high Cd doping level are formed by an electron or a proton beam.

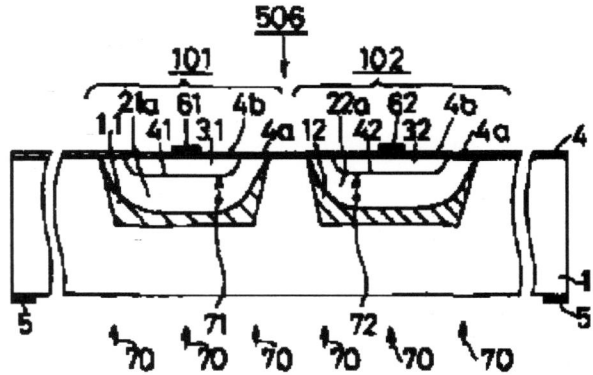

Fig. 2.2.206 (GB-A-2284930 fig. 7)

Pixel regions 31, 32 are separated by regions 11,12 having an high Cd doping level. The regions 11, 12 are formed by irradiating the HgCdTe substrate 1 with an electron or a proton beam to evaporate some of the Hg. In other embodiments the pixel regions are separated by open or insulation-filled grooves, buffer regions with defects, regions of high resistivity semiconductor or a lattice of base semiconductor material. In further arrangements, separation means between pixels are replaced by lenses and electrodes on the face opposite to the regions 31,32 to restrict the region where carriers are produced.

*

Passivation and Leakage Current Preventing Measures

A process for the production of photodiodes by a planar technique is presented in US-A-3988774 (Societe Anonyme de Telecommunication, France, 26.10.76). A thin transition layer is formed between the substrate and a dielectric masking layer necessary for local diffusion to increase the shunt resistance of the photodiodes. The shunt resistance is largely dependent on surface states of the material at the level of transition zone of n-type and p-type.

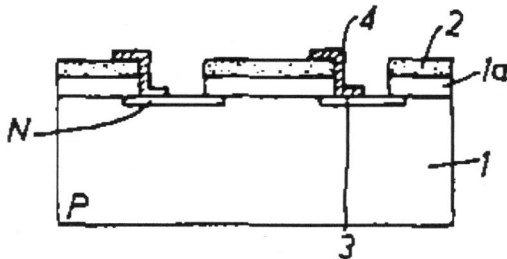

Fig. 2.2.207 (US-A-3988774 fig. 5)

An thin intermediate layer 1a is deposited on a HgCdTe substrate. The intermediate layer is formed from an HgCdTe layer having a mercury concentration equal to or below that of the substrate or zero (CdTe). A dielectric layer 2 is deposited on the intermediate layer. Next, heat treatment is performed to bring about diffusion between the intermediate layer which recrystallizes and the substrate. Windows are opened in the dielectric layer and the intermediate layer and doped areas 3 are formed by diffusion. Finally, metallization 4 is carried out to obtain electrical contacts. The heat treatment makes it possible to obtain an intermediate layer with a composition gradient.

*

To improve the resolution of an imager the distance between adjacent photodiodes is reduced. A problem occurs when the distance reaches the same dimensions as the diffusion length of photogenerated charge carriers. In EP-A-0024970 (Thomson-CSF, France, 11.03.81) an insulating opaque film is introduced between the photodiodes allowing a close spacing of the photodiodes.

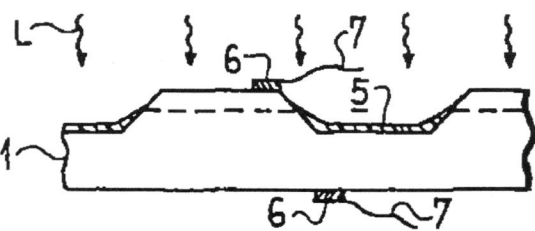

Fig. 2.2.208 (EP-A-0024970 fig. 5)

Mesa shaped photodiodes are formed in an HgCdTe substrate 1. A thin film 5 of an opaque insulating material is formed at regions surrounding the photodiodes. Electrical connections 7 are made to connection pads 6 which are connected to each photodiode and to the substrate.

* *

The structure shown above in EP-A-0024970 is further improved in US-A-4665609 (Thomson-CSF, France, 19.05.87) in which an anodic oxide is applied on the surface of the detector before an opaque metal film is formed at regions surrounding the photodiodes. This anodic oxide stabilizes the pn-junctions.

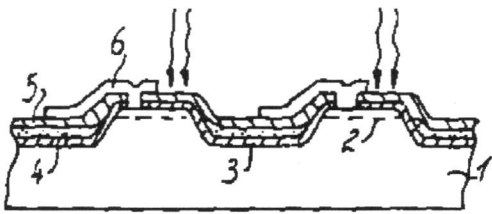

Fig. 2.2.209 (US-A-4665609 fig. 1e)

PN-junctions 2 are formed in a substrate 1. An oxide is grown by anodic oxidation over the whole surface of the substrate. A metallic layer 4 is deposited on the anodic oxide. This layer may consist of a first layer which has a good adhesion to the oxide and a second layer which is opaque to infrared radiation. The metallic layer is shaped by photolithography in order to cover only zones surrounding the detector regions. A dielectric layer 5 is deposited covering the whole surface and the photodiodes are connected by metal connections 6. An embodiment with a planar structure is also shown.

*

The detector elements of JP-A-61268075 (Fujitsu Ltd, Japan, 27.11.86) are formed in mesas where the mercury to cadmium ratio is increased towards the surface of the mesas. The design provides an imager with a low level of surface leakage-currents.

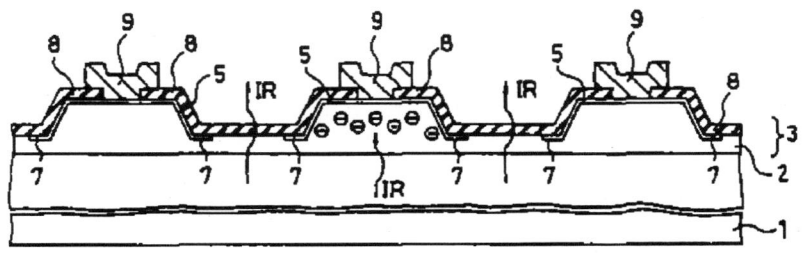

Fig. 2.2.210 (JP-A-61268075 fig. 3)

An $Hg_{1-x}Cd_xTe$ layer 2 is grown epitaxially on a CdTe substrate 1. The value of x has a high value at the interface with the CdTe substrate and decreases as the layer is grown. Mesas are formed in the layer 2 and n-type regions 5 are formed in the surface layer of the mesas thereby forming pn-junctions. Electrodes 9 are connected to the n-type regions.

*

An opaque conductive shield is positioned around photodiodes of an imager presented in JP-A-62090986 (NEC Corp., Japan, 25.04.87). An electrical potential is applied to the shield. The arrangement reduces leakage-currents in the device.

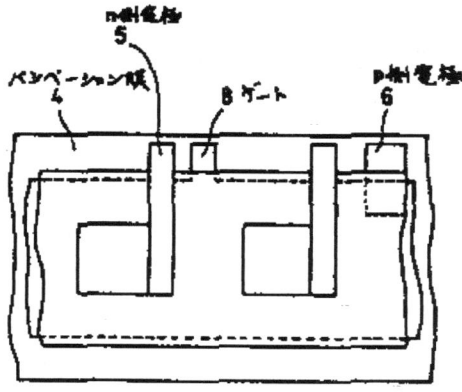

Fig. 2.2.211 (JP-A-62090986 fig. 1)

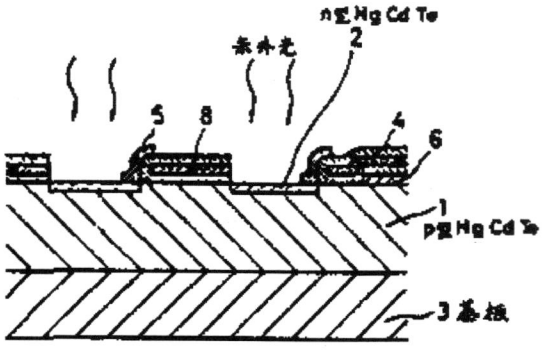

Fig. 2.2.212 (JP-A-62090986 fig. 2)

A p-type layer 1 of HgCdTe is formed on a substrate 3. The HgCdTe layer is connected to an electrode 6. N-type regions 2, forming photodiodes, are provided in the HgCdTe layer. The n-type regions are connected to electrodes 5. An opaque aluminium shield 8, 500 nm thick, is formed on an insulating layer having openings corresponding to the n-type regions. The leakage currents of the device are reduced because the shield is opaque and because an electrical potential is applied to the shield.

*

The surface leakage-current of the imager presented in JP-A-62165973 (Fujitsu Ltd, Japan, 22.07.87) is reduced by providing a layer of HgCdTe with a large bandgap at peripheral regions of the photodiodes of the imager.

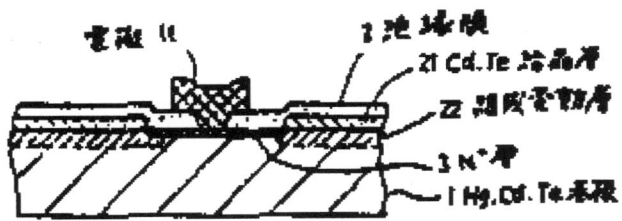

Fig. 2.2.213 (JP-A-62165973 fig. 1)

Photodiodes are created by forming n^+-type regions 3 in a p-type substrate of $Hg_{1-x}Cd_xTe$. A CdTe layer 21 is formed at peripheral regions to the photodiodes. An $Hg_{1-y}Cd_yTe$ layer 22, y > x, is formed under the CdTe layer by diffusion. An insulating film 2 and electrodes 11 are provided. Since the peripheral regions to the photodiodes have a bandgap which is larger than the bandgap of the n^+-type regions and the substrate, the surface leakage currents at the peripheries of the photodiodes are reduced.

*

The conversion efficiency of individual photodiode elements in an imaging array is a function of the area of the pn-junction. As the surface area of an n-type material in contact with a p-type material increases so does the conversion efficiency of infrared radiation to electrical current. However, by increasing the area of the n-type region, the amount of thermal leakage-current, or dark current, generated by the diode also increases, with a consequent reduction in the diode signal-to-noise ratio. Furthermore, the n-type regions within the diode array are typically formed by ion implantation techniques. This implantation process causes damage to the crystalline lattice thereby increasing the thermally induced diode leakage-current.

These problems are addressed in the invention of US-A-4751560 (Santa Barbara Research Center, USA, 14.06.88). A field plate and a guard plate are formed which, when correctly biased, generate an inversion layer at a semiconductor surface surrounding each photodiode.

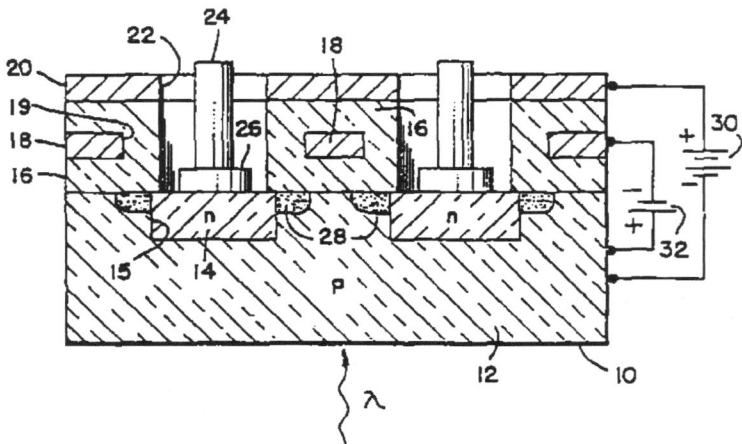

Fig. 2.2.214 (US-A-4751560 fig. 2)

N-type well regions 14 are formed within an upper surface of a p-type substrate 12 of HgCdTe creating pn-junctions which define diodes 15. An insulating layer 16 of silicon dioxide or ZnS deposited by a chemical vapor deposition process overlies the upper surface except for the regions 14. A metal layer 18 of titanium having openings 19 forms a guard plate. The openings 19 are of a larger area than the underlying regions 14. The layer 16 of insulating dielectric is further built up by a chemical vapor deposition process. A second metal layer 20 of titanium is formed on the dielectric layer. This layer has openings 22 and serves as a field plate. The areas of the openings 22 are slightly smaller than the areas of the underlying regions 14. The inequality of opening areas between the field plate and the guard plate permits the development of an inversion layer 28 in the surface of the substrate 12 laterally adjacent to the regions 14, which surface region underlies the field plate but not the guard plate. Diode contacts 24 of indium connect the photodiodes to a second semiconductor device comprising a multiplexer.

The inversion layer operates electrically to fill up the surface energy states of the substrate 12 with electrons. This results in a reduction of the leakage, or thermal current, thereby increasing the signal-to-noise ratio. Furthermore, an n-type layer is created within the surface of substrate 12. This n-type layer effectively enlarges the area of the n-type regions 14, resulting in a larger pn-diode, hence an increased conversion efficiency of radiation to electrical current. A reduction in the amount of dark current results because the fractional area of the n-type diode junction produced by the ion implant is reduced.

*

The surface leakage-current of the imager presented in JP-A-62224982 (Fujitsu Ltd, Japan, 02.10.87) is improved by forming anodic sulphide films at surface regions overlying the pn-junctions of the photodiodes of the imager.

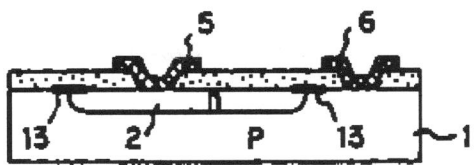

Fig. 2.2.215 (JP-A-62224982 fig. 2d)

Anodic sulphide films 13 are provided at surface regions where junctions are made between a p-type HgCdTe substrate and an n-type region 2. An insulating layer of silicon dioxide is formed and the p-type substrate and the n-type region are contacted by electrodes 5 and 6.

*

JP-A-62224983 (Fujitsu Ltd, Japan, 02.10.87) also refers to forming anodic sulphide films at the surface of an HgCdTe substrate. A metal grid, covered with silicon dioxide, is formed before the film is formed.

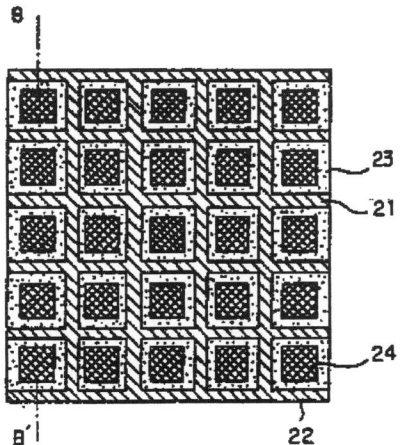

Fig. 2.2.216 (JP-A-62224983 fig. 1a)

An electrode is formed as a metal grid 22 on a surface of an HgCdTe substrate 21. The metal grid is covered by an insulating film 23. Openings corresponding to detector element regions are formed in the insulating film and an anodic sulphide film 24 is formed in the openings.

*

An imager having mesa-shaped photodiodes overlayed with a layer of passivation which contains a fixed charge is presented in WO-A-8707083 (Santa Barbara Research Center, USA, 19.11.87). The charge creates an inversion layer within the surface of the mesa walls, thereby enlarging the pn-junction and reducing the diode leakage-current of each photodiode.

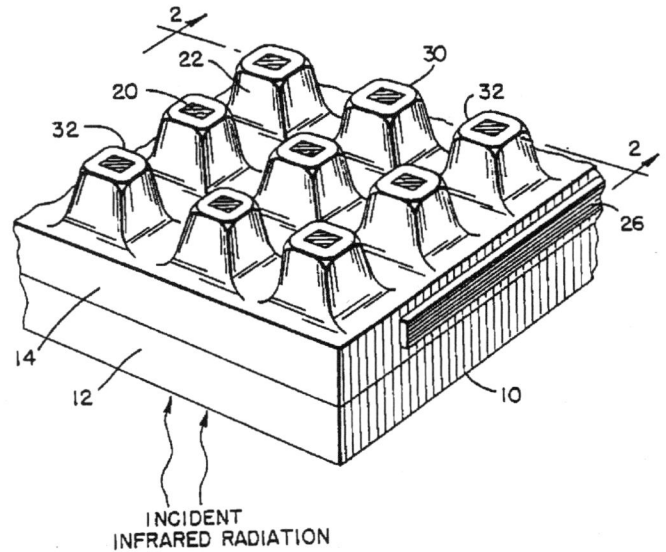

Fig. 2.2.217 (WO-A-8707083 fig. 1)

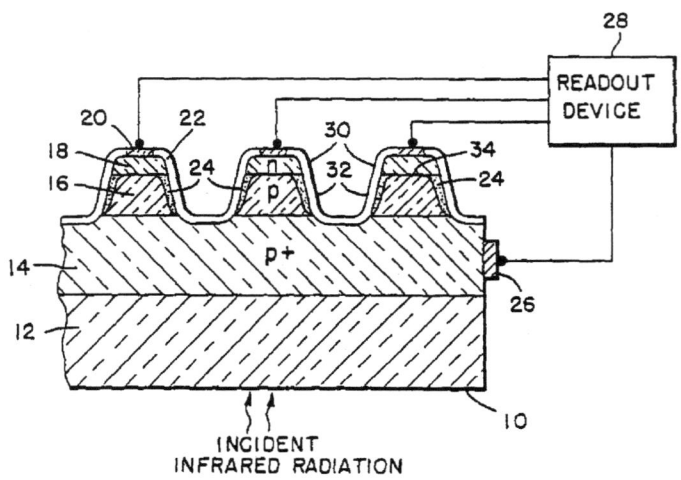

Fig. 2.2.218 (WO-A-8707083 fig. 2)

A photodiode array 10 comprises a substrate 12 of CdTe or CdZnTe, an epitaxially grown transparent buffer layer 14 of p-type $Hg_{0.55}Cd_{0.45}Te$, an epitaxially grown base layer 16 of p-type $Hg_{0.7}Cd_{0.3}Te$ and an epitaxially grown capping layer 18 of n-type $Hg_{0.7}Cd_{0.3}Te$. U-shaped grooves 32 are etched through the base layer and partly into the buffer layer, thereby forming mesas 30. A layer of passivation 22, composed of silicon dioxide, is formed over the mesa surfaces. It contains a fixed positive charge which interacts with the underlying surface of the base layer to create an inversion layer 24. The inversion layer causes an overall decrease in the surface leakage effects of the array. Furthermore, the inversion layer acts to enlarge the n-type capping layer, thus creating a larger pn-junction 34. Electrical contact is made by diode contact metallization areas 20 and a common contact metallization area 26 which is in contact with the buffer layer. Radiation which has passed through the transparent substrate and the buffer layer is absorbed in the base layer.

*

In a structure which is fabricated by etching a plurality of mesas to isolate individual photodiodes, the resulting non-planar surface is difficult to passivate. Furthermore, the pn-junctions intersect the surface of the device which gives a high leakage-current due to surface states. Another disadvantage is that the device is removed from a growth chamber in order to etch the mesas thereby increasing the susceptibility to surface contamination. These are some of the drawbacks of the mesa structure which are acknowledged in US-A-5189297 (Santa Barbara Research Center, USA, 23.02.93). Instead, a structure having a planar surface is proposed. Isolation junctions are formed by a thermally driven process converting regions in a top layer to an opposite type of material. A method to fabricate the structure without removing it from an epitaxial reactor is also disclosed.

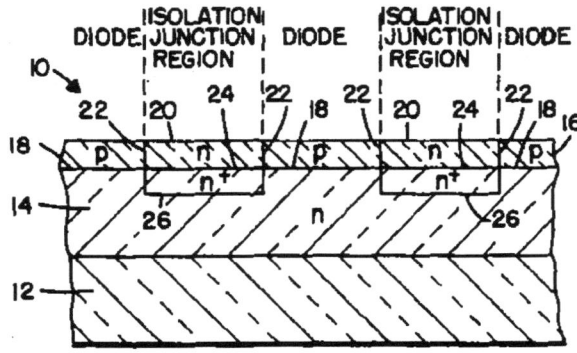

Fig. 2.2.219 (US-A-5189297 fig. 1)

An n-type base layer 14 of HgCdTe overlies a substrate layer 12 of CdZnTe. A thin p-type HgCdTe collector layer 16 having a wider bandgap than the base layer is formed on the base layer. Isolation junctions 20 are formed by type-converting the p-type collector layer to n-type material by a thermally driven process. PN-homojunctions 22 are formed at the edges of the isolation junctions which isolate the individual photodiodes. Heterojunction 24 and homojunction 26 are created, which together reflect excess minority charge carriers away from the surface of the device as well as from neighboring photodiodes.

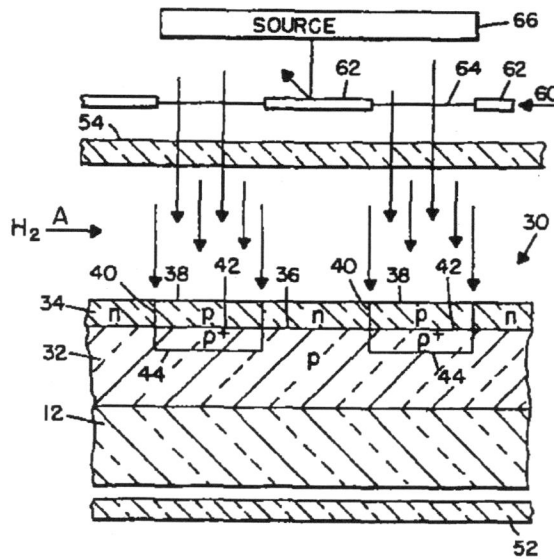

Fig. 2.2.220 (US-A-5189297 fig. 3a)

A method of manufacturing of the array is shown above. The array is positioned within a MOCVD chamber having reactor walls 52 and 54. A base layer 32 and a collector layer 34 are grown on a substrate 12. A mask 60 having transparent regions 64 is provided external to the reactor. A pulsed laser or an arc lamp is used as a high-intensity source 66 of radiation. The result of the selective absorption of the radiation is to heat the isolation junctions without significantly heating the photodiode regions.

*

The imager of EP-A-0406696 (Santa Barbara Research Center, USA, 09.01.91) is made up of a compositionally graded p-type base layer and a compositionally graded cap layer in which pn-junctions are formed. The structure has a small amount of leakage current and a large diode impedance.

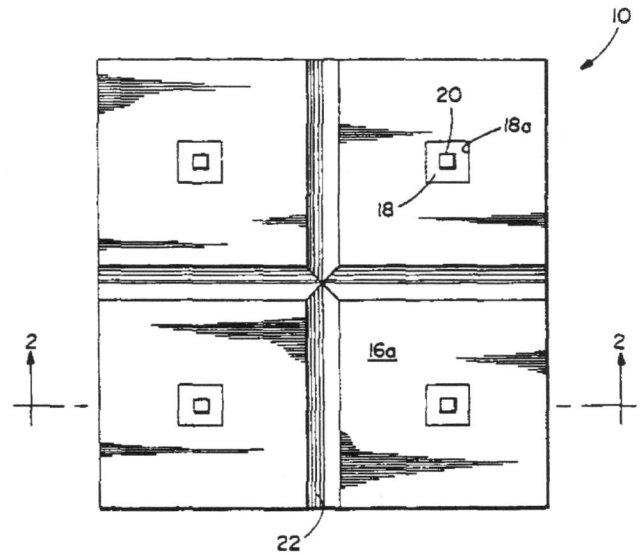

Fig. 2.2.221 (EP-A-0406696 fig. 1)

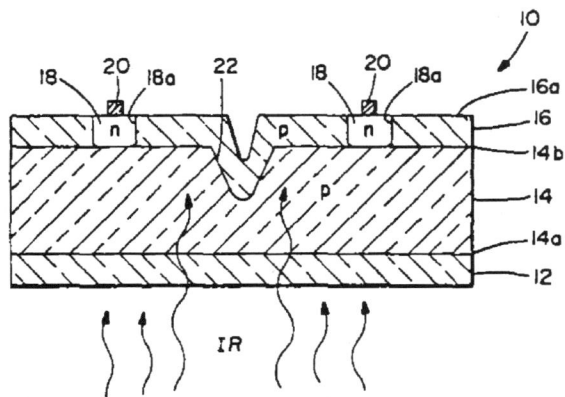

Fig. 2.2.222 (EP-A-0406696 fig. 2)

A substrate 12 of CdZnTe is overlied with a radiation absorbing base layer 14 having a lower surface 14a in contact with the substrate 12 and an upper surface layer 14b. The base layer is comprised of p-type $Hg_{1-x}Cd_xTe$ where x has a value of 0.6 to 0.8 at the surface 14a and is graded in value such that at the surface 14b x has a value of 0.2, 0.3 and 0.4 for long-, mid-, and short-wavelength infrared radiation, respectively. The energy band gap of the base layer nonlinearly decreases in width as the surface 14b is approached. The base layer has a thickness of 15 microns. Overlying the base layer is a graded composition $Hg_{1-x}Cd_xTe$ cap layer 16 that is compositionally graded to wider bandgap material comprised of CdTe, $Hg_{1-x}Cd_xTe$, HgCdZnTe, $Cd_{1-y}Zn_yTe$ or combinations thereof at the upper surface 16a. The energy band gap of the cap layer increases in width as the surface 16a is approached. The cap layer, with a wider band gap relative to the upper portion of the layer 14, functions as a layer of high quality surface passivation. The array includes n-type regions 18 within the cap layer for forming photodiode pn-junctions 18a. Contact metalization 20 is formed to couple the photodiodes to an external read-out circuit. A channel 22 which is etched before the deposition of the cap layer is formed partly into the base layer inorder to prevent cross-talk between individual photodiodes.

*

The imager presented in GB-A-2246907 (Mitsubishi Denki Kabushiki Kaisha, Japan, 12.02.92) includes a transistion layer formed between an HgCdTe light absorbing layer and a CdTe substrate. The energy bandgap of the transistion layer varies through its thickness. The structure includes only one grown layer since the transistion layer is a natural product of the growth process when the light absorbing layer is grown. Moreover, the amount of surface leakage-current is low in the structure because the pn-junctions of the imager are not exposed at any surface.

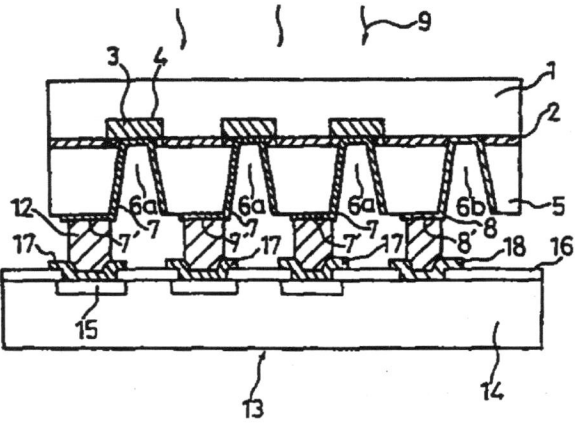

Fig. 2.2.223 (GB-A-2246907 fig. 5)

A transition layer 2 is formed between a CdTe substrate 5 and an $Hg_{0.8}Cd_{0.2}Te$ light absorbing layer 1. N-type regions 3 are formed within parts of the transition layer 2 and the light absorbing layer 1. Recesses 6a extend through the substrate to corresponding doped regions 3. An electrode 7 is disposed on the substrate within each of the recesses and is in contact with a corresponding doped region 3. Another recess 6b extends through the substrate and the transition layer to the light absorbing layer. An electrode 8 is disposed in the recess 6b in contact with the light absorbing layer. Each of the electrodes 7 and 8 includes a planar portion

7', 8' that extends onto the front side of the photo-detector. These planar portions are employed to provide bases for indium masses 12 that are used as bump electrodes interconnecting the photo-detector to a CCD 13.

*

The photodiodes of the imager in JP-A-4318979 (NEC Corp., Japan, 10.11.92) are created by first forming an n^+-type region by ion-implantation and then carrying out an annealing step. The pn-junctions of the photodiodes will be situated away from the regions which have been damaged by the ion implantation. The structure shows a high sensitivity.

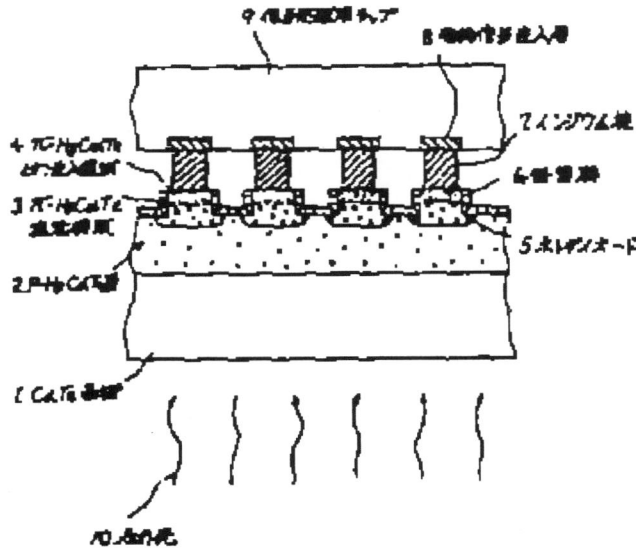

Fig. 2.2.224 (JP-A-4318979 fig. 1)

N^+-type HgCdTe regions 4 are formed by ion-implantation in a p-type HgCdTe layer 2 which has been formed on a CdTe substrate 1. The individual detector elements, defined by the n^+-type regions, are separated by mesa etching. Thereafter, the structure is heated to form n^--type regions by diffusion.

*

Contact metals often used for photovoltaic HgCdTe sensors include gold (Au) with a nickel (Ni) overcoat for individual p-type contacts, and palladium (Pd) with a nickel overcoat for n-type common contacts. However, during high temperature storage both Au and Pd diffuse into HgCdTe, causing a high density of dislocations in the case of Pd, and shorting out the pn-junction in the case of Au, thereby degrading the performance of the device. Furthermore, the Au/Ni and Pd/Ni metal systems each have a coefficient of thermal expansion that differs from the coefficient of thermal expansion of HgCdTe. In US-A-5296384 (Santa Barbara Research Center, USA, 22.03.94) a fabrication process is presented which combines passivation and contact metallization of the p and n sides of a photodiode junction in such a way that a contact metal is applied before a passivation film is formed and before a high temperature anneal is carried out. The method also includes the fabrication of a layer of dielectric overglass to eliminate the occurrence of undesirable chemical reactions during a hybridization process, and also to prevent Hg from diffusing through the passivation during high temperature storage. The fabrication process is claimed in US-A-5296384 and the corresponding device is claimed in US-A-5401986 (Santa Barbara Research Center, USA, 28.03.95).

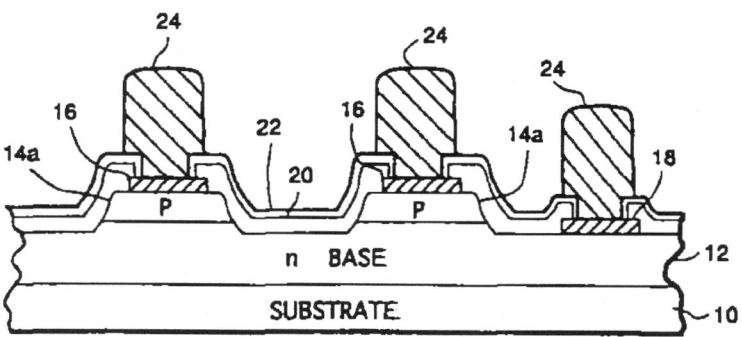

Fig. 2.2.225 (US-A-5296384 fig. 2g)

An n-type HgCdTe base layer 12 is grown on an insulating substrate 10 of CdZnTe. A p-type HgCdTe cap layer is epitaxially grown upon the base layer. Individual photodiodes, patterned by a mesa-etch process, are formed in regions 14a of the cap layer. Metal contact pads 16 and 18 of molybdenum (Mo) are formed. Mo has a low diffusion characteristic at elevated temperatures and a more closely matched coefficient of thermal expansion to HgCdTe compared to the contact metals mentioned above. The structure is first overcoated with a layer 20 of wide bandgap material such as CdTe, CdZnTe or CdTeSe, and then annealed in Hg vapor at a first temperature and then at a second lower temperature. One result of the anneal is to interchange Hg and Cd at cation sites along an interface between the layers 12 and 14a and the layer 20. Next, windows are opened through the layer 20 to expose the Mo contacts 16 and 18. A dielectric layer 22 is formed over the structure and a mask is used to open windows to expose the Mo contacts 16 and 18. Next, indium bumps 24 are applied to the Mo contact pads. Only the indium bumps are exposed, with the underlying Mo contact metallization buried beneath the dielectric layer 22. The dielectric layer is also substantially impervious to the passage of Hg, and thus prevents the out-diffusion of Hg from the base layer 12 and the regions 14a.

*

A structure with low amounts of crosstalk and dark current is presented in JP-A-6163969 (Fujitsu Ltd, Japan, 10.06.94).

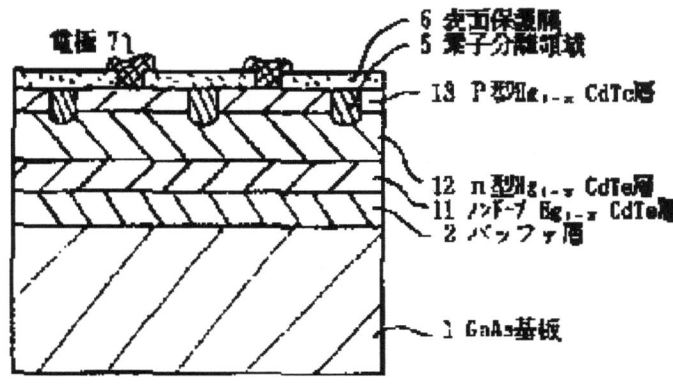

Fig. 2.2.226 (JP-A-6163969 fig. 1a)

A buffer layer 2 of CdTe is grown formed on a GaAs substrate 1. An HgCdTe layer 11, an n-type HgCdTe layer 12 and a p-type HgCdTe layer 13 are formed on the CdTe layer. The n-type layer and the p-type layer form an heterojunction. The individual detector elements are separated by n-type insulation regions 5 which penetrate through the p-type layer. Electrodes 7 connect the p-type regions of the detector elements.

*

The leakage current of the imager presented in JP-A-6237005 (Fujitsu Ltd, Japan, 23.08.94) is reduced. N$^+$-type photodetector regions which have been formed in a p-type substrate are surrounded by n-type regions at the surface portion of the substrate.

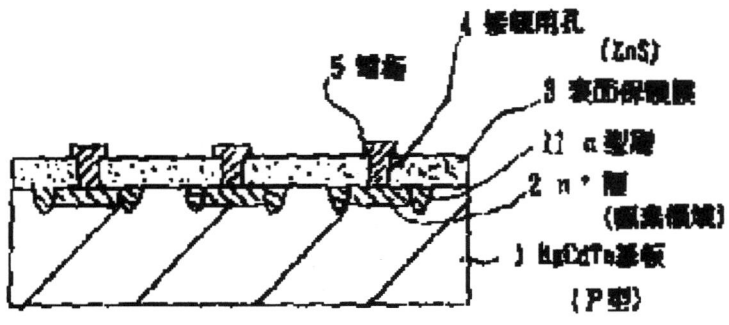

Fig. 2.2.227 (JP-A-6237005 fig. 1a)

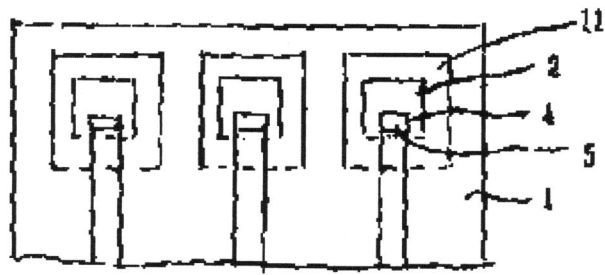

Fig. 2.2.228 (JP-A-6237005 fig. 1b)

N⁺-type regions 8 are formed in a p-type HgCdTe substrate 1. N-type regions 11 having a doping concentration which is lower than the doping concentration of the regions 8 are formed at the outer edges of the regions 8 at the surface of the substrate. The regions 8 are connected by connectors 4 and electrodes 5.

*

Chapter 2.3

Flip-Chip Arrangements

Flip-chip arrangements for connecting a mercury cadmium telluride chip to a read-out chip are presented in this chapter.

Summary

At the time US-A-3808435 was filed, a conventional infrared detector system had an array of infrared detectors and each detector had an amplifier for amplifying its output. The amplified output was connected to a multiplexer scan converter. Such a system required a large number of amplifiers. It is acknowledged in US-A-3808435 that an improved system would be obtained if an imager with a charge transfer device (CTD) could be used. Two problems, however, were encountered in this technique. The first problem was the fabrication of a CTD on an infrared sensitive material and the second the fact that the thermal background radiation at room temperature is many orders of magnitude greater in the infrared than in the visible (41 orders of magnitude greater at 10 µm than at 500 nm).

- The problems mentioned above are overcome in US-A-3808435 by a.c. coupling detector signals of an infrared detector array to a CTD array. The detector array and the CTD array are formed in an HgCdTe chip and a silicon chip respectively and are connected by placing the chips on top of each other with intermediate conductor elements. This technique of connecting two chips is generally called the flip-chip technique.

- In US-A-4369458 the contact array of an infrared detector includes a substantially greater number of contacts than the contact array of a signal processor. At the flip-chip bonding operation it is not a prerequisite to precisely align the two integrated circuits in a one-to-one contact relationship.

- A method to manufacture an imager having a flip-chip structure is given in JP-A-57073984. The detector substrate is thinned and a slot is used to facilitate this operation.

- The photodiodes of the imagers presented in US-A-4566024 are formed in a p-type HgCdTe substrate by diffusing n-type impurities from two opposite faces, a front face and a rear face, of the substrate. The photodiodes can be connected to a read-out device by a flip-chip bonding process on the rear face of the substrate and still be illuminated from its front face.

- The imager presented in JP-A-59112652 comprises a detector chip which is bonded to a first side of a silicon read-out chip by a flip-chip technique. Connection is made through the silicon chip to a read-out CCD.

- N-type regions corresponding to photodiodes, are formed in JP-A-62011265 by diffusing indium into a p-type HgCdTe layer. The indium is diffused from straps which also connect the formed n-type regions to a read-out device comprised in a silicon substrate. No alignment between the detector layer and the read-out device is needed which would be the case if the n-type regions were formed before the bonding of the detector array and the read-out device takes place.

- A cold shield layer which is formed on a silicon substrate is introduced in JP-A-62272564. The substrate is bonded by a flip-chip process to an HgCdTe detector substrate.

- In JP-A-63170961 a shape-memory alloy structure is used to release mechanical stress generated in connection bumps during a flip-chip process comprising pressure bonding.

- The resistivity of a common contact to an array of photodiodes can be reduced by the introduction of a highly doped HgCdTe layer as is shown in JP-A-63300559.

- In JP-A-1050560 an indium layer is applied over an n-type layer of HgCdTe which covers a p-type layer of HgCdTe. Individual bumps and individual photodiodes are formed in the same etch step.

- When two substrates, each provided with connection bumps, are joined together, a problem arises in positioning the substrates with respect to one another. The invention of JP-A-1227472 gives a solution to this problem. An metal bump on one of the substrates meets a corresponding contact on the other substrate in recessed regions which are formed in one of the two substrates.

- A difficulty arises in filling the space between a detector substrate and a read-out substrate with epoxy. The epoxy tends to compress the two substrates together upon hardening and in doing so, while providing good electrical contact and mechanical strength, it also causes the bumps, which are often made of indium, to penetrate into the detector material itself which often causes unreliable operation of the device. This problem is solved in WO-A-9001802 where areas surrounding the detectors are not filled with epoxy.

- The indium bumps on the detector substrate and the indium bumps on the read-out substrate in US-A-4935627 have the same geometric shape but are oriented in a nonparallel orientation to each other, or alternatively have the same geometric shape but have different surface areas. These designs facilitate the alignment when several detector arrays are butted together. Furthermore, indium bumps of the same size are unstable against lateral motion during bump bonding operations.

- A method for aligning detector arrays on the surface of a signal processing module is shown in US-A-5075201.

- A detector substrate and a read-out substrate are bonded together in JP-A-4293240 by using two indium alloy bumps for each connection. The two indium alloy bumps have different melting points and the regions where the bumps are joined will have a higher melting point than the indium alloy having the lower melting point.

- Sets of bumps are provided in JP-A-7050330 in addition to the bumps which are used to connect detector elements in a detector substrate to input regions in a read-out substrate. The additional bumps are used when the two substrates are joined to improve the bonding process.

Summary - Thermal Stress Preventing Measures

When the flip-chip technique is used to connect two chips of different materials having different thermal expansion coefficients, the connection will be subjected to mechanical stress to a degree dependent on the the thermal history of the array.

- In JP-A-55150279 the connection between an HgCdTe detector chip and a silicon read-out chip comprises bent buffers which absorb such mechanical stress and which prevent the fragile detector chip being damaged during the connection.

- When a silicon read-out chip and an HgCdTe detector chip are bonded together by a flip-chip bonding process, the chips are exposed to a mechanical stress which may lead to damages, especially of the fragile detector chip. Photo-resist films are used in JP-A-61059771 to form a spacer. The spacer reduces the mechanical stress of the semiconductor chips during the bonding process.

- In JP-A-63268271 metal rods are used to provide spacers between a detector substrate and a read-out substrate.

- The problem of cross-talk between adjacent photodiodes and the problem of different thermal expansion coefficients between a silicon substrate and an HgCdTe substrate are approached in JP-A-63281460. A detector is made up of HgCdTe wells, comprising photodiodes, formed in a CdTe substrate. The detector is bonded to a silicon substrate by a flip-chip process, and finally the CdTe substrate is etched away.

- A similar structure to the structure discussed in JP-A-63281460 above is disclosed in JP-A-63296272. In the latter, the HgCdTe wells are not connected by indium bumps but by Au films supported by an epoxy resin.

- The problem of difference in thermal expansion coefficients between an HgCdTe substrate and a silicon read-out substrate is addressed in JP-A-2214159 where a design similar to the design of JP-A-63296272 is presented.

- Cross-talk is reduced in US-A-5376558 by forming a three layered light absorption structure in the areas between the photodiodes of the array. The second layer has a refractive index which is greater than that of the first layer and the third layer is a metal layer.

- To reduce the effect of difference in thermal expansion coefficients between a silicon read-out substrate and an HgCdTe detector layer, which is bump bonded to the silicon substrate, it is proposed in JP-A-1061056 to affix a second silicon substrate to the HgCdTe detector layer with an intermediate layer of GaAs.

- The problem of difference in thermal expansion coefficients between a silicon read-out substrate and an HgCdTe detector substrate is approached in US-A-4783594 by filling the space between the two substrates with a resilient electrically insulating polymeric material and thereafter separating the detector elements from each other by removing a layer of the HgCdTe detector substrate.

- The invention of US-A-5365088 approaches the problem of stresses generated by the mismatch of the thermal expansion coefficients of HgCdTe and silicon by including a buffer layer of sapphire. The characteristic thermal expansivity of sapphire is more similar to the thermal expansivity of HgCdTe than that of silicon.

- When indium bumps are used to connect two chips having different thermal expansion coefficients and the chips are exposed to repeated temperature cycling, the mechanical stress in the bumps is reduced when the bumps are made taller. If the indium bumps are formed using vapor deposition through a photo-reduced mask pattern, the height of the bumps are limited to 6-9 microns. A fabrication process for making taller indum bumps is disclosed in EP-A-0405865.

- Mechanical stress in connection bumps formed between a read-out circuit in a silicon substrate and detectors in an HgCdTe substrate due to a difference in thermal expansion coefficients of the two substrates is reduced in US-A-4943491 by bonding a layer of a material, having a greater thermal expansion coefficient than the coefficient of the two substrates, to the silicon substrate.

- An imager in which the optical energy is incident on a p-type doped region prior to being incident on a bulk n-type doped region is presented in EP-A-0485115. The short distance between the surface of the p-type region and the pn-junction causes efficient detectivity including at the short wavelength end of the spectrum of interest. The description is made for InSb detectors. HgCdTe is mentioned.

- Mechanical stress due to temperature variations is reduced in the imager of JP-A-4199877 by growing two layers of HgCdTe of different compositions on each side of a silicon substrate. One of the layers comprises the photodiodes and the second is transparent to the infrared radiation of interest.

- A detector layer of an imager presented in US-A-5264699 is thinned to allow the detector to act like a flexible membrane and to elastically respond to thermal mismatch resulting from different coefficients of thermal expansion between the detector and a semiconductor read-out circuit.

- In JP-A-6163865 the mechanical stress of indium bumps connecting a silicon read-out substrate and an HgCdTe detector substrate is reduced by bonding the two substrates together at a temperature which is between the operational and the non-operational temperature of the detector.

- During the bonding process in JP-A-7111323 a silicon read-out substrate and an HgCdTe detector substrate are held at different temperatures. The amount of stress in the connector bumps will be reduced when the device is cooled from room temperature to an operating temperature of 77 K.

- In WO-A-9417557 an integrated circuit assembly is presented which includes a silicon thin film circuit bonded to a substrate of a material selected to provide the assembly with an effective thermal expansion characteristic that approximately matches that of an HgCdTe detector array.

- An imager with insulator walls formed between connection bumps is shown in JP-A-6236981. The structure reduces the detrimental effects due to a difference in the coefficient of thermal expansion between a detector substrate and a read-out substrate.

Flip-Chip Arrangements

At the time US-A-3808435 (Texas Instruments Incorporated, USA, 30.04.74) was filed, a conventional infrared detector system had an array of infrared detectors and each detector had an amplifier for amplifying its output. The amplified output was connected to a multiplexer scan converter. Such a system required a large number of amplifiers. It is acknowledged in US-A-3808435 that an improved system would be obtained if an imager with a charge transfer device (CTD) could be used. Two problems, however, were encountered in this technique. The first problem was the fabrication of a CTD on an infrared sensitive material and the second the fact that the thermal background radiation at room temperature is many orders of magnitude greater in the infrared than in the visible (41 orders of magnitude greater at 10 µm than at 500 nm). These problems are overcome in US-A-3808435 by a.c. coupling detector signals of an infrared detector array to a CTD array. The detector array and the CTD array are formed in an HgCdTe chip and a silicon chip respectively and are connected by placing the chips on top of each other with intermediate conductor elements. This technique of connecting two chips is generally called the flip-chip technique.

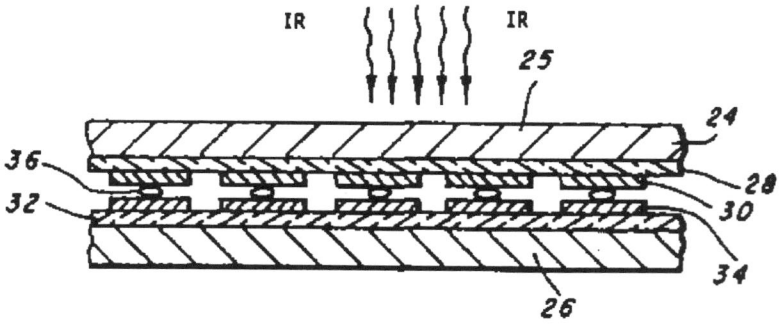

Fig. 2.3.1 (US-A-3808435 fig. 3)

The hybrid system includes an infrared detector chip 24 of HgCdTe, comprising metal-insulator-semiconductor (MIS) detectors, sandwiched with a CCD multiplexing chip 26. The thickness of the layer 24 is less than the diffusion length of charge carriers generated by the infrared radiation. A thin insulating layer 28 is formed over a surface of the semiconductor 24. An array of conductive electrodes 30 completes the detector structure. The CCD multiplexer array includes a silicon substrate 26, a silicon dioxide insulating layer 32 and an array of electrodes 34. The electrodes 34 correspond to the electrodes 30 of the detectors. The electrodes 30 and 34 are connected by ball bonds 36.

The background radiation is eliminated by utilization of a quantum differential detector (QDD). The QDD requires a chopper which is held at background temperature. The signal from the QDD is proportional to the difference between the background temperature and the scene. By removing the background, the requirements on detector uniformity and dynamic range of the associated electronics are greatly reduced.

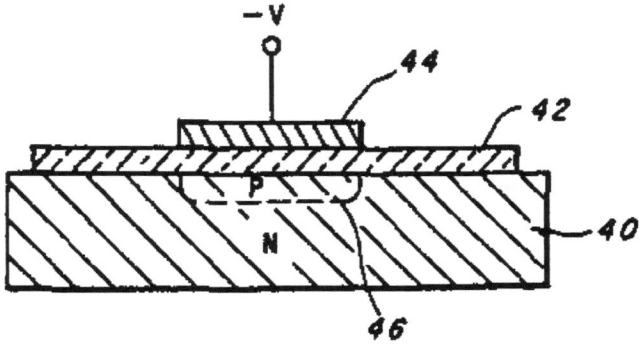

Fig. 2.3.2 (US-A-3808435 fig. 6)

Shown above is a MIS photocapacitor which can be used as a QDD. The MIS photocapacitor, when biased into inversion, is a photodiode in series with a capacitance. A substrate 40 of n-type HgCdTe is covered by an insulating layer 42. A conductor 44 is in turn formed on the insulator. An inversion region, enclosed by dashed line 46, is formed when the conductor is suitably biased with a negative voltage.

Instead of the MIS photocapacitor, it is possible to form a photodiode having a capacitor plate placed above either the p-type or the n-type region of the photodiode.

*

In US-A-4369458 (Westinghouse Electric Corp., USA, 18.01.83) the contact array of an infrared detector includes a substantially greater number of contacts than the contact array of a signal processor. At the flip-chip bonding operation it is not a prerequisite to precisely align the two integrated circuits in a one-to-one contact relationship.

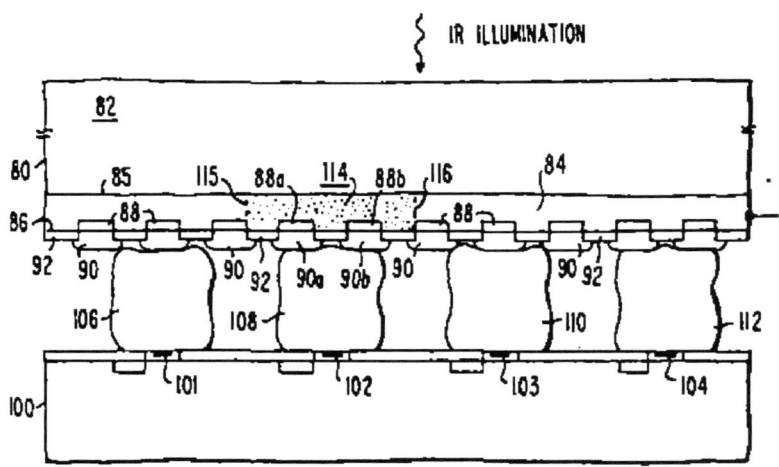

Fig. 2.3.3 (US-A-4369458 fig. 4)

A photovoltaic radiation detector 80 comprises an infrared transparent substrate 82 of CdTe. A layer 84 of p-type HgCdTe is formed over a surface 85 by a liquid phase epitaxy technique. A high density array of regions 88 doped with an n-type impurity is formed on the surface 86 of the layer 84. Metallic contacts 90 of indium contact the doped regions 88 and a dielectric layer 92 insulates the contacts 90 from one another. A silicon chip 100 comprising a CCD with input signal injection sites 101-104 is prepared. Metallic cell contacts 106, 108, 110 and 112 of indium contact the input sites 101-104, respectively. In operation, photogenerated charge carriers denoted by the dots surrounding the diodes 88a and 88b will be collected in the depletion regions of these diodes thereby defining an elemental region 114. Those diodes which have unbonded contacts are not capable of collecting photogenerated charge carriers.

*

A method to manufacture an imager having a flip-chip structure is given in JP-A-57073984 (Fujitsu Ltd, Japan, 08.05.82). The detector substrate is thinned and a slot is used to facilitate this operation.

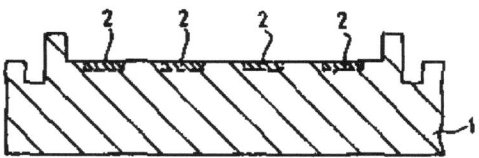

Fig. 2.3.4 (JP-A-57073984 fig. 5)

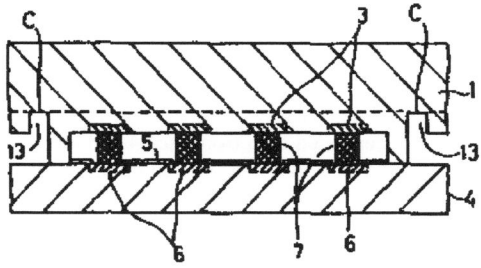

Fig. 2.3.5 (JP-A-57073984 fig. 6)

A slot 13 and a ridge is formed at each of the outer edges of an HgCdTe chip 1. Photodiodes 2 are formed between the ridges. The HgCdTe chip is connected to a silicon read-out chip 4 having input regions 2 by the aid of indium metal bumps 7.

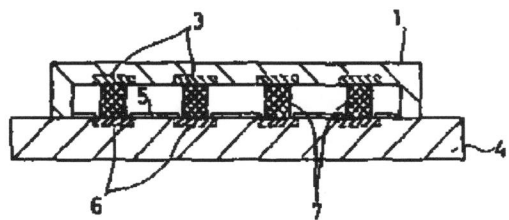

Fig. 2.3.6 (JP-A-57073984 fig. 7)

The ridges make contact with the silicon chip. The HgCdTe substrate is polished to a thickness corresponding to the depth of the slots.

*

The photodiodes of the imagers presented in US-A-4566024 (Societe Anonyme de Telecommunications, France, 21.01.86) are formed in a p-type HgCdTe substrate by diffusing n-type impurities from two opposite faces, a front face and a rear face, of the substrate. The photodiodes can be connected to a read-out device by a flip-chip bonding process on the rear face of the substrate and still be illuminated from its front face.

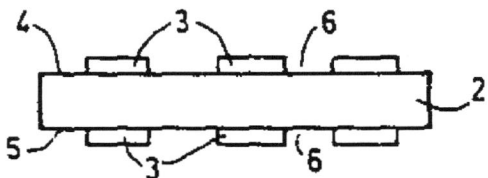

Fig. 2.3.7 (US-A-4566024 fig. 3)

A p-type substrate of HgCdTe is polished to obtain a wafer 2 with a thickness of from 20-30 µm. A masking layer of ZnS is deposited on a front face 4 and a rear face 5 of the wafer, and openings are made to form windows 6 in a mosaic configuration separated by masking bands 3. The windows of the two faces are respectively opposite one another.

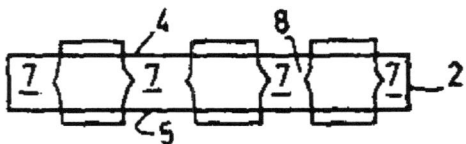

Fig. 2.3.8 (US-A-4566024 fig. 4)

An n-type diffusion is performed from the two faces 4 and 5 to obtain zones 7 extending from face 4 to face 5 of the wafer.

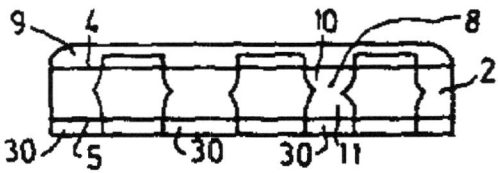

Fig. 2.3.9 (US-A-4566024 fig. 5)

The front face 4, which will be exposed to radiation, is covered with a protection layer 9. An indium layer is then deposited in the windows 6 of the rear face in order to make studs 30.

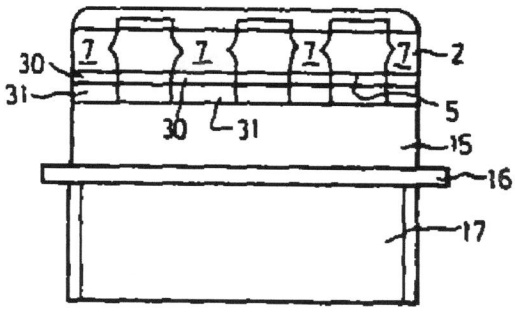

Fig. 2.3.10 (US-A-4566024 fig. 6)

The indium studs 30 are used for connection of the n-type zones by cold welding with corresponding input studs 31 of a silicon integrated circuit formed in a silicon wafer 15. The two wafers 2 and 15 are disposed on a cooled surface 16 in a cryogenic enclosure 17.

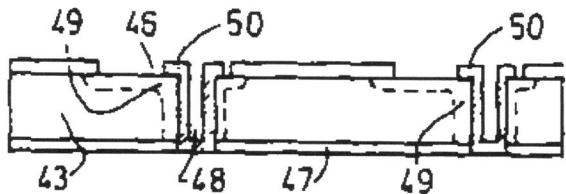

Fig. 2.3.11 (US-A-4566024 fig. 12)

In a second embodiment, a masking layer having windows 46 is formed on a thinned wafer of p-type HgCdTe. Holes 48 are etched through the wafer in the regions inside the windows by ion etching. N-type zones 49 are developed under the windows and along the walls of the holes. The n-type zones are connected to input studs of a silicon circuit via a deposited metal layer 50. The holes may be filled with a reinforced epoxy resin to perform the same function of connection.

*

The imager presented in JP-A-59112652 (Fujitsu KK, Japan, 29.06.84) comprises a detector chip which is bonded to a first side of a silicon read-out chip by a flip-chip technique. Connection is made through the silicon chip to a read-out CCD.

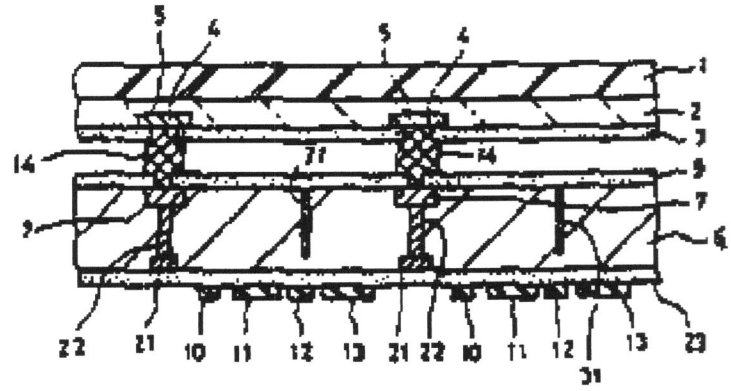

Fig. 2.3.12 (JP-A-59112652 fig. 2)

A p-type HgCdTe film 2 is epitaxially grown on a PbTe substrate 1. Photodiodes are prepared by forming n-type regions 4 in the HgCdTe film. A p-type silicon substrate 6 is prepared on a first side with n-type regions 7 forming input diodes. Plasma etching is used to form holes 22 which are filled with n-type polysilicon. The polysilicon contacts n-type regions 21 which are formed on an opposite side to the first side. Channel stops 31, gate electrodes 10-12 and transfer electrodes 13 of a CCD are formed.

A similar imager is presented in US-A-4761681 (Texas Instruments Incorporated, USA, 21.04.87) and US-A-4660066 (Texas Instruments Incorporated, USA, 02.08.88) where a silicon chip has mesas covered with metal contacts to which a detector chip is connected by a flip-chip technique. The read-out circuit is either formed on the mesa side of the silicon chip or on the opposite side. In the latter case, connections are made via conductive elements extending through the silicon substrate. These documents do not specify the material of the detector chip.

*

N-type regions corresponding to photodiodes, are formed in JP-A-62011265 (Toshiba Corp., Japan, 20.01.87) by diffusing indium into a p-type HgCdTe layer. The indium is diffused from straps which also connect the formed n-type regions to a read-out device comprised in a silicon substrate. No alignment between the detector layer and the read-out device is needed which would be the case if the n-type regions were formed before the bonding of the detector array and the read-out device takes place.

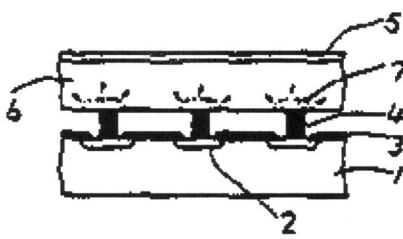

Fig. 2.3.13 (JP-A-62011265 fig. 1d)

Indium straps 4 are formed by evaporation on n-type regions formed in a p-type silicon substrate 1. A p-type HgCdTe layer 6 is formed on a CdTe substrate 5 and the layer 6 is placed to make contact with the straps 4. When the assembly is heated to 150°C in an N_2 atmosphere, indium is diffused into the p-type HgCdTe layer, thereby forming n-type regions 7. A similar method of manufacturing an imager is disclosed in JP-A-62013085 (Fujitsu Ltd, Japan, 21.01.87).

*

A cold shield layer which is formed on a silicon substrate is introduced in JP-A-62272564 (Fujitsu Ltd, Japan, 26.11.87). The substrate is bonded by a flip-chip process to an HgCdTe detector substrate.

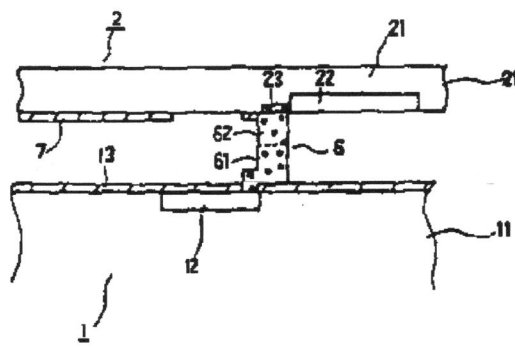

Fig. 2.3.14 (JP-A-62272564 fig. 1)

An HgCdTe substrate 11, covered with a non-reflecting film 13, comprises a photo-detector 12 which is connected by a conductor post 6 to a signal processor 22 formed in a silicon substrate 21. The silicon substrate, which is transparent to infrared radiation, is equiped with an opaque cold shield layer 7 of aluminium except for a region corresponding to the photo-detector 12.

*

In JP-A-63170961 (Fujitsu Ltd, Japan, 14.07.88) a shape-memory alloy structure is used to release mechanical stress generated in connection bumps during a flip-chip process comprising pressure bonding.

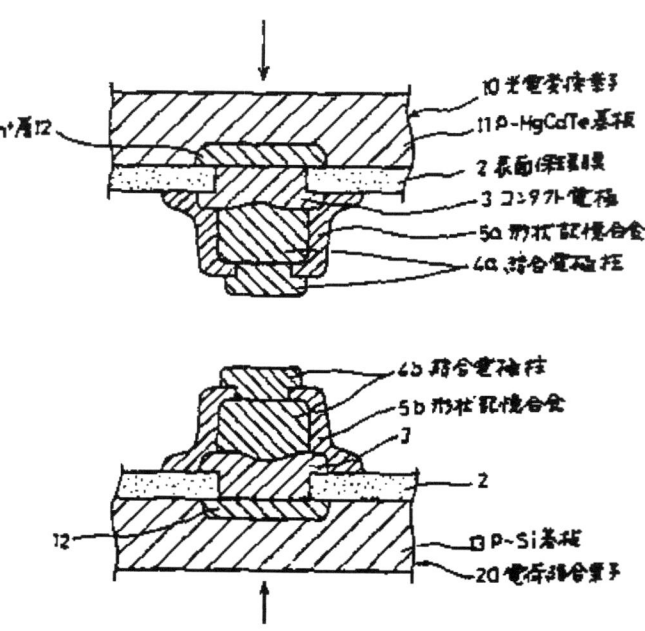

Fig. 2.3.15 (JP-A-63170961 fig. 1)

A photoelectric conversion device 10 is formed in a p-type HgCdTe substrate 11 and a CCD 20 is formed in a p-type silicon substrate 13. Indium electrodes 4a and 4b, which are connected to the photoelectric conversion device and the CCD, respectively, are covered on their sides with shape-memory metal layers 5a and 5b. Pressure bonding is used to couple the photoelectric conversion device to the CCD. Thereafter, the temperature is raised to return the shape-memory metal to its original shape, at the same time releasing mechanical stress in the indium electrodes.

*

The resistivity of a common contact to an array of photodiodes can be reduced by the introduction of a highly doped HgCdTe layer as is shown in JP-A-63300559 (NEC Corp., Japan, 07.12.88).

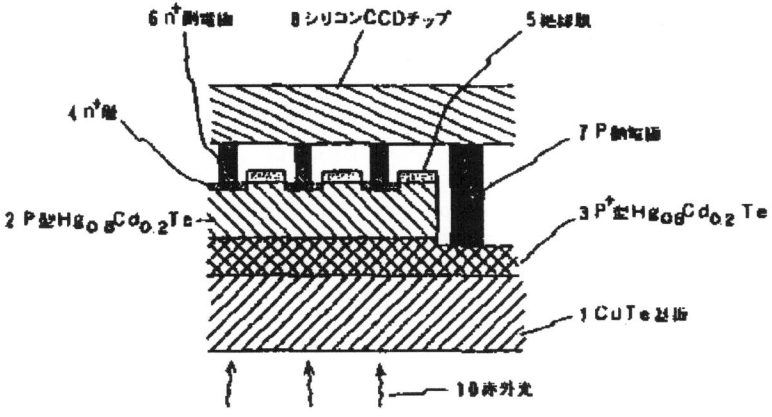

Fig. 2.3.16 (JP-A-63300559 fig. 1)

A p^+-type layer 3 is introduced between a CdTe substrate 1 and a p-type $Hg_{0.8}Cd_{0.2}Te$ layer 2. Photodiodes are created in the layer 2 by forming n-type regions 4 which are connected to a CCD 8 via columns 6. The p-type side of the photodiodes are commonly connected to the CCD by the aid of a column 7. The resistance of the common connection is reduced by the p^+-type layer 3.

*

In JP-A-1050560 (Fujitsu Ltd, Japan, 27.02.89) an indium layer is applied over an n-type layer of HgCdTe which covers a p-type layer of HgCdTe. Individual bumps and individual photodiodes are formed in the same etch step.

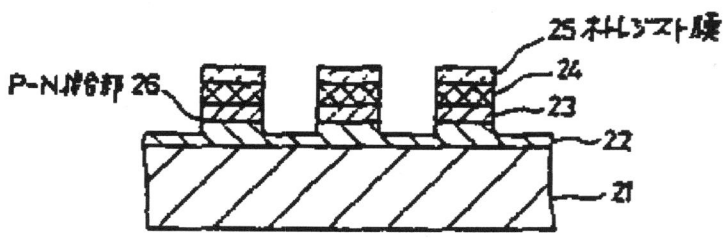

Fig. 2.3.17 (JP-A-1050560 fig. 1c)

A p-type layer 22 of HgCdTe, an n-type layer 23 of HgCdTe and an indium layer 24 are successively provided on a CdTe substrate 21. A photo-resist film 25 is patterned and mesas are formed by etching through the layers 23 and 24 and into the layer 22 until the pn-junction 26 is exposed.

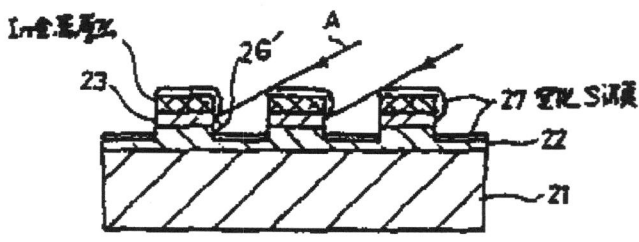

Fig. 2.3.18 (JP-A-1050560 fig. 1d)

An ECR-plasma CVD method is used to provide a protective film 27 on the mesas except for an exposed part of the pn-junction 26'.

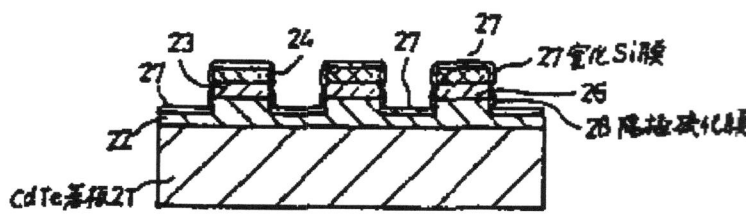

Fig. 2.3.19 (JP-A-1050560 fig. 1e)

An anodic sulphide film 28 is applied on the exposed part of the pn-junction 26' by the use of the protective film 27.

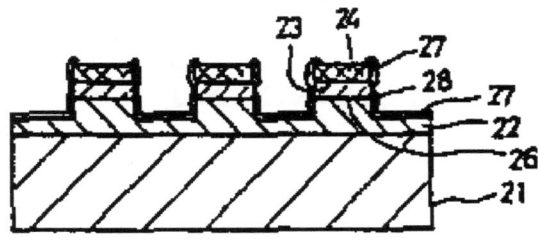

Fig. 2.3.20 (JP-A-1050560 fig. 1f)

Next, the protective film 27 is removed from a top surface of the indium layer 24.

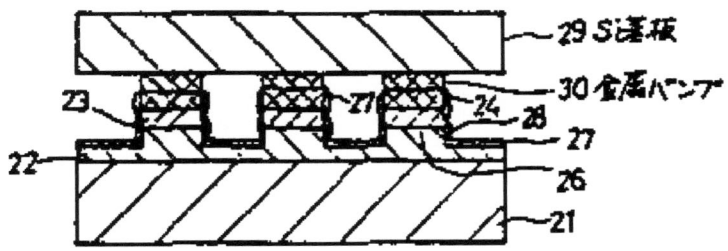

Fig. 2.3.21 (JP-A-1050560 fig. 1g)

A silicon substrate 29 equipped with connection bumps 30 is connected to the photodiodes by the use of a flip-chip process.

*

When two substrates, each provided with connection bumps, are joined together, a problem arises in positioning the substrates with respect to one another. The invention of JP-A-1227472 (Fujitsu Ltd, Japan, 11.09.89) gives a solution to this problem. An metal bump on one of the substrates meets a corresponding contact on the other substrate in recessed regions which are formed in one of the two substrates.

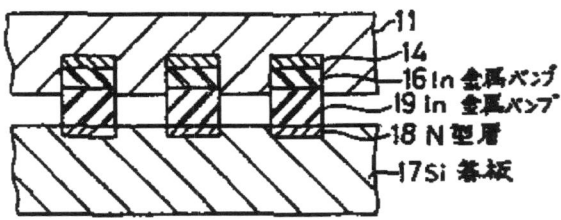

Fig. 2.3.22 (JP-A-1227472 fig. 1c)

N-type layers are formed in recessed regions formed in a p-type HgCdTe substrate. Indium bumps 16 are formed in the recessed regions and on corresponding input regions of a silicon substrate 17. The two substrates are joined in a flip-chip process in which the corresponding bumps will meet in the recessed regions.

*

A difficulty arises in filling the space between a detector substrate and a read-out substrate with epoxy. The epoxy tends to compress the two substrates together upon hardening and in doing so, while providing good electrical contact and mechanical strength, it also causes the bumps, which are often made of indium, to penetrate into the detector material itself which often causes unreliable operation of the device. This problem is solved in WO-A-9001802 (Honeywell Inc., USA, 22.02.90) where areas surrounding the detectors are not filled with epoxy.

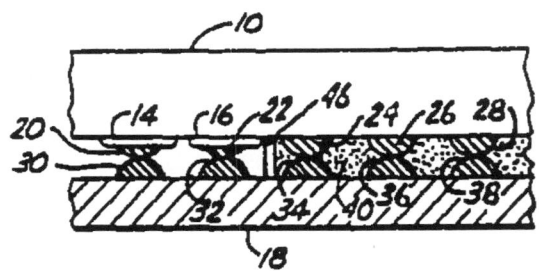

Fig. 2.3.23 (WO-A-9001802 fig. 3)

A substrate 10 of HgCdTe having detectors 14 and 16 and bumps 20, 22, 24, 26 and 28 is shown. The substrate is bump bonded to a second read-out substrate 18 having bumps 30, 32, 34, 36 and 38 thereon. A dam member 46 extends vertically between the two substrates so as to prevent the flow of epoxy 40 from entering into the region around detectors 14 and 16. The epoxy, when hardening, will draw the substrates closer together in the areas near the detectors but not in the areas of the detectors.

*

The indium bumps on the detector substrate and the indium bumps on the read-out substrate in US-A-4935627 (Honeywell Inc., USA, 19.06.90) have the same geometric shape but are oriented in a nonparallel orientation to each other, or alternatively have the same geometric shape but have different surface areas. These designs facilitate the alignment when several detector arrays are butted together. Furthermore, indium bumps of the same size are unstable against lateral motion during bump bonding operations.

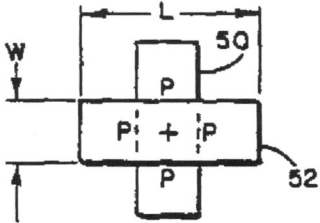

Fig. 2.3.24 (US-A-4935627 fig. 2a)

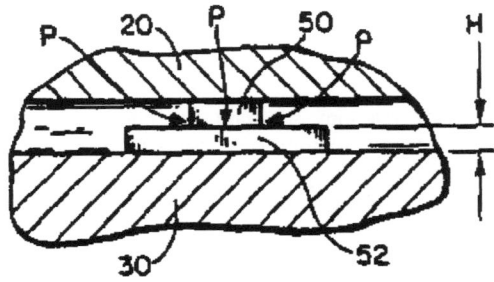

Fig. 2.3.25 (US-A-4935627 fig. 2b)

Indium bumps 50 and 52 connect a detector substrate 20 of HgCdTe with a read-out substrate 30. The two bumps have a rectangular cross-section and are rotated 90° with respect to each other. Alternatively, the bumps 50 and 52 have a round shape with two different diameters.

*

A method for aligning detector arrays on the surface of a signal processing module is shown in US-A-5075201 (Grumman Aerospace Corporation, USA, 24.12.91).

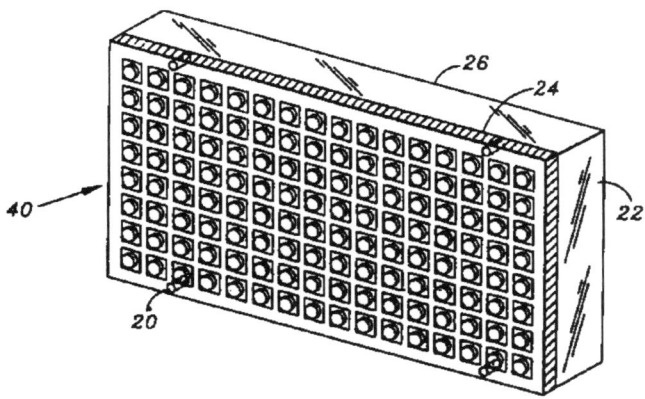

Fig. 2.3.26 (US-A-5075201 fig. 2)

A support substrate 22 of CdTe is attached to the backside of an infrared detector array of HgCdTe having four fiducial through holes 20 which provide a means for visibly observing an indication of the position of the detector elements from the back side 26 of the support substrate 22. The hole 20 is used in the alignment of the detector array to indices formed upon a signal processing module.

*

A detector substrate and a read-out substrate are bonded together in JP-A-4293240 (Mitsubishi Electric Corp., Japan, 16.10.92) by using two indium alloy bumps for each connection. The two indium alloy bumps have different melting points and the regions where the bumps are joined will have a higher melting point than the indium alloy having the lower melting point.

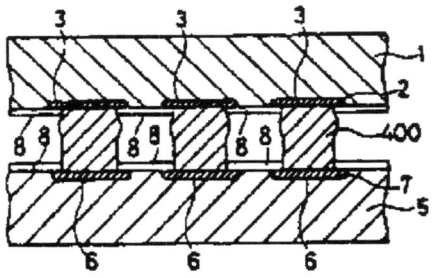

Fig. 2.3.27 (JP-A-4293240 fig. 1c)

A first set of indium alloy bumps, having a first melting point, are connected to input regions 6 of a CCD formed in a silicon substrate 5. A second set of indium alloy bumps, having a different, second melting point, are connected to detector elements 3 which are formed in an HgCdTe substrate 1. The two sets of indium alloy bumps are fused together to form a connection between the detector elements and the CCD. The regions where the two indium alloy bumps are joined will have a melting point at a value between the first and the second melting points.

*

Sets of bumps are provided in JP-A-7050330 (Fujitsu Ltd, Japan, 21.02.95) in addition to the bumps which are used to connect detector elements in a detector substrate to input regions in a read-out substrate. The additional bumps are used when the two substrates are joined to improve the bonding process.

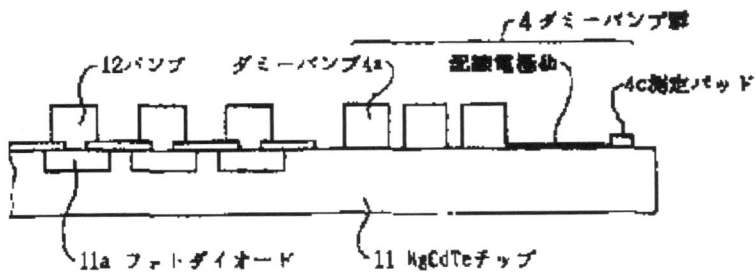

Fig. 2.3.28 (JP-A-7050330 fig. 2)

Bumps 12 are connected to detector elements 11a which are formed in an HgCdTe substrate 11. Sets of bumps 4 comprising bumps 4a are formed in addition to the bumps 12. Each set of bumps is connected to a measuring pad 4c by an electrode 4b. In a second embodiment the bumps 4a are separated in groups, each group having a characteristic height.

*

Thermal Stress Preventing Measures

When the flip-chip technique is used to connect two chips of different materials having different thermal expansion coefficients, the connection will be subjected to mechanical stress to a degree dependent on the the thermal history of the array. In JP-A-55150279 (Fujitsu Ltd, Japan, 22.11.80) the connection between an HgCdTe detector chip and a silicon read-out chip comprises bent buffers which absorb such mechanical stress and which prevent the fragile detector chip being damaged during the connection.

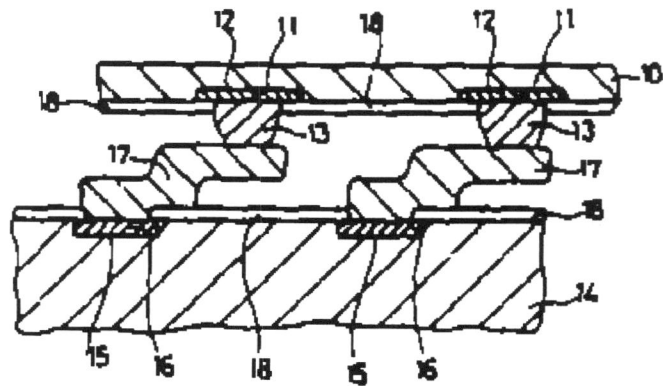

Fig. 2.3.29 (JP-A-55150279 fig. 2)

A detector chip 10 of HgCdTe comprising detector elements 11 is connected via bumps 13 to bent buffers 17 of gold. The buffers 17 are in turn connected to input regions 15 of a CCD formed in a silicon substrate 14. A method to form the buffers is also provided.

*

When a silicon read-out chip and an HgCdTe detector chip are bonded together by a flip-chip bonding process, the chips are exposed to a mechanical stress which may lead to damages, especially of the fragile detector chip. Photo-resist films are used in JP-A-61059771 (Fujitsu Ltd, Japan, 27.03.86) to form a spacer. The spacer reduces the mechancial stress of the semiconductor chips during the bonding process.

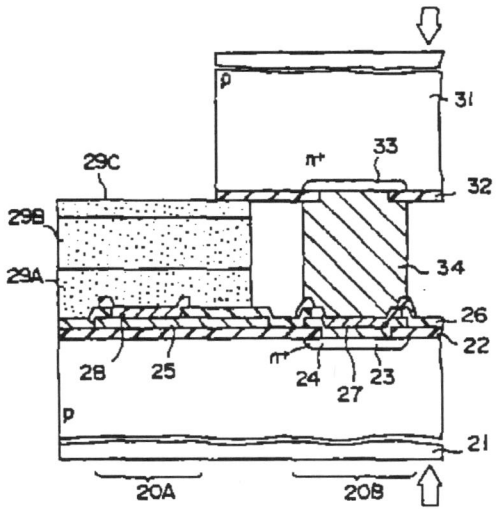

Fig. 2.3.30 (JP-A-61059771 fig. 5)

A read-out device, comprising an input region 23, is formed in a silicon substrate 21. A positive photo-resist film 29A, a negative photo-resist film 29B and a second positive photo-resist film 29C are patterned into a spacer. Connecting bumps 34 are applied and a detector chip 31, comprising photodiodes 33, is bonded to the silicon chip by pressing the two chips together. Finally, the photo-resist films 29A, 29B and 29C are removed.

* *

In JP-A-63268271 (Fujitsu Ltd, Japan, 04.11.88) metal rods are used to provide spacers between a detector substrate and a read-out substrate.

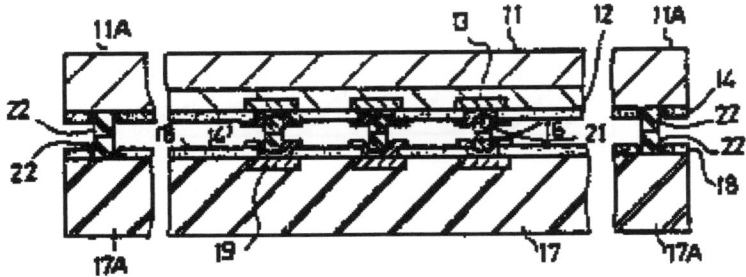

Fig. 2.3.31 (JP-A-63268271 fig. 2)

Photodiodes are made in an HgCdTe layer 12 formed in a CdTe substrate 11. A silicon substrate 17, comprising read-out circuits, is prepared. Indium rods 22 are formed at the ends of the substrates and metal rods 16 and 21 of Wood's metal are connected to the photodiodes and input regions of the read-out circuits, respectively. The indium rods 22 are made to be slightly lower than the rods 16 and 21. The two substrates are pressure-welded together. Thereafter, the metal rods 16 and 21, which have a lower melting point than the indium rods, are fused together in a heating treatment.

*

The problem of cross-talk between adjacent photodiodes and the problem of different thermal expansion coefficients between a silicon substrate and an HgCdTe substrate are approached in JP-A-63281460 (Fujitsu Ltd, Japan, 17.11.88). A detector is made up of HgCdTe wells, comprising photodiodes, formed in a CdTe substrate. The detector is bonded to a silicon substrate by a flip-chip process, and finally the CdTe substrate is etched away.

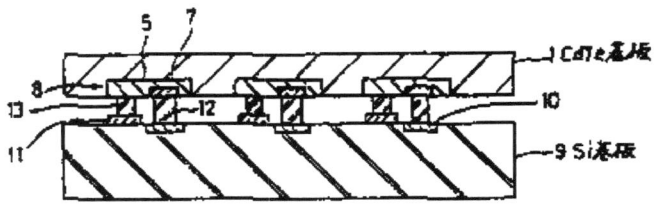

Fig. 2.3.32 (JP-A-63281460 fig. 6)

HgCdTe wells 5, comprising photodiodes 7, are formed in a CdTe substrate 1. The anodes and cathodes of the photodiodes are connected by indium columns 12 and 13 to a silicon substrate 9 by a flip-chip bonding process.

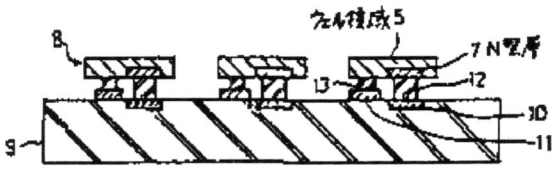

Fig. 2.3.33 (JP-A-63281460 fig. 1)

Next, the CdTe substrate is removed by dipping the device in an etching liquid comprising hydrofluoric acid (HF), nitric acid (HNO_3), acetic acid (CH_3COOH) and water (H_2O) with a weight ratio of 2-5:3-5:6:6.

*

A similar structure to the structure shown in JP-A-63281460 above is disclosed in JP-A-63296272 (Fujitsu Ltd, Japan, 02.12.88). In the latter, the HgCdTe wells are not connected by indium bumps but by Au films supported by an epoxy resin.

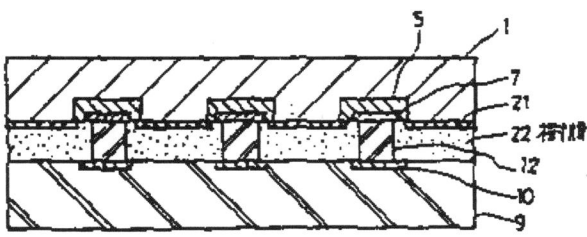

Fig. 2.3.34 (JP-A-63296272 fig. 3)

P-type HgCdTe well regions 5 are formed in a CdTe substrate 1. Photodiodes are created by forming n-type regions in the p-type wells. The p-type well regions 5 are connected to each other by an Au film 21 forming a common electrode. The n-type regions are connected to input diodes 10 prepared in a silicon substrate 9 by indium columns 12 in a flip-chip process. The space between the CdTe substrate and the silicon substrate is filled with an epoxy resin 22.

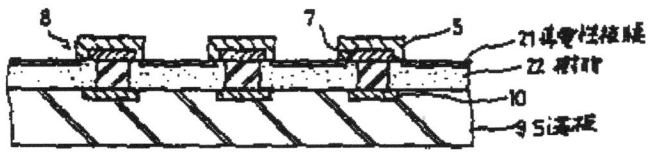

Fig. 2.3.35 (JP-A-63296272 fig. 4)

Next, the CdTe substrate is etched away by using the same etching liquid as is used in JP-A-63281460 above.

* *

The problem of difference in thermal expansion coefficients between an HgCdTe substrate and a silicon read-out substrate is addressed in JP-A-2214159 (Mitsubishi Electric Corp., Japan, 27.08.90) where a design similar to the design of JP-A-63296272 is presented.

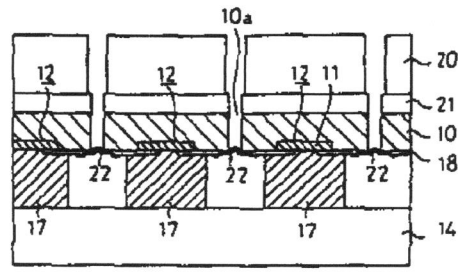

Fig. 2.3.36 (JP-A-2214159 fig. 1)

N-type regions 11 are formed in a p-type substrate 10 of HgCdTe thereby forming detector elements 12 which are separated by grooves 10a except for metal electrodes 22 which bridge the grooves. The metal electrodes connect the p-type regions of all detector elements. The detector elements are connected to a silicon read-out substrate 14 via electrodes 17.

* *

Cross-talk is reduced in US-A-5376558 (Fujitsu Limited, Japan, 27.12.94) by forming a three layered light absorption structure in the areas between the photodiodes of the array. The second layer has a refractive index which is greater than that of the first layer and the third layer is a metal layer.

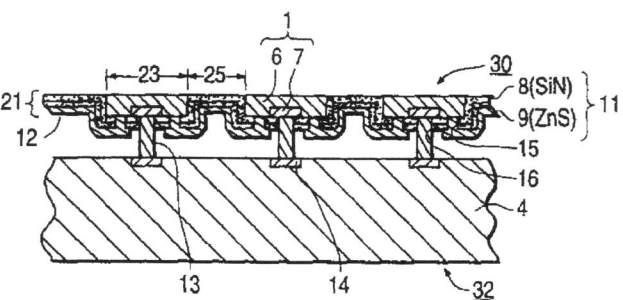

Fig. 2.3.37 (US-A-5376558 fig. 2)

Detector elements 1 comprising p-type HgCdTe regions 6 and n-type regions 7 are insulated from neighboring detector elements by a double insulation layer 11 composed of a silicon nitride layer 8 (refraction index approx. 2.0) and a ZnS layer 9 (refraction index approx. 2.2). A metal layer 12 of aluminium covers the insulation layer 11. The metal layer and the insulation layer form a light shield layer 21 which covers an outer surface of the p-type regions 7 except at light input ports 23. Holes 15 and 16 are formed through the insulation layer, and the metal layer is in contact with the p-type regions through holes 15, and a bump is formed through holes 16. The photo-sensitive array 30 is assembled with a signal processing substrate 32 using bumps 13.

Infrared rays which are incident to regions 25 penetrate into the silicon nitride layer and ZnS layer are thereby partly attenuated and reflected, and are finally reflected from the metal layer. The incident rays and the reflected rays cause a complex interference with each other, and an apparent overall reflectance of the light shield layer is reduced if proper thickness and refraction index are chosen for each layer of the insulation layer. Preferable, the refractive index of the first layer 8 is less than the refractive index of the second layer 9.

A method of manufacturing the structure is also disclosed. In short, islets of HgCdTe are formed from an HgCdTe layer which has been formed on a CdTe substrate. Next, the double insulation layer and the metal layer are formed. The bumps are provided before the CdTe substrate is removed by etching.

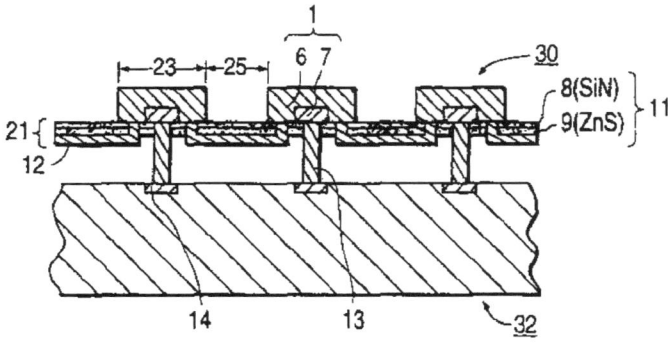

Fig. 2.3.38 (US-A-5376558 fig. 3)

In a second embodiment, the detector elements 1 are protruding from a flat light shield layer 21. Bumps 13 are used to connect the array to a signal processing substrate 32. The light shield layer comprises the same layers as in the first embodiment.

A method of manufacturing the structure of the second embodiment is also disclosed. In short, buried HgCdTe regions are formed from the surface of a CdTe substrate. Next, the double insulation layer and the metal layer are formed. The bumps are provided before the CdTe substrate is removed leaving the buried HgCdTe regions protruding from the from the light shield layer.

*

To reduce the effect of difference in thermal expansion coefficients between a silicon read-out substrate and an HgCdTe detector layer, which is bump bonded to the silicon substrate, it is proposed in JP-A-1061056 (NEC Corp., Japan, 08.03.89) to affix a second silicon substrate to the HgCdTe detector layer with an intermediate layer of GaAs.

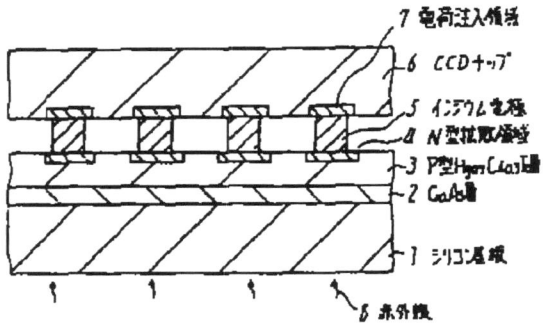

Fig. 2.3.39 (JP-A-1061056 fig. 1)

A p-type GaAs buffer layer 2 and a p-type HgCdTe layer 3 are epitaxially grown on a p-type semi-insulating silicon substrate 1. Photodiodes are created by forming diffused n-type regions 4 in the p-type HgCdTe layer. The n-type regions are connected to a silicon substrate via indium electrodes 5 which are connected to input regions 7 of a CCD.

*

The problem of difference in thermal expansion coefficients between a silicon read-out substrate and an HgCdTe detector substrate is approached in US-A-4783594 (Santa Barbara Research Center, USA, 08.11.88) by filling the space between the two substrates with a resilient electrically insulating polymeric material and thereafter separating the detector elements from each other by removing a layer of the HgCdTe detector substrate.

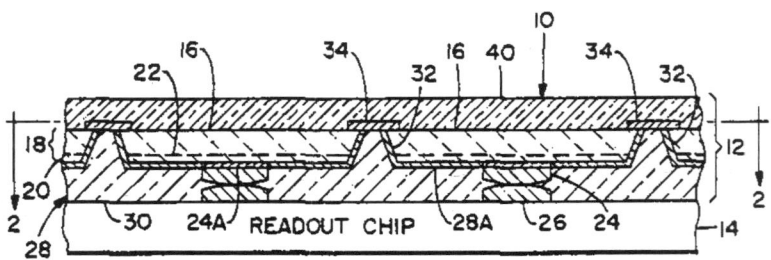

Fig. 2.3.40 (US-A-4783594 fig. 1)

A detector array 12, which is composed of detectors 16 arranged in rows and columns, is disposed on a silicon read-out chip 14. Each detector includes two HgCdTe layers 18 and 20, which form a pn-junction. A set of upper contacts 24 and a set of lower contacts 26 of indium are provided for connecting the detectors to the read-out chip. A resilient support 28 of a resilient polymer, such as silicone elastomer, envelops and holds the detectors. An insulating layer 28A of silicon dioxide separates the detectors from the support and passivates the surfaces of the HgCdTe layers to prevent development of electric charge by interaction with polymer material of the support. Electrical connection through the insulating layer 28A is made by contact metal 24A. An electrically conductive grid 34 is disposed on top of the array to form an electrically conductive ohmic contact with each detector followed by a radiation-transmissive coating 40 of ZnS. The process for constructing the assembly 10 is shown below.

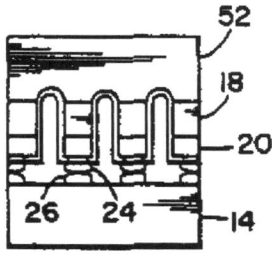

Fig. 2.3.41 (US-A-4783594 fig. 5h)

A p-type layer 18 and an n-type layer 20 of HgCdTe are grown epitaxially on a substrate 52 of cadmium zinc tellurium. A set of through holes are formed which extend through the layers 18 and 20 up to the surface of the substrate. A silicon dioxide layer is applied and at each detector a window is etched. A contact metal 24A is applied within each window. Contacts 24 and 26 are constructed on the detectors and on the read-out chip 14. The chip and the substrate 52 are pushed together to compress the contacts 24 and 26 against each other which cold welds them to each other.

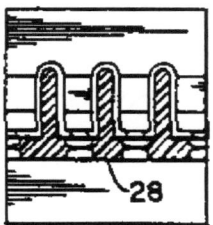

Fig. 2.3.42 (US-A-4783594 fig. 5i)

The spaces between the welded sets of contacts are filled with a polymer material in liquid form. The polymer material is then allowed ot cure and solidify to provide the support 28.

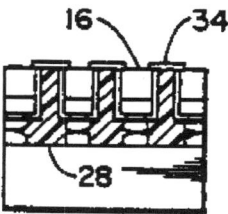

Fig. 2.3.43 (US-A-4783594 fig. 5k)

The substrate 52 is removed and a grid 34 is deposited followed by an anti-reflection coating. Due to the increased resilience of the support relative to the chip, any differential expansion or contraction of the support relative to the chip due to changes in temperature can occur without the development of excessive stress and strain in the contacts. Furthermore, the structure eliminates a significant amount of cross-talk between signals of neighboring detectors.

*

The invention of US-A-5365088 (Santa Barbara Research Center, USA, 15.11.94) approaches the problem of stresses generated by the mismatch of the thermal expansion coefficients of HgCdTe and silicon by including a buffer layer of sapphire. The characteristic thermal expansivity of sapphire is more similar to the thermal expansivity of HgCdTe than that of silicon.

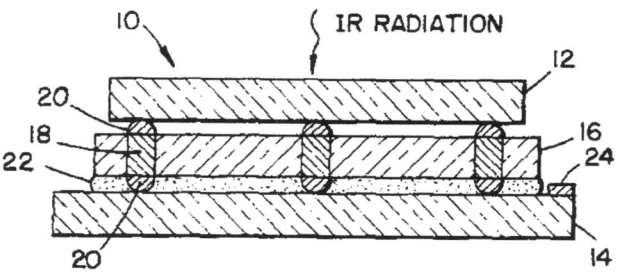

Fig. 2.3.44 (US-A-5365088 fig. 1)

The hybrid circuit 10 comprises a buffer structure 16 which is comprised of a material which accommodates the difference in thermal expansion coefficients of the HgCdTe detector array 12 and the silicon read-out chip 14. The buffer layer is made of sapphire which also has good thermal conductivity properties. The buffer structure has laser drilled vias 18 which are formed in registration with unit cells of the detector array and the read-out circuit. Each of the vias is provided with indium bumps 20 at opposing ends thereof. The buffer structure is interposed between the detector array and the read-out chip. Cold weld indium bump technology is employed to couple the bumps 20 to the buffer structure. The buffer structure is further coupled to the read-out circuit by an epoxy layer 22 which is wicked in between the buffer structure and the read-out circuit. A single buffer layer may be employed with a plurality of detector arrays and a plurality of read-out chips.

*

When indium bumps are used to connect two chips having different thermal expansion coefficients and the chips are exposed to repeated temperature cycling, the mechanical stress in the bumps is reduced when the bumps are made taller. If the indium bumps are formed using vapor deposition through a photo-reduced mask pattern, the height of the bumps are limited to 6-9 microns. A fabrication process for making taller indum bumps is disclosed in EP-A-0405865 (Hughes Aircraft Company, USA, 02.01.91).

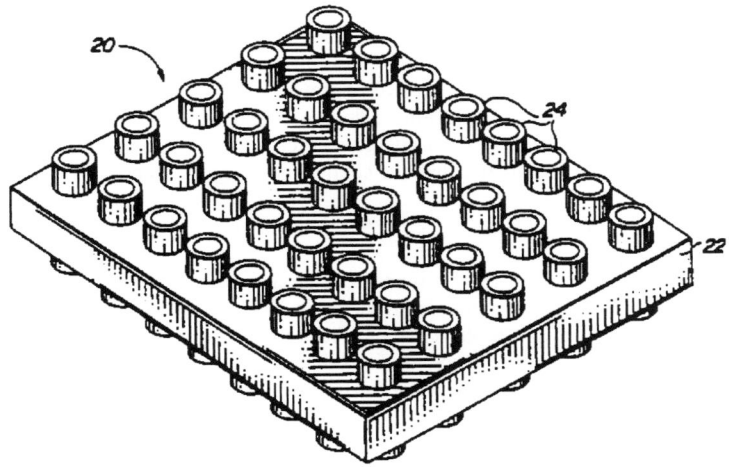

Fig. 2.3.45 (EP-A-0405865 fig. 2)

An integrated circuit connector pad, consisting of a polymer film 22 acting as a carrier for an array of metal tubes 24 which extend through the film and protrude from each side, is continuously dipped in molten indium until the tubes are filled with indium by capillary action. The indium-filled tubes are affixed on a read-out chip. Next, the original metal tubes are etched away and the carrier is lifted off. Finally, the detector chip is aligned in position with the indium columns and the read-out chip and the detector chip are joined by pressure welding.

*

Mechanical stress in connection bumps formed between a read-out circuit in a silicon substrate and detectors in an HgCdTe substrate due to a difference in thermal expansion coefficients of the two substrates is reduced in US-A-4943491 (Honeywell Inc., USA, 24.07.90) by bonding a layer of a material, having a greater thermal expansion coefficient than the coefficient of the two substrates, to the silicon substrate.

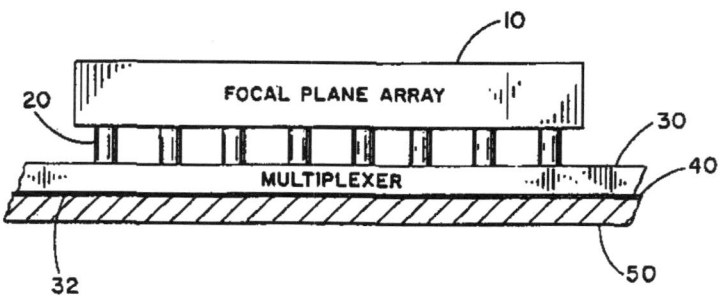

Fig. 2.3.46 (US-A-4943491 the figure)

A focal plane array is formed in a substrate 10 of HgCdTe. The substrate is bonded by indium bumps 20 to a multiplexer circuit formed in a silicon substrate 30. A copper layer 50 is bonded to the silicon substrate by means of an adhesive layer 40. The copper layer has a thermal expansion coefficient which is greater than the coefficient of HgCdTe and silicon.

*

An imager in which the optical energy is incident on a p-type doped region prior to being incident on a bulk n-type doped region is presented in EP-A-0485115 (Cincinnati Electronics Corporation, USA, 13.05.92). The short distance between the surface of the p-type region and the pn-junction causes efficient detectivity including at the short wavelength end of the spectrum of interest. The description is made for InSb detectors. HgCdTe is mentioned.

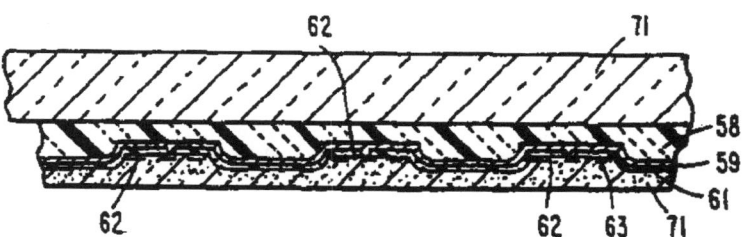

Fig. 2.3.47 (EP-A-0485115 fig. 6)

P-type regions 62 are formed in an n-type substrate 71. A metal grid 59, which connects the p-type regions, is formed over an insulating layer 61. The substrate is bonded to a dielectric plate 57 by an epoxy glue layer 58, which overlays the metal grid. Next, the thickness of the substrate is reduced.

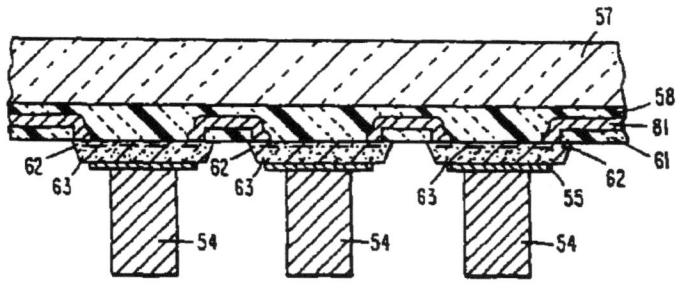

Fig. 2.3.48 (EP-A-0485115 fig. 8)

Trenches are formed in the n-type substrate, thereby separating the photodiodes on n-type islands 63.

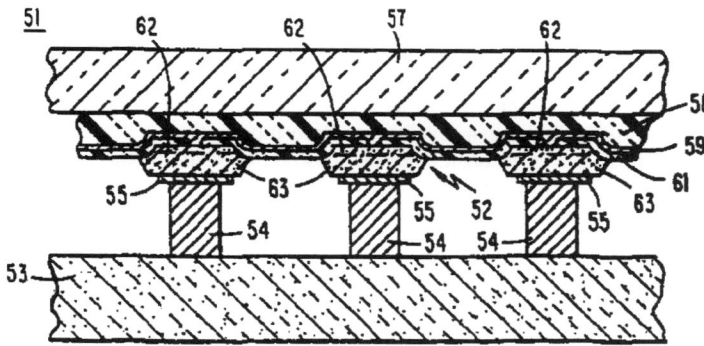

Fig. 2.3.49 (EP-A-0485115 fig. 3)

Metal ohmic contact pads 55 are formed on the n-type islands and the detector is bonded to a multiplexer integrated circuit substrate 53 via indium columns 54.

*

Mechanical stress due to temperature variations is reduced in the imager of JP-A-4199877 (Mitsubishi Electric Corp., Japan, 21.07.92) by growing two layers of HgCdTe of different compositions on each side of a silicon substrate. One of the layers comprises the photodiodes and the second is transparent to the infrared radiation of interest.

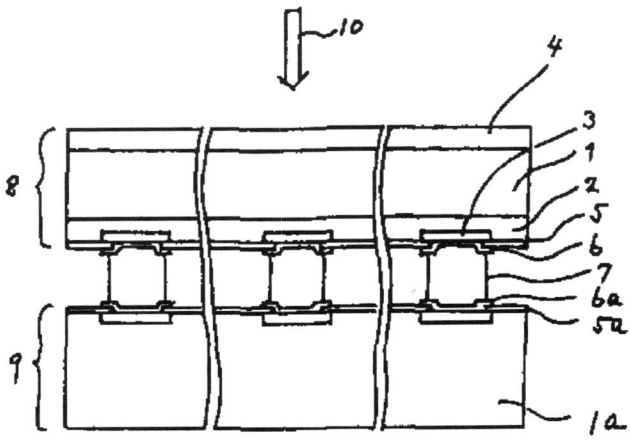

Fig. 2.3.50 (JP-A-4199877 fig. 1)

An $Hg_{1-x}Cd_xTe$ layer 2, in which photodiodes 3 are formed, is grown epitaxially on a silicon substrate 1. An $Hg_{1-y}Cd_yTe$ layer 4, having a composition with $y > x$ and thereby a larger bandgap than layer 2, is formed on the silicon substrate opposite to the layer 2. The larger bandgap allows infrared radiation 10 to pass through the layer 4. The photodiodes are connected to substrate via columns 7.

*

A detector layer of an imager presented in US-A-5264699 (Amber Engineering Inc., USA, 23.11.93) is thinned to allow the detector to act like a flexible membrane and to elastically respond to thermal mismatch resulting from different coefficients of thermal expansion between the detector and a semiconductor read-out circuit.

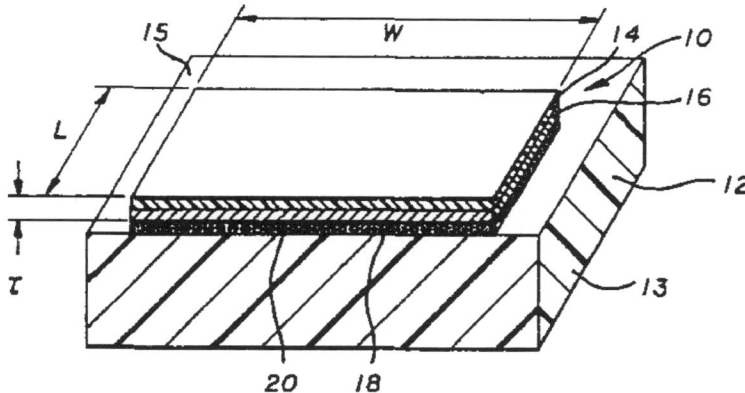

Fig. 2.3.51 (US-A-5264699 fig. 3)

A thin detector array portion 10 is bonded to a read-out circuit 12. The detector array employs a thin transparent substrate 14 of CdTe and a thin detector layer 16 of HgCdTe. After the detector array has been bonded to the read-out circuit, the detector layer and the transparent substrate are thinned to a thickness of 25 to 400 µm by using lapping and polishing techniques.

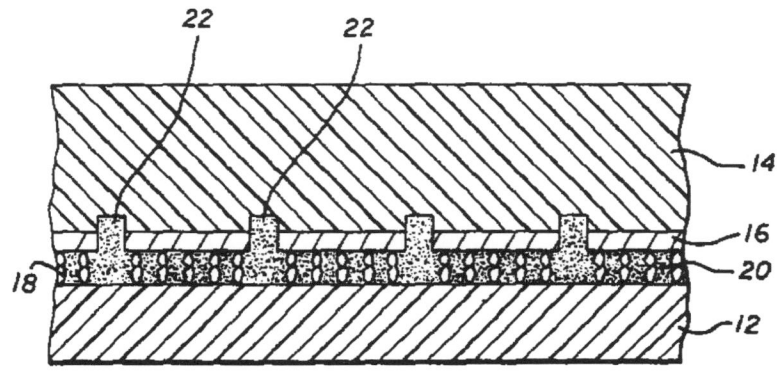

Fig. 2.3.52 (US-A-5264699 fig. 6c)

In an alternative embodiment, channels 22 are etched through the detector layer 16 and into the transparent substrate 14 to a depth slightly exceeding the required thickness. The detector array portion and the silicon circuit 12 are bonded together by the aid of indium bumps 18, and an epoxy material 20 is back-filled into the space between the detector array portion and the read-out chip.

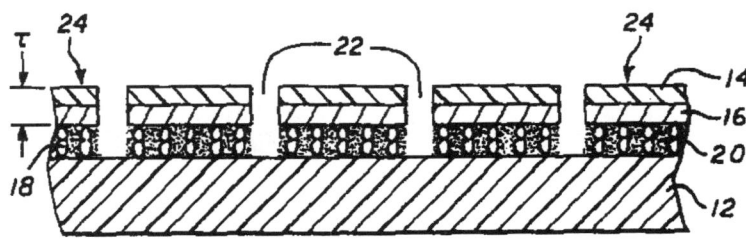

Fig. 2.3.53 (US-A-5264699 fig. 6d)

The detector array portions are lapped and polished to a thickness less than the depth of the channels 22, thereby dividing the detector array portions into sub-arrays 24, each including a number of photodiodes. Finally, the channels 22 are cleared of epoxy.

*

In JP-A-6163865 (Fujitsu Ltd, Japan, 10.06.94) the mechanical stress of indium bumps connecting a silicon read-out substrate and an HgCdTe detector substrate is reduced by bonding the two substrates together at a temperature which is between the operational and the non-operational temperature of the detector.

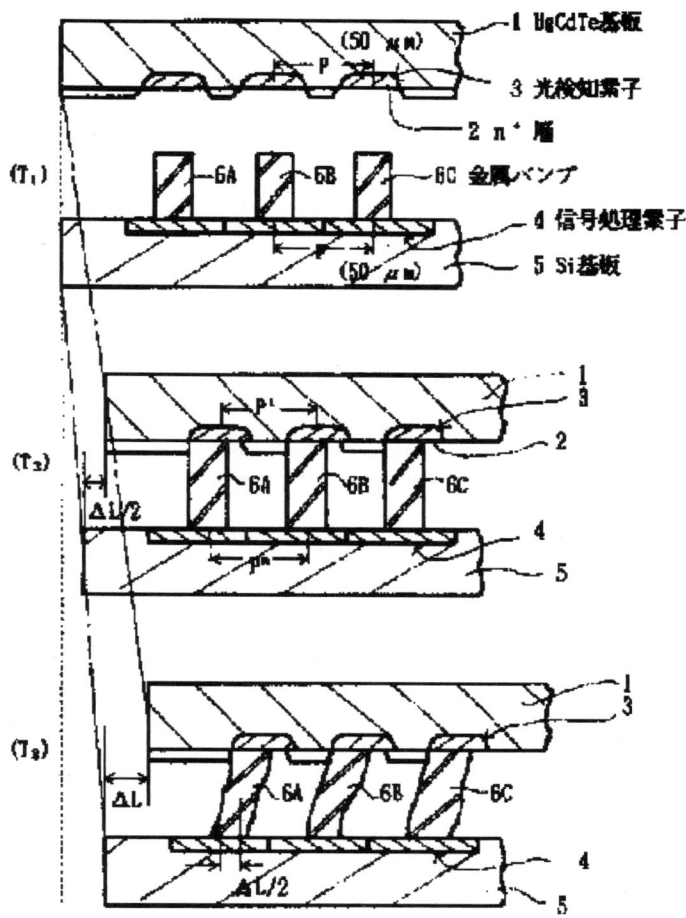

Fig. 2.3.54 (JP-A-6163865 fig. 1)

An HgCdTe detector substrate 1 having photodiodes 3 is bonded to a silicon read-out substrate 5 by means of indium bumps 6A to 6C at a temperature T3 which is between the operational temperature of the detector array T2 and the non-operational temperature of the detector array T1 (room temperature).

* *

During the bonding process in JP-A-7111323 (NEC Corp., Japan, 25.04.95) a silicon read-out substrate and an HgCdTe detector substrate are held at different temperatures. The amount of stress in the connector bumps will be reduced when the device is cooled from room temperature to an operating temperature of 77 K.

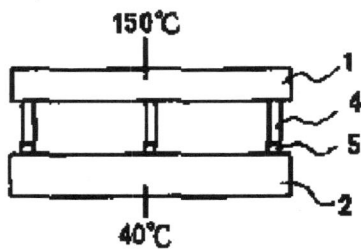

Fig. 2.3.55 (JP-A-7111323 fig. 2b)

The connector bumps 4 and 5 are formed on a read-out substrate 1 and an HgCdTe detector substrate, respectively. The pitch between the bumps 4 and the pitch between the bumps 5 are chosen such that the two sets of bumps are aligned when the read-out substrate is held at 150°C and the detector substrate is held at 40°C. At those temperatures the two substrates are bonded together.

*

In WO-A-9417557 (Hughes Aircraft Company, USA, 04.08.94) an integrated circuit assembly is presented which includes a silicon thin film circuit bonded to a substrate of a material selected to provide the assembly with an effective thermal expansion characteristic that approximately matches that of an HgCdTe detector array.

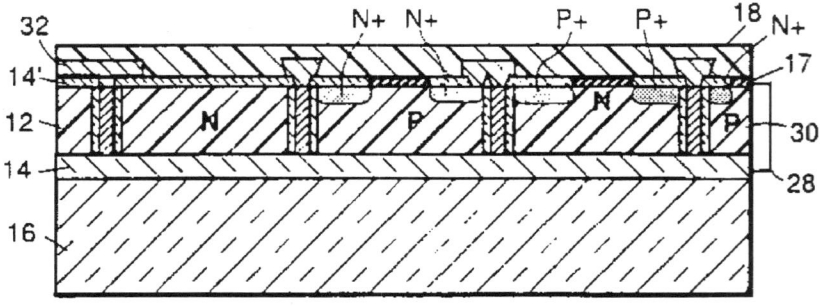

Fig. 2.3.56 (WO-A-9417557 fig. 1c)

A 0.2 to 10 μm thick single crystal silicon film 12, a SiO_2 layer 14 and a silicon substrate 16 are formed together. Electrical feedthroughs 20 are formed by etching trenches through the film 12. The walls of the trenches are oxidized and the holes are filled with doped polysilicon which may be used as a conductive material. Read-out circuits are formed in the film 12. This is schematically shown as n-type and p-type regions and gate electrodes 17. A further layer of SiO_2 14' is formed to bury the gate electrodes followed by an insulating overglass layer 18.

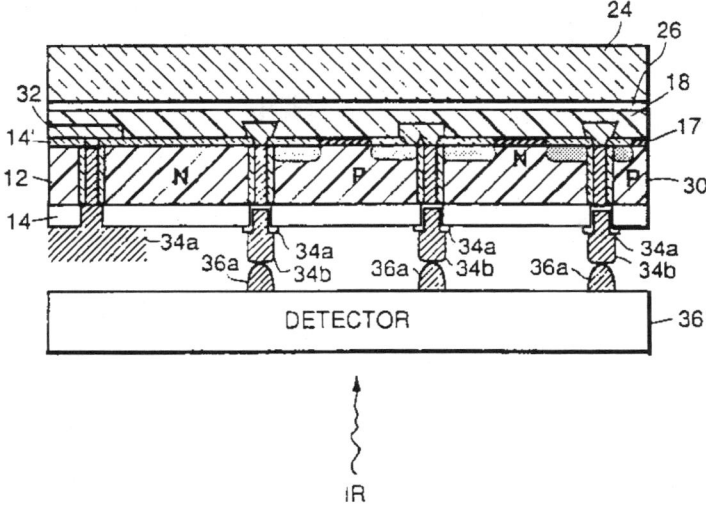

Fig. 2.3.57 (WO-A-9417557 fig. 1d)

The structure is bonded to a substrate 24 which is chosen to have a coefficient of thermal expansion that is selected for providing the resultant read-out chip assembly with an effective coefficient of thermal expansion that is approximately the same as an HgCdTe detector array 36. The substrate material may be GaAs (4.5-5.9 x 10^{-6} m/mK), CdTe, Ge (5.5-6.4 x 10^{-6} m/mK), and a-plane sapphire (3.5-7.5 x 10^{-6} m/mK) where the coefficients of thermal expansion are given in parentheses. The coefficients of thermal expansion for silicon, HgCdTe and epoxy are 1.2 x 10^{-6} m/mK, 3.8-4.5 x 10^{-6}m/mK and 30-50 x 10^{-6} m/mK, respectively. Next, the substrate 16 is removed and aluminium pads 34a are formed. Indium bumps 34b are cold welded to corresponding indium bumps 36b.

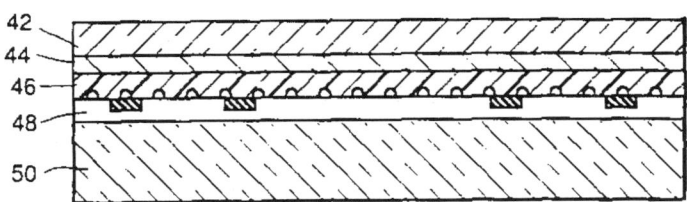

Fig. 2.3.58 (WO-A-9417557 fig. 2b).

In a second embodiment, a silicon substrate 42, a layer of silicon dioxide 44, and a thinned silicon layer 46 which includes read-out circuitry are attached to a temporary carrier substrate 50 with a layer of adhesive or wax 48.

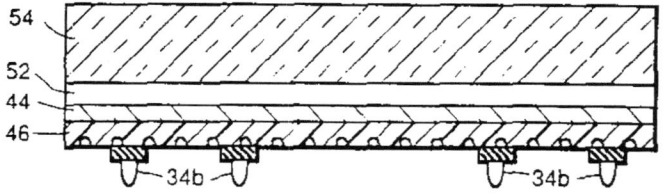

Fig. 2.3.59 (WO-A-9417557 fig. 2e)

The silicon substrate is removed while the temporary substrate provides mechanical support. Next, the substrate is bonded to a final carrier substrate 54 with a layer 52 of epoxy adhesive. The bonding layer 48 is removed, which also removes the temporary carrier substrate 50. The final carrier substrate is chosen according to the same criteria as described in the first embodiment. Processing continues by forming indium interconnects 34b for later hybridization of the read-out integrated circuit assembly with a detector array.

*

An imager with insulator walls formed between connection bumps is shown in JP-A-6236981 (Fujitsu Ltd, Japan, 23.08.94). The structure reduces the detrimental effects due to a difference in the coefficient of thermal expansion between a detector substrate and a read-out substrate.

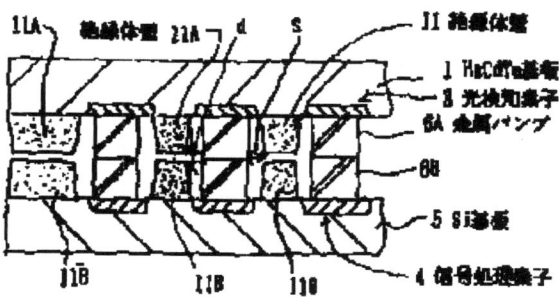

Fig. 2.3.60 (JP-A-6236981 fig. 1a)

A detector substrate 1 of HgCdTe comprising photodiodes 3 is connected to a silicon read-out substrate 5 by means of metal bumps 6A and 6B. Insulator walls 11A and 11B are formed at regions surounding the metal bumps.

*

Chapter 2.4

Z-Technology Arrangements

Z-technology arrangements for mercury cadmium telluride imagers are presented in this chapter.

Summary

The use of the Z-technology [1-3] results in a very compact detector unit. The detector unit is made by connecting a two-dimensional detector array to an end of a module formed by a layered structure. The layered structure comprises means for reading-out information from the detector array.

- In US-A-3852714 a two-dimensional electro-optical detector array is presented which has coupling conductors between the detectors and their respective amplifiers packaged three-dimensionally on multilayered modules.

- The photo-detector array module disclosed in US-A-4304624 comprises a focal plane array of detectors which is directly mounted, without a separate substrate, on an end of a module formed by a layered structure. Electronic components are mounted in "wells" in the layered structure. The method of fabricating the structure is claimed in the patent and the photo-detector mosaic array module is claimed in US-A-4354107.

- Placing detectors on a focal plane face of a supporting module is very demanding. The goal is to end up with thin diode chips attached to the end of the module and in electric contact with lead points on the module. In US-A-4290844 a wafer of a detector material is secured

---

[1] J. Y. Wong and J. P. Rode, "The Advent of Three-Dimensional Imaging Array Architectures: A Perspective", *Proc. Soc. Photo-Opt. Instrum. Eng.*, 501, 128-135, 1984.
[2] J. C. Carson, "Applications of Advanced "Z" Technology Focal Plane Architectures", *Proc. Soc. Photo-Opt. Instrum. Eng.*, 930, 164-182, 1988.
[3] P. I. Zappella, W. L. Robinson, J. W. Slemmons, P. J. Redmond and F. J. Woolston, "Implementation of a Hybridization Station for Mating Grumman Z-Plane Modules with HgCdTe Arrays", *Proc. Soc. Photo-Opt. Instrum. Eng.*, 1097, 117-125, 1989.

to the end of the module and ion-milling is used to pattern separate detector islands. This method permits a very high density of detectors to be formed on the focal plane.

- The packing density of a detector module is further improved in US-A-4551629 in which silicon chips are stacked to create a focal plane to which photo-detectors are mounted.

- The primary deterrent to reducing the detector center-to-center distance for the structure presented above in US-A-4551629 is the required thickness of each silicon chip. If the chips are too thin, they will be very fragile, and will therefore be difficult to handle during fabrication of the modules. In US-A-4403238 the distance between adjacent detector rows can be less than the thickness of a given chip, by orienting the detector rows along lines which extend diagonally with respect to the planes in which the silicon chips extend.

- In US-A-4618763 a detector module is disclosed which is formed of stacked multi-channel integrated circuits, a detector array and a module header interface. The detector array and the module header interface are disposed transverse to the plane of the integrated circuits on opposite edge portions of the integrated circuits. Each integrated circuit is formed in a semiconductor material which has been deposited upon a thin sapphire wafer.

- A detector mosaic which is bonded to a multilayer alternating thick film and thin film interconnect pattern is disclosed in EP-A-0281026.

One problem which arises when a detector array is attached to the face of a multi-layer module is the inability of the detector material to absorb forces generated by a mismatch of coefficient of thermal expansion between the detector array material and the module. Furthermore, it is difficult to isolate a fault that may be attributable to either the detector elements, module wiring or processing elements.

- A buffer board is introduced in WO-A-8807764 which facilitates electrical communication between the detector elements and the module and conductive patterns formed on the module layers, and also enhances the structural characteristics and separate testability of the system components.

- When a detector array is connected to the side edges of a plurality of thin silicon substrates, which comprise signal processing electronics, there is a problem in forming interconnect bumps on the side edges of the silicon substrates. This problem is solved in US-A-5081063 by an electroplating technique.

Other problems, which occur when an array of detectors is connected to the side edges of a plurality of thin silicon integrated circuit layers, are the problem of aligning the detector array and the integrated circuit layers and the problem of replacing a single integrated circuit layer after the detector array is connected to the integrated circuit layers.

- The problems stated above are solved in US-A-4992908 by bevelling the edges of the silicon layers and connect the layers to a contact board which has a beveled surface formed to receive and support the integrated circuit layers.

- When ceramic substrates, having thin film micro circuits printed thereupon, are laminated using thin thermosetting adhesive sheets to form a module to which a detector array is bonded, the coefficient of expansion between the ceramic layers and the thermoset adhesive sheets leads to thermal stress when the module is subjected to large range of thermal cycling. The thermal stress is transmitted to the detectors and can cause the detector array to crack. Non-uniformity in the thickness of the laminating adhesives is one of several problems mentioned in US-A-5128749. Instead of the thermosetting adhesive sheets, it is proposed to use a glass binding material which has a coefficient of expansion approximately equal to the coefficient of expansion of the substrates.

Z-Technology Arrangements

In US-A-3852714 (Eocom Corporation, USA, 03.12.74) a two-dimensional electro-optical detector array is presented which has coupling conductors between the detectors and their respective amplifiers packaged three-dimensionally on multilayered modules.

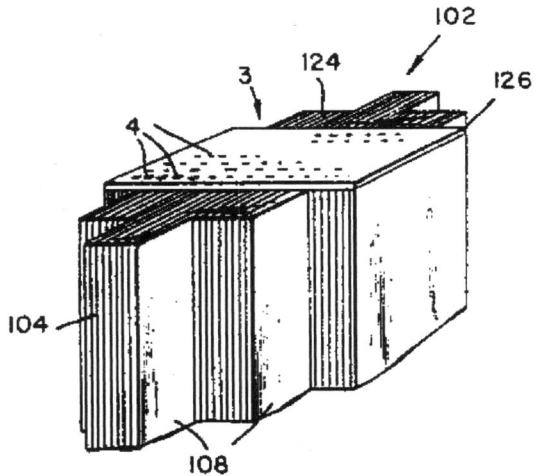

Fig. 2.4.1 (US-A-3852714 fig. 6)

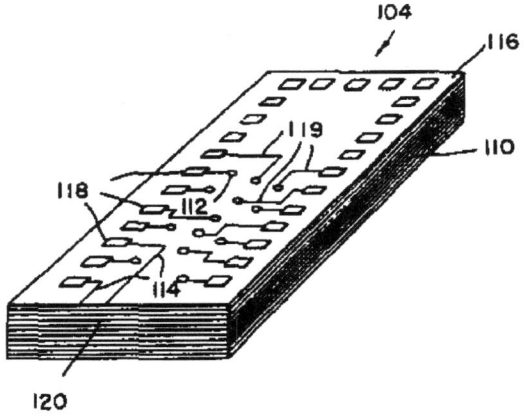

Fig. 2.4.2 (US-A-3852714 fig. 7)

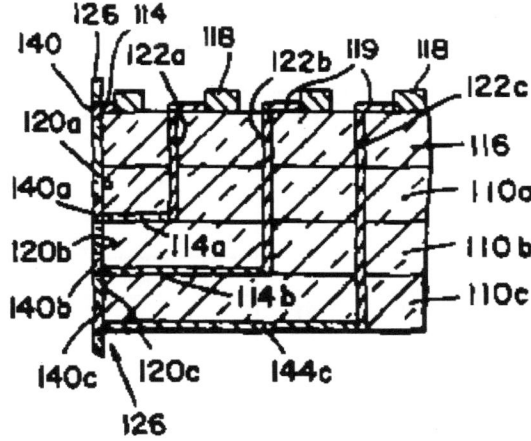

Fig. 2.4.3 (US-A-3852714 fig. 15)

A sensor module 102 is shown comprising a set of multilayered boards 104 of different widths stacked to form a shelved structure with multiple shelves 108. Each board is comprised of uniform alumina wafers 110 stacked together. Each board also comprises a top wafer 116, the edges of which comprise the shelves. Each wafer has a pattern of holes 112 therethrough, a series of terminal pads 118 and conductors 119 each coupled between a hole 112 and a pad 118 thereon. The conductors 114 on each wafer 110 extend from and in a plane perpendicular to that of an end 120 thereof to the various metallized holes 112 therethrough. Each wafer 116 also has a series of conductors 114 extending from an end 120 thereof to those terminal pads 118 thereon not coupled to any holes 112 therethrough. The wafers 110 are stacked such that the various holes 112 of each wafer 110 are aligned with the holes 112 of the wafer above and below, whereby the metal in the aligned holes 112 form electrically conducting vias 122 perpendicular to the wafers 110 and 116. Each via 122 connects a conductor 114 on a wafer 110 to a conductor 119 on the top wafer 116 of its board 104.

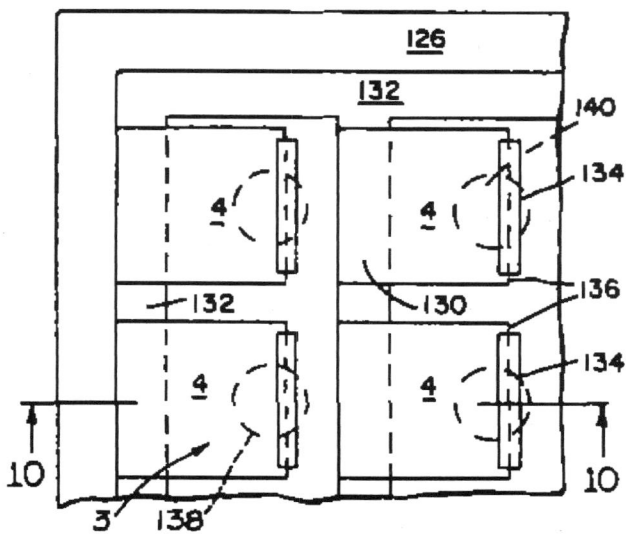

Fig. 2.4.4 (US-A-3852714 fig. 9)

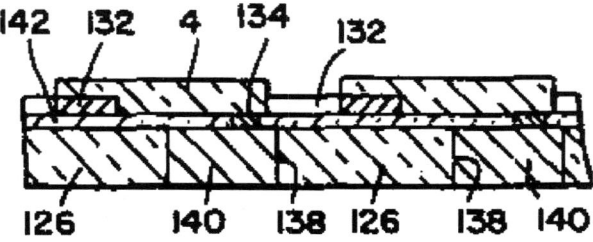

Fig. 2.4.5 (US-A-3852714 fig. 14)

The module further comprises a flat end surface 124 to which a substrate 126 may be bonded. The substrate is of alumina or sapphire and has disposed thereon a mosaic array 3 of electro-optical detectors 4. The detectors are formed by etching or laser cutting of a mercury cadmium telluride layer which has been deposited on the substrate. Each detector is connected at one end 130 to a common reference terminal 132 and at the other end 136 to terminal 134. The substrate further comprises a pattern of through holes 138 and metal dots 140 such that each signal terminal 134 contacts a dot 140. The substrate 126 is aligned on the end surface 124 such that the metal dots 140 in each hole 138 contacts a conductor 114. As a result, each detector 4 is coupled to a pad 118.

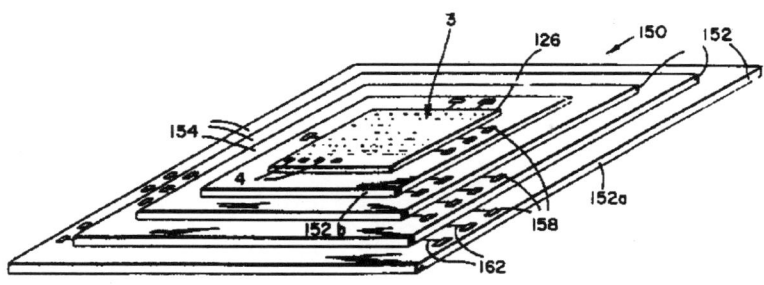

Fig. 2.4.6 (US-A-3852714 fig. 16)

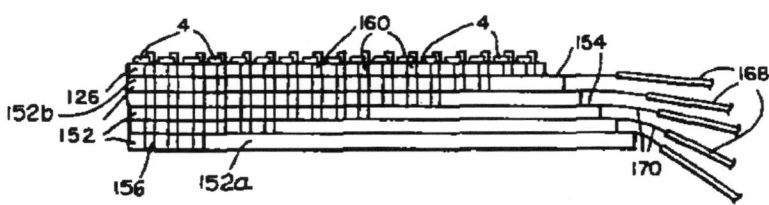

Fig. 2.4.7 (US-A-3852714 fig. 17)

The module of another embodiment uses a set of various sized wafers 152 stacked to form a mesa structure with the edges of the wafers comprising shelves 154. Each wafer has a pattern of metallized holes 156 therethrough, and a series of terminal pads 158 on the shelves. The method of manufacturing the imager is claimed in US-A-3970990 (Grumman Aerospace Corporation, USA, 20.07.76).

*

The photo-detector array module disclosed in US-A-4304624 (Irvine Sensors Corporation, USA, 08.12.81) comprises a focal plane array of detectors which is directly mounted, without a separate substrate, on an end of a module formed by a layered structure. Electronic components are mounted in "wells" in the layered structure. The method of fabricating the structure is claimed in the patent and the photo-detector mosaic array module is claimed in US-A-4354107 (Irvine Sensors Corporation, USA, 12.10.82).

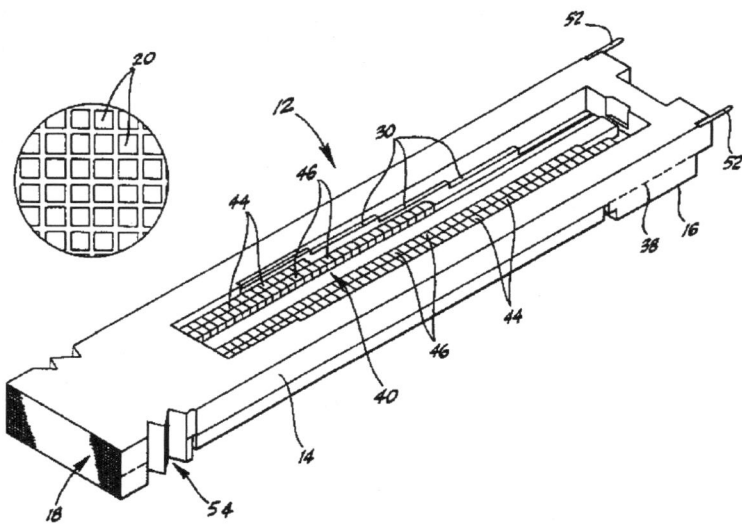

Fig. 2.4.8 (US-A-4304624 fig. 1)

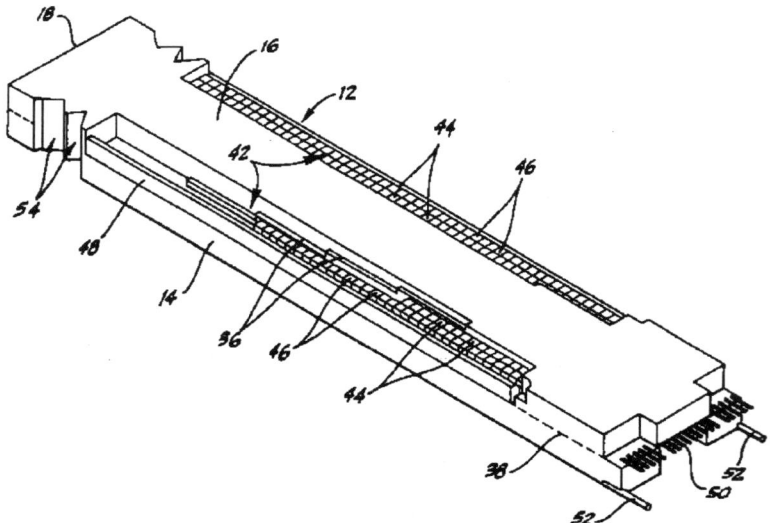

Fig. 2.4.9 (US-A-4304624 fig. 2)

The figures show opposite sides of a completed module. The module consists of an "O-shaped" sub-module 14 and an "I-shaped" module 16. The photo-detector array, which comprises individual infrared photo-detector elements 20 (see the "blown-up" view), is located on the end 18 of the module. Each of the sub-modules is composed of a number of thin layers. Each of the thin layers constitute a ceramic insulating and supporting layer which supports a layer of thin film conductors. The thin layers are glued together. The "O-shaped" and "I-shaped" sub-modules are also glued together to form the complete module. A dotted line 38 is used to show the location of the interface between the two sub-modules. The interface provides the surface to support electronic components, 44 and 46, which are located in the wells, 40 and 42, provided at the center of the "O-shaped" sub-module and at the sides of the "I-shaped" sub-module. When one of the thin layers is in the complete module, detector elements mounted on the focal plane face 18 of the layer are connected individually by the thin film conductors to the electronic components. The complete module also comprises a control bus pallet 48, connector pins 50, guide pins 52 and alignment and cooling rod cutouts 54.

*

Placing detectors on a focal plane face of a supporting module is very demanding. The goal is to end up with thin diode chips attached to the end of the module and in electric contact with lead points on the module. In US-A-4290844 (Carson Alexiou Corporation, USA, 22.09.81) a wafer of a detector material is secured to the end of the module and ion-milling is used to pattern separate detector islands. This method permits a very high density of detectors to be formed on the focal plane.

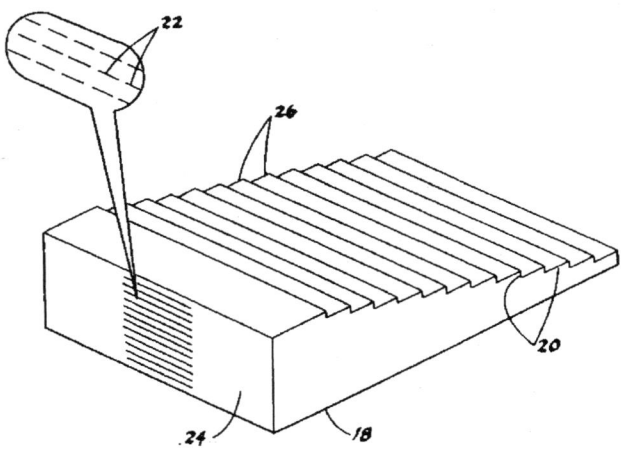

Fig. 2.4.10 (US-A-4290844 fig. 1)

A multi-layer module 18 consisting of thin insulating wafers 20 formed from a ceramic material is prepared. The wafers have a large number of thin film electrically-conducting leads on their surfaces. The leads end at points 22 located on the focal plane surface 24 of the module. The lengths of the ceramic wafers are staggered to give access to lead pads 26 on the face of each wafer.

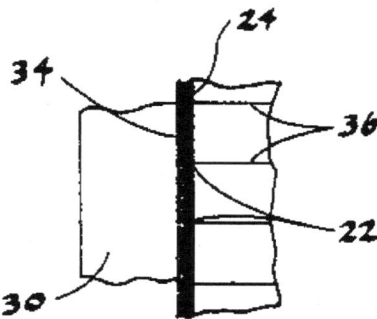

Fig. 2.4.11 (US-A-4290844 fig. 3)

A wafer 30 of HgCdTe is cut from an ingot and lapped to remove irregularities. The wafer is then cut into strips which are suitable for mounting on the focal plane face 24. In order to improve the conductivity between the wafer and the contact points 22, a p^+-type region is formed at the "back" of the wafer on top of which is plated a gold layer. The wafer 30 is secured to the face 24 by a thin layer 34 of silver epoxy, which also provides electrical continuity to the contacts 22. The epoxy is cured and the thickness of the wafer is reduced by lapping and polishing. The final thinning is produced by etching. Next, titanium is sputtered on the surface of the wafer in order to provide an ion beam milling mask. The mosaic detector pattern is etched into the titanium in alignment with the contacts 22.

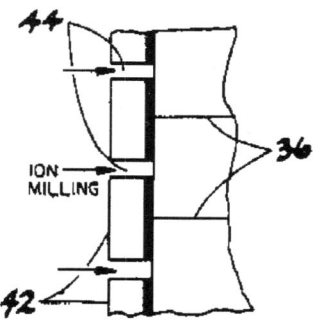

Fig. 2.4.12 (US-A-4290844 fig. 5)

Ion beam milling is carried out to cut through the HgCdTe wafer 30 and the epoxy layer 34, so as to electrically isolate the individual HgCdTe detector elements 42 by channels 44. Each element is in electrical contact with an individual contact point 22 and its electrical lead 36. Next, the titanium mask is removed by etching and the surface is cleaned.

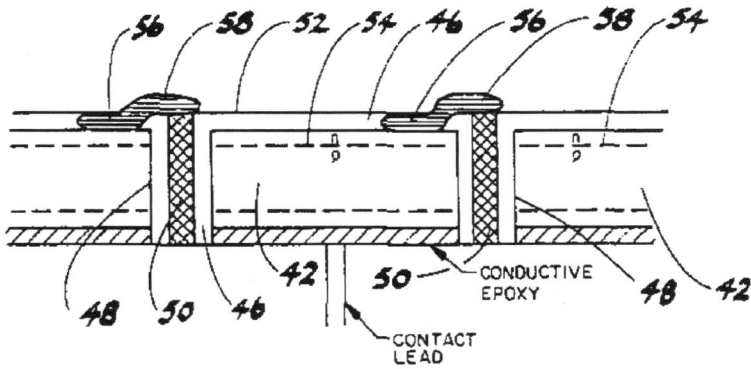

Fig. 2.4.13 (US-A-4290844 fig. 10)

A ZnS coating 46 is used to encapsulate the detectors. A dielectric filler is deposited in the channels between the detector elements to provide a supporting surface for a common electrode and to provide lateral mechanical support for the detector elements. Next, diode junctions 54 of the detectors are created by ion-implantation of boron ions. Indium contact pads 56 are formed in holes formed in the coating 46, and a common indium electrode 58 is formed on top of the dielectric material 50.

*

The packing density of a detector module is further improved in US-A-4551629 (Irvine Sensors Corporation, USA, 05.11.85) in which silicon chips are stacked to create a focal plane to which photo-detectors are mounted.

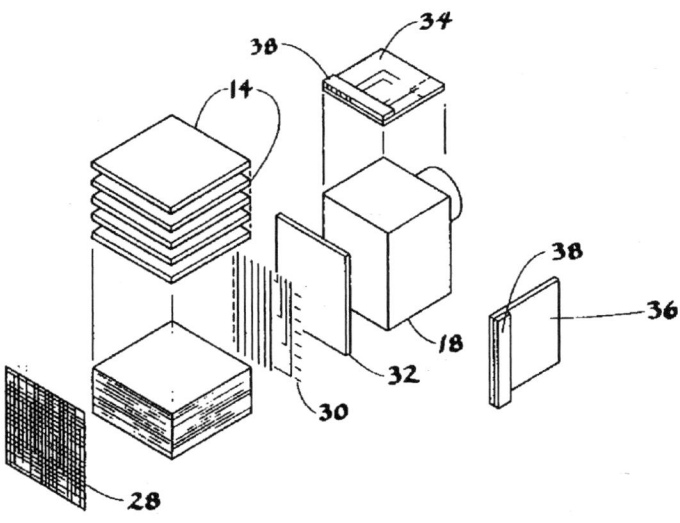

Fig. 2.4.14 (US-A-4551629 fig. 2)

The module, shown above in an exploded view, comprises a stack of silicon chips 14 which are secured together by an adhesive material. Each chip carries an integrated circuit having electrical leads terminating at a focal plane where they are individually in electrical contact with separate very closely spaced photo-detectors. The stack of chips is mounted on a supporting block 18 of molybdenum, which also functions as a cooling structure and which has a mounting stud 20 used when the module and its supporting block are secured in a larger assembly. A detector "mosaic" 28 is provided on the front plane of the silicon stack according to the method presented in US-A-4290844, *supra*, and a back plane wiring 30 is provided on the rear plane of the silicon stack. The supporting block is insulated from the back plane of the silicon stack by a flat insulating board 32. Two additional insulating boards 34 and 36 are secured to two sides of the supporting block, and are arranged to carry lead-out conductors from the back plane wiring to ribbon cables. A process for the purpose of providing an insulating surfaces on the front and rear planes which do not cover the ends of the electrical leads but otherwise fully cover the end surfaces of the stacked silicon chips is also disclosed. This aspect is also described in US-A-4525921 (Irvine Sensors Corporation, USA, 02.07.85). Alternative processes are disclosed in EP-A-0317083 (Grumman Aerospace Corporation, USA, 24.05.89), EP-A-0317084 (Grumman Aerospace Corporation, USA, 24.05.89), US-A-4877752 (The United States of America as represented by the Secretary of the Army, USA, 31.10.89) and US-A-4956695 (Rockwell International Corporation, USA, 11.09.90). A method and related fixtures which permit formation of the stack of chips are presented in WO-A-8603338 (Irvine Sensors Corporation, USA, 05.06.86). The method includes the steps of measuring the thickness of separate chips and selecting groups of chips having appropriate thickness.

*

The primary deterrent to reducing the detector center-to-center distance for the structure presented above in US-A-4551629 is the required thickness of each silicon chip. If the chips are too thin, they will be very fragile, and will therefore be difficult to handle during fabrication of the modules. In US-A-4403238 (Irvine Sensors Corporation, USA, 06.09.83) the distance between adjacent detector rows can be less than the thickness of a given chip, by orienting the detector rows along lines which extend diagonally with respect to the planes in which the silicon chips extend.

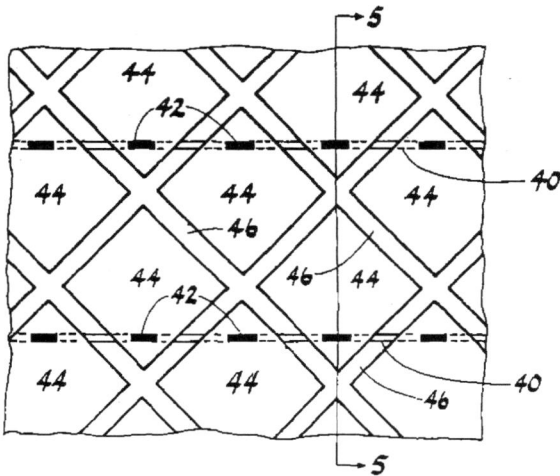

Fig. 2.4.15 (US-A-4403238 fig. 4)

The figure shows the photo-detectors 44 in place on top of the focal plane. Each photo-detector is spaced from adjacent photo-detectors by channels 46, thereby isolating the individual photo-detectors electrically. The photo-detectors, which are square in shape, are arranged in rows which extend diagonally across the focal plane. As a result, the thickness of each chip is equal to the diagonal of a square having its side equal to the distance between the centers of adjacent detectors. Furthermore, the electrical leads 42 make contact with the detectors 44 near the corners of the detectors.

*

In US-A-4618763 (Grumman Aerospace Corporation, USA, 21.10.86) a detector module is disclosed which is formed of stacked multi-channel integrated circuits, a detector array and a module header interface. The detector array and the module header interface are disposed transverse to the plane of the integrated circuits on opposite edge portions of the integrated circuits. Each integrated circuit is formed in a semiconductor material which has been deposited upon a thin sapphire wafer.

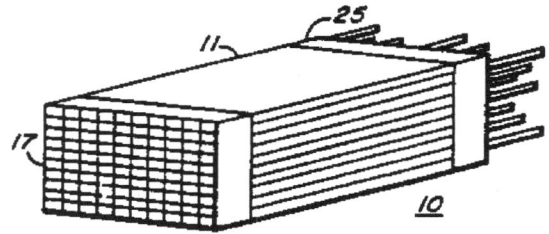

Fig. 2.4.16 (US-A-4618763 fig. 4)

An infrared array segment is bonded to a front end 17 of stacked integrated circuits 11, transverse to the orientation of the integrated circuit layers. The detector array segment is formed of a plurality of discrete detector elements. Each element is connected to a dedicated channel on one of the integrated circuits through connections between the detector element connectors and metallized pads. Each integrated circuit is formed by epitaxially depositing semiconductor material upon a thin insulating sapphire wafer. Conductors leading to a back edge of the integrated circuits are exposed and provided with metallized pads to which a header pin matrix interface 25 is bonded.

*

A detector mosaic which is bonded to a multilayer alternating thick film and thin film interconnect pattern is disclosed in EP-A-0281026 (Honeywell Inc., USA, 07.09.88).

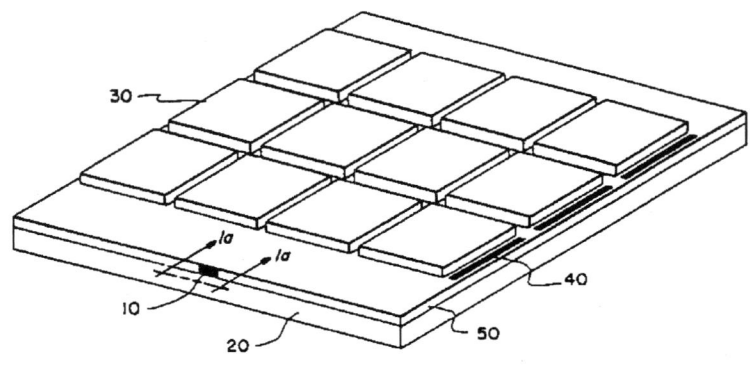

Fig. 2.4.17 (EP-A-0281026 fig. 1)

The assembly comprises a detector mosaic 10, a substrate 20, a plurality of integrated circuit devices 30, IC read-out pads 40 and a multilayer alternating thick film and thin film interconnect pattern 50. The detector mosaic comprises rows and columns of detectors 14. The figure below shows the cross-section taken along 1a-1a.

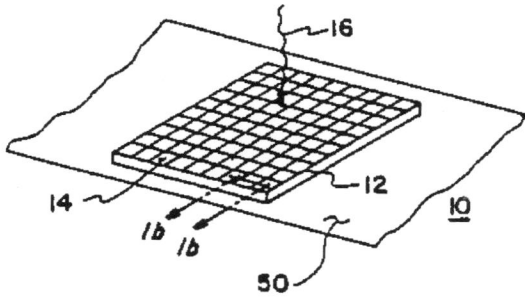

Fig. 2.4.18 (EP-A-0281026 fig. 1a)

Below a cross-section along 1b-1b is shown which shows the bonding construction of a detector 14.

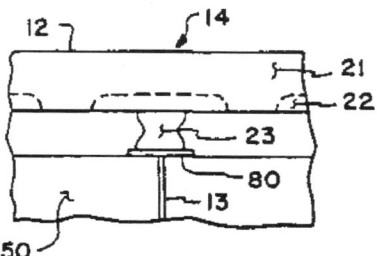

Fig. 2.4.19 (EP-A-0281026 fig. 1b)

Each detector 14 includes a detector crystal region 21 having a junction 22 which is connected to an indium bump 23. The indium bump bonds to the bonding pad 80 which is, in turn, bonded to a metallized trace 13. The sensitive area plane 12 is parallel to the plane formed by the edge of the supporting substrate and the multilayer thick/thin film interconnect pattern 50.

*

One problem which arises when a detector array is attached to the face of a multi-layer module is the inability of the detector material to absorb forces generated by a mismatch of coefficient of thermal expansion between the detector array material and the module. Furthermore, it is difficult to isolate a fault that may be attributable to either the detector elements, module wiring or processing elements. A buffer board is introduced in WO-A-8807764 (Grumman Aerospace Corporation, USA, 06.10.88) which facilitates electrical communication between the detector elements and the module and conductive patterns formed on the module layers, and also enhances the structural characteristics and separate testability of the system components.

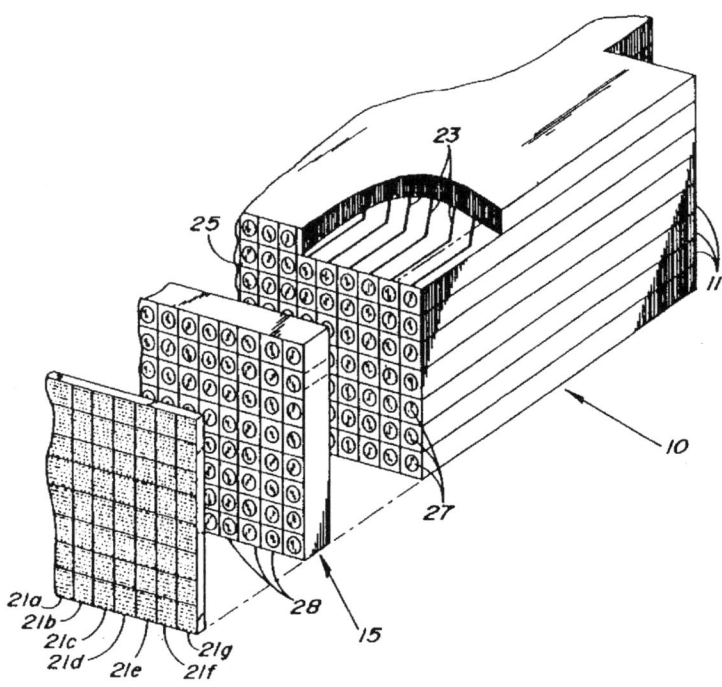

Fig. 2.4.20 (WO-A-8807764 fig. 1)

A module 10 is formed by a plurality of stacked circuit layers 11. Each of the circuit layers is provided with conductors 23 extending to a first edge portion 25. Bump bond connectors 27 are formed on the first edge portion of each of the circuit layers. The bump bond connectors facilitate electrical contact between the conductors 23 and connectors formed on a buffer board 15. A detector array segment is formed of discrete detector elements 21a-21g. Each detector element is connectable to a dedicated conductor 23 through dedicated connectors 28 formed on the buffer board, and a dedicated connector 27 formed on the edge portions of the circuit layers. The buffer board is disposed intermediate the detector array segment and the module. Two buffer boards may be used to facilitate separate testing of the detector array and the module.

*

When a detector array is connected to the side edges of a plurality of thin silicon substrates, which comprise signal processing electronics, there is a problem in forming interconnect bumps on the side edges of the silicon substrates. This problem is solved in US-A-5081063 (Harris Corporation, USA, 14.01.92) by an electroplating technique.

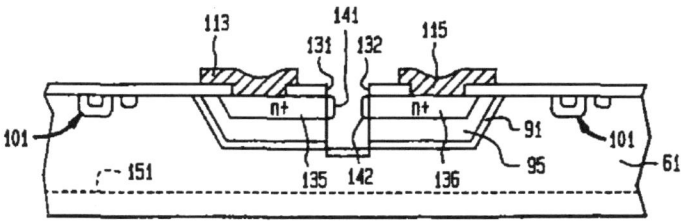

Fig. 2.4.21 (US-A-5081063 fig. 10)

A depression is anisotropically etched in a silicon substrate 61 and a silicon dioxide layer 91 is formed. An n$^+$-type doped polysilicon layer 95 is deposited to fill the depression. A pair of signal processors are formed in device regions 101. Impurities are introduced into the top surface of the polysilicon layer to form a heavily doped n$^+$-type surface region, to which a pair of electrodes 113 and 115 are contacted. Next, a groove is cut or scribed into the top surface of the substrate so as to cut through the bottom floor of the polysilicon layer 95 and expose a pair of side portions 131 and 132 of the heavily doped surface region which now forms region 135 and 136, respectively. The exposed surfaces of the substrate are immersed in an electroplating solution, such as gold cyanide, and each of the electrodes 113 and 115 is connected to the negative potential electrode of an electroplate power supply, the positive potential electroplate electrode of which is immersed in the electroplate solution, so as to cause raised land portions or bumps 141 and 142. After completion of the electroplating process the wafer is backlapped until the the material removal plane intersects the groove and thereby separates the substrate into separate dice.

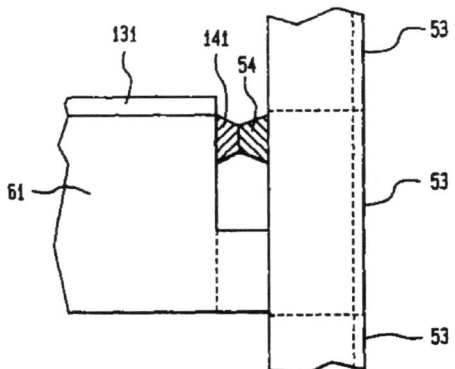

Fig. 2.4.22 (US-A-5081063 fig. 12)

The substrate may be partially lapped and then scribed to the bottom of the groove so as to leave a portion of the substrate as a projection beneath the bump, thereby facilitating location and joining of the silicon substrate and the detector substrate comprising photodiodes 53.

*

Other problems, which occur when an array of detectors is connected to the side edges of a plurality of thin silicon integrated circuit layers, are the problem of aligning the detector array and the integrated circuit layers and the problem of replacing a single integrated circuit layer after the detector array is connected to the integrated circuit layers. These problems are solved in US-A-4992908 (Grumman Aerospace Corporation, USA, 12.02.91) by bevelling the edges of the silicon layers and connect the layers to a contact board which has a beveled surface formed to receive and support the integrated circuit layers.

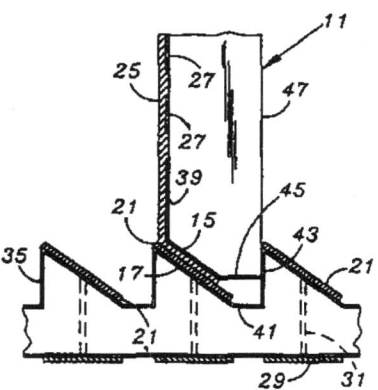

Fig. 2.4.23 (US-A-4992908 fig. 2b)

The integrated circuit layers 11 are formed on (100) oriented single crystalline silicon material. An anisotropic etch selectively etches away atoms of a silicon until it reaches the portion of the silicon crystal lattice defined by the (111) plane. The etching results in a beveled edge, the angle of which is determined by the silicon crystal lattice. The layers also comprise coductive pads connected with conductive pattern 25. The pads are in electrical connection with contact board pads 21, which are formed to accommodate the bevel of the layers. The contact board is provided with conductive pads 29 which are connected to pads 21 by conductive vias 31. The integrated circuit layers may be supported between two contact boards to avoid the need for adhesive layers to provide structural integrity of the module. In such a case the removal of an individual layer is possible without disturbing adjacent layers. This document does not explicitly specify the detector material which is connected to the contact board.

*

When ceramic substrates, having thin film micro circuits printed thereupon, are laminated using thin thermosetting adhesive sheets to form a module to which a detector array is bonded, the coefficient of expansion between the ceramic layers and the thermoset adhesive sheets leads to thermal stress when the module is subjected to large range of thermal cycling. The thermal stress is transmitted to the detectors and can cause the detector array to crack. Non-uniformity in the thickness of the laminating adhesives is one of several problems mentioned in US-A-5128749 (Grumman Aerospace Corporation, USA, 07.07.92). Instead of the thermosetting adhesive sheets, it is proposed to use a glass binding material which has a coefficient of expansion approximately equal to the coefficient of expansion of the substrates.

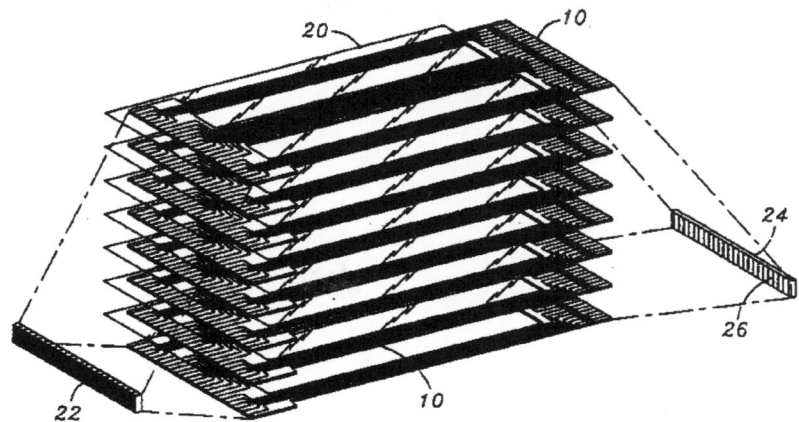

Fig. 2.4.24 (US-A-5128749 fig. 3)

A plurality of substrate layers 10 have electronic devices mounted thereon and interconnected by conductive conduits formed upon the surface thereof. An array of detector elements 22 are flip-chip bump bonded to the edges of the substrate layers. A dielectric or glass binding material layer 20, which has a melting point between 400 and 550°C, is disposed adjacent to the layers 10. The dielectric layer may comprise materials such as lead borosilicate glass, glass-rich ceramics, green ceramics or lead borate glass. Glass binding layers can typically be formed to more uniform and precise thickness than thermosetting adhesive layers. Other advantages are greater mechanical strength, higher stiffness, less outgassing and better chemical and solvent resistance.

*

Chapter 2.5

Detector Arrays Directly Contacting the Read-Out Chip

Imagers having detector arrays directly contacting the read-out chip are presented in this chapter. The imagers having electrical connections made by through hole technologies are presented in a separate section.

Summary

- An imager having HgCdTe detectors inlaid in a silicon semiconductor substrate and the manufacturing process thereof is disclosed in GB-A-1337968.

- In US-A-3842274 an infrared semiconductor is capacitively coupled to a silicon CCD register array.

- An imager having an element packing density of 90% is disclosed in US-A-4104674. Infrared photovoltaic detectors of mercury cadmium telluride are mounted on a silicon substrate. Electrical contacts are made by thin-film metallizations.

- A detector structure which may be fabricated by planar processing and which permits fabrication of thin infrared detectors is shown in US-A-4197633.

- An improved process for manufacturing the device disclosed in US-A-4197633, discussed above, is taught in EP-A-0050512.

- A multicolour imager in which detectors having different wavelength sensitivities are stacked in a piggy-back fashion on top of one another and in which thin film interconnectors are used to connect the detectors with an integrated circuit, forming the detector substrate, is presented in US-A-4137625.

A problem which arises when a read-out chip of for example silicon is attached to a detector chip of mercury cadmium telluride is the mechanical damage which may occur when the array is cooled to cryogenic temperatures for operation. The stress is due to a mismatch in the coefficients of thermal expansion between the two materials.

- The problem stated above is solved in FR-A-2484705 by removing the detector material except for islands which form the detector elements.

- The difference of thermal expansion coefficient between an HgCdTe detector material and a silicon read-out material is dealt with in FR-A-2494910 where detector elements in the shape of islands are connected on one side to a silicon read-out chip and on the other side to a substrate having a similar thermal expansion coefficient as the silicon read-out chip.

- The problem of electrical and mechanical coupling is the basis of the invention in US-A-4311906. A first and a second array of metallic electrodes connect detector elements formed as a mesa-structure in a first substrate by capacitive coupling to a read-out circuit in a second substrate.

- An imager which uses Metal-Insulator-Semiconductor gate Static Induction Transistors (MISSIT), formed in silicon, for read-out and refresh and which has a photo-detecting layer of HgCdTe is presented in EP-A-0038697.

- The imager of EP-A-0038697, discussed above, has an complex structure and hence is difficult to produce. In EP-A-0042218 a simplified structure is presented. In all the embodiments the refresh transistor present in most of the embodiments of EP-A-0038697 has been eliminated.

- An array which is compact and easy to manufacture is disclosed in JP-A-57031170. A MOSFET with a polysilicon gate electrode is formed in a silicon substrate. A thin HgCdTe film composed of an n-type layer and a p-type layer is prepared on the source.

- In GB-A-2113467 a manufacturing process of an imager is disclosed using the step of converting a polycrystalline layer of HgCdTe to a monocrystalline layer.

- A multi-colour imager formed by multi-level HgCdTe layers is disclosed in EP-A-0087842.

- The imager of FR-A-2593642 comprises a silicon read-out substrate on which a detector substrate is glued. A general problem is addressed in forming reliable electrical connectors which are formed from a region on the read-out substrate, over the edge of the detector substrate, to a region on the detector substrate.

- An imager with a similar process of manufacturing to that disclosed in FR-A-2593642, discussed above, is presented in US-A-4532699.

- Operation of photo-conductive detectors requires current bias. A steady bias current produces a standing DC output. The bias pedestal is difficult to compensate for accurately for each detector of an imager. An imager designed to be operated by a current bias which is cyclic in time and has a waveform and frequency to produce zero net average signal both in dark and in presence of radiation of uniform intensity is presented in GB-A-2128019.

- An imager comprising detectors formed in an HgCdTe layer which has been epitaxially grown on a semiconductor layer of a different material, the semiconductor layer comprising a charge transfer element, is disclosed in US-A-4553152.

- In MIS-type detectors, radiation generated charge is stored in potential-wells in a semiconductor under MIS-gates. When MIS-gates are formed by depositing metal and patterning the metal by a photolithographic process, the device is exposed to potentially damaging mechanical, thermal and ultraviolet radiation stresses. A different approach to fabricate a MIS-type detector is presented in EP-A-0173074.

- An imager which has a detector substrate and a silicon read-out substrate attached on different sides of a sapphire substrate is presented in JP-A-61128564.

- An imager similar to the imager presented in JP-A-61128564 above is described in JP-A-63147366 in which HgCdTe photodiodes are formed on one side of a CdTe substrate and a GaAs substrate comprising a read-out CCD is formed on the other side.

- In JP-A-62022474 vapor deposition of metal electrodes is used to connect an HgCdTe layer with a silicon read-out substrate. The metal electrodes form Schottky diodes or generate pn-junctions depending on the type of metal used.

- An imager is presented in JP-A-62071270 in which HgCdTe layers are grown directly on a read-out substrate by an epitaxial process.

- Molecular Beam Epitaxy, MBE, is used to grow infrared detectors on a silicon read-out substrate in US-A-5003364.

- In JP-A-63116459 an imager is made up of photodiodes formed in an HgCdTe substrate. The photodiodes are placed in columns and rows. The number of connections between the HgCdTe substrate and a silicon substrate, comprising a read-out device, is reduced by connecting the anodes of the photodiodes in each row to each other and connecting the cathodes of the photodiodes in each column to each other.

- In JP-A-63310165 infrared ray detector elements of HgCdTe are formed on a CdTe substrate in which a read-out CCD is integrated.

- The lattice mismatch between silicon and mercury cadmium telluride makes it difficult to grow an epitaxial crystalline layer of mercury cadmium telluride on silicon. In EP-A-0343738 an imager is formed by growing mercury cadmium telluride on a sapphire substrate at openings formed in a silicon layer which has been grown on the sapphire substrate (silicon-on-sapphire, SOS) at an earlier stage.

- GaAs, CdTe and HgCdTe are deposited epitaxially on a silicon substrate in US-A-4910154. If islands of individual detector elements are formed, the deposited layers will have a combined thickness which makes it difficult to interconnect the detector elements with corresponding silicon circuitry due to the steep sides of the detector elements. It is therefore proposed to grow the GaAs, CdTe and HgCdTe layers in recesses formed in the silicon substrate, and interconnect with a planar technology.

- Manufacturing of an imager, such as the imager presented in US-A-4197633 above, comprises the steps of bonding a mercury cadmium telluride body via a first surface by an adhesive layer to a substrate on which a conductor pattern is present, removing portions of the body thoughout its thickness and providing electrical connections from the substrate to regions of the body. A problem is that the conductor pattern is exposed in the removal step. Furthermore, pn-junctions formed in the body adjacent to a second surface, opposite to the first surface, can be degraded by heat treatments. In GB-A-2231199 the body is adhered to the substrate via the second surface thereby protecting the pn-junctions against material degradation during heat treatments.

- When electrical connections from a detector array to processor circuits are formed by an array of small diameter vias or electrical conduits through a wafer substrate, a problem is that sub-micron vias can be formed by either etching or laser drilling only as long as they do not extend more than a few microns in length. The imager of US-A-5315147 comprises an integrated circuit wafer having a signal processing circuitry formed on a first surface thereof and detector elements formed upon a surface opposite to the first surface. The wafer has an array of waffle-like hollows, each hollow has a dense array of small diameter vias which extend through the wafer. Conductive conduits are formed through the hollows and vias to connect the detectors to associated signal processing circuitry.

- The imager disclosed in WO-A-9202959 comprises thin film transistors deposited on a substrate, a planarization layer deposited on the transistors followed by a mercury cadmium telluride layer and a top electrode layer.

- Each detector element of the imager disclosed in JP-A-5315580 comprises a photodiode formed from HgCdTe layers which have been grown in a recess in a silicon substrate having a surface orientation selection layer and a buffer layer formed in between.

- In JP-A-6021419 HgCdTe layers are formed on a silicon substrate with an high-resistance compound semiconductor in between.

- The imager of JP-A-6089991 comprises HgCdTe detector regions which have been grown on a silicon substrate. A method to clean the silicon surface before the HgCdTe regions are grown thereon is disclosed.

- The detector elements of the imager presented in JP-A-6326342 are separated from each other by providing amorphous material only at regions between the detector elements and removing this amorphous material in a wet etching process.

Summary - Connections made by Through Hole Technologies

A different approach to the problem of connecting terminals of individual detectors formed in an HgCdTe layer with corresponding terminals of a read-out device integrated in a semiconductor substrate is to provide connections via holes formed through the thickness of the HgCdTe layer. One particular technique is known as the loophole technique. The characteristic features of this technique are that an HgCdTe substrate is adhered to a semiconductor substrate which comprises a read-out circuit and that holes are drilled through the HgCdTe substrate where the pn-junctions are formed and the detector terminals are connected to input terminals of the read-out circuit by filling the apertures with a metallization layer. The technique gives rise to a robust device which is rather easy to manufacture and which may have a small size of the individual detector elements.

- An imager and a method of manufacturing an imager using the loophole technique described above is presented in EP-A-0061803.

- In GB-A-2208256 detector elements are provided in an hexagonal arrangement whereby each detector in the array is surrounded by six neighbouring detector elements. Compared to a square X-Y matrix arrangement, in which each detector is surrounded by eight neighbouring detector elements, the detector elements can have increased detectivity and lower leakage currents. Furthermore, the array can have a higher spatial resolution for a given detecting area and a higher probability of detecting small objects of less than one element subtense.

- It is conventional to cool mercury cadmium telluride photodiodes to about 80 K for 8-14 μm waveband detection material and to about 200 K for 3-5 μm waveband detection material. By designing the photodiodes so as to incorporate a minority-carrier extraction region and to inhibit minority-carrier injection, the minority-carrier concentration in the photosensitive region can be depressed to produce low noise and high sensitivity characteristics similar to those as are obtained by cooling. The resulting photodiodes for 3-5 μm waveband detection may be operated at ambient temperature, about 295 K, and those for the 8-14 μm waveband at about 190 K. An imager comprising photodiodes designed in this manner is presented in GB-A-2239555.

- An anodized film is used in JP-A-2272765 to suppress cross-talk between detector elements.

- Cross-talk may also be reduced by forming trenches between individual detectors. In JP-A-4267562 and JP-A-4269869 gate electrodes are also provided in the trenches.

- Two or more HgCdTe layers having different bandgaps are formed as different levels on a read-out substrate in EP-A-0475525 to provide different wavelength response. The individual detectors are connected by the same method as described in EP-A-0061803 above.

- The shorter-wavelength photodiode 10 and the longer-wavelength photodiode 20 of the imager presented above in EP-A-0475525 are placed side-by-side and not on top of each other. The signals from the photodiodes will therefore not refer to the same source from a scene projected on the imager. This problem is solved in EP-A-0481552 by the use of infrared lenses placed on the imager.

- EP-A-0561615 teaches how to form an electrical connection on a side-wall, and only on the side-wall, of a step structure. This technique is used to form a compact two-wavelength imager.

- A detector involving an ambipolar drift of radiation-generated free minority carriers using the loophole technique is disclosed in GB-A-2207802. (Ambiplar drift field imagers are presented in Chapter 1.2.)

- In US-A-4720738 an HgCdTe layer is attached to a read-out substrate. Detector elements are connected to aluminium buses which are connected to bonding pads on the substrate via conductors that occupy holes through the HgCdTe layer.

- An imager connecting pn-junction detectors or MIS detectors to connection pads of a silicon read-out chip via through holes is presented in EP-A-0137988. The through holes are created by first ion-milling small holes and then enlarge the same by spray etching.

- The storage gate of EP-A-0137988, presented above, is formed of nickel or chromium. There is a tradeoff between transmittance and electrical conductivity of the gate. This problem is further analysed in EP-A-0416299. It is proposed to use bismuth (Bi), antimony (Sb) or titanium oxynitride (TiN_xO_y).

- A method of passivation of HgCdTe by anodic selenidization is disclosed in US-A-4726885.

- A process for forming a ZnS insulative layer on an HgCdTe material characterized by the absence of an intermediate native oxide layer is taught in EP-A-0366886.

- As discussed in EP-A-0366886 above, a sulphide passivation layer may be formed without an intermediate oxide layer. If this technique is used to form the insulation layer of the MIS detector array discussed above in EP-A-0137988, channel stops can not be provided by trapped positive charge in the passivation layer. An approach to solve this problem is to etch away the sulphide passivation layer in channel stop areas and grow anodic oxide to regain the fixed positve charge, but this has the problem of extra wet processing steps. In EP-A-0416320 n-type channel stops are provided by lattice damage implants using boron or by ion-milling using argon ions.

- During the operation of a MIS detector of the kind presented in EP-A-0137988 above, a voltage is applied to the gate to form a depletion region in the HgCdTe. Photons penetrate the gate and create electron-hole pairs. The holes accumulate in the HgCdTe at the interface with the gate insulator and form an inversion layer which reduces the size of the depletion region and lowers the absolute value of the gate potential. An increase in the magnitude of the gate voltage implies a larger depletion region so that more holes can be collected. However, too large a gate voltage leads to breakdown in the HgCdTe. In US-A-5144138 the photocapacitors include a heterojunction. This structure allows increased potential well capacity, reduced dark current and detection of two colours. Other features of this document are presented in chapter 1.5.

- Each pixel of the imager presented in GB-A-2197984 comprises a homojunction formed in parallel with a heterojunction.

- In the invention disclosed in JP-A-1061055 a read-out chip is attached to an HgCdTe layer and connection is made via holes through the silicon chip.

- A problem encountered when aluminium is used as gate metal for a MIS structure is the significant etching of the aluminium by the bromine solution which is used to form vias or contacts in overlaying ZnS insulator films. To solve this problem, it is proposed in EP-A-0407062 to use refractory metals as the metallization layers of an infrared detector.

- A structure having low amount of surface leakage currents and a high quantum efficiency is presented in JP-A-5343729. The HgCdTe detecting layer of the presented structure is overlayed on both sides by HgCdTe layers having a larger bandgap than the bandgap of the detecting layer.

- The electrical connection between detector elements and an input pad of a read-out circuit is improved in JP-A-6204448 by providing an indium film on the detector substrate corresponding to the input pad prior to forming a through hole.

- In US-A-5318666 photodiodes and vias are formed simultaneously by employing a dry reactive etching process. Portions of an HgCdTe body adjacent to the vias are thereby type converted.

- A metal interconnect fabrication process is disclosed in US-A-5384267. A metal layer and a photoresist layer are formed on an array of HgCdTe detectors. The photoresist layer is patterned to form a positive mask and the metal interconnect is formed by using a dry etching technique.

Detector Arrays Directly Contacting the Read-Out Chip

An imager having HgCdTe detectors inlaid in a silicon semiconductor substrate and the manufacturing process thereof is disclosed in GB-A-1337968 (Selenia Industrie Elettroniche Associate S.P.A., Italy, 21.11.73).

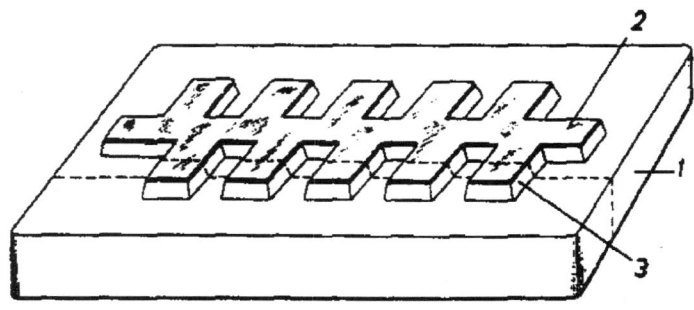

Fig. 2.5.1 (GB-A-1337968 fig. 2)

A monocrystalline chip 1 of HgCdTe is polished and a mask 2 is patterned according to the desired geometry of the array. The chip is chemically etched by HCL and HNO_3 in the areas not covered by the mask, so as to form a projecting structure.

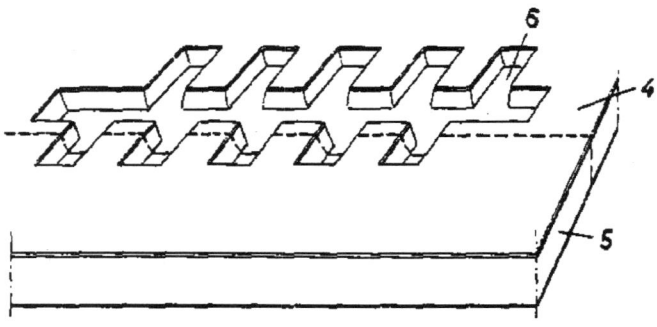

Fig. 2.5.2 (GB-A-1337968 fig. 4)

An analogous mask 4, with a complementary geometry with respect to that on the chip 1, is applied to a slice of monocrystalline silicon. The substrate is chemically etched to form a recess 6.

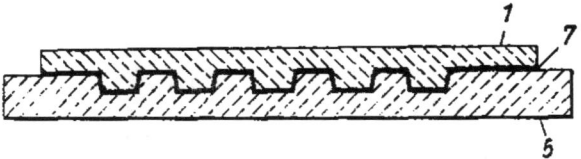

Fig. 2.5.3 (GB-A-1337968 fig. 5a)

The chip 1 and the substrate 5 are placed face to face with the structure 3 fitted into the recess 6. The chip and the substrate are cemented by the use of an epoxy resin base cement 7.

Fig. 2.5.4 (GB-A-1337968 fig. 6a)

The photosensitive material standing proud of the substrate is removed by mechanical lapping or by chemical etching until the free surface of the substrate is exposed.

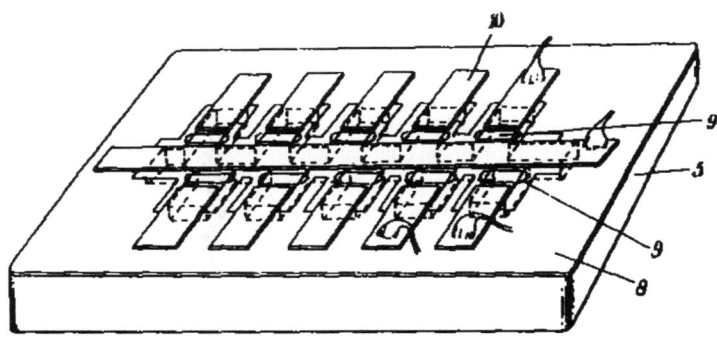

Fig. 2.5.5 (GB-A-1337968 fig. 7)

The necessary electrical contacts are then formed. Electronic read-out elements and preamplifying elements may be incorporated in the silicon substrate.

*

In US-A-3842274 (The United States of America as represented by the Secretary of the Navy, USA, 15.10.74) an infrared semiconductor is capacitively coupled to a silicon CCD register array.

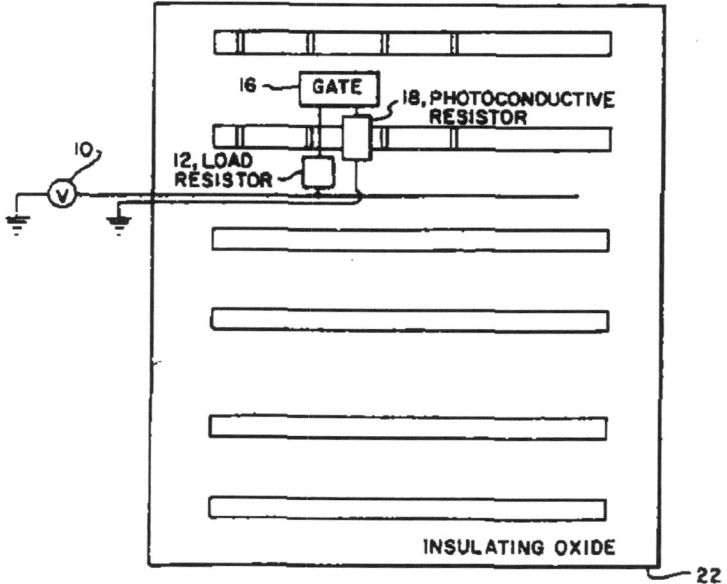

Fig. 2.5.6 (US-A-3842274 fig. 3)

A voltage source 10 is applied to a photo-conductive circuit comprising a load resistor 12 and a polycrystalline or amorphous HgCdTe semiconductor element 18. The photo-conductive circuit is connected to a gate 16 of a conventional silicon charge coupled device. The potential on the gate 16 will control the amount of carriers transferred from the source bucket to the drain bucket of the CCD creating a signal charge in the drain bucket corresponding to the incident infrared radiation. Multiple configurations of two or more photo-conductive resistors, sensitive to different infrared wavelengths, are also proposed.

*

An imager having an element packing density of 90% is disclosed in US-A-4104674 (Honeywell Inc., USA, 01.08.78).

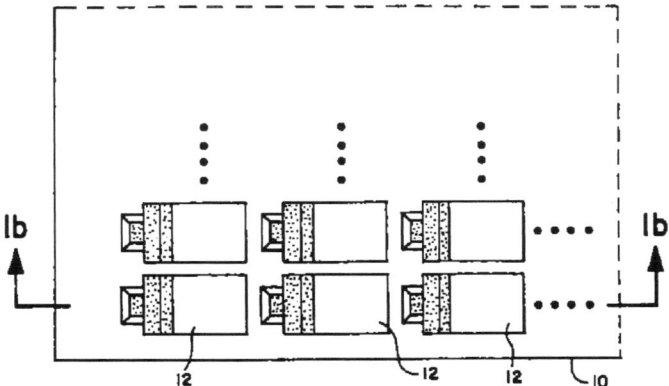

Fig. 2.5.7 (US-A-4104674 fig. 1a)

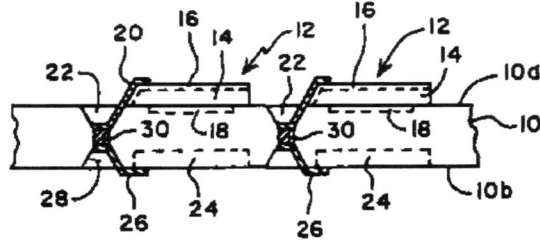

Fig. 2.5.8 (US-A-4104674 fig. 1b)

Infrared photovoltaic detectors 12 of mercury cadmium telluride are mounted on a silicon semiconductor substrate 10 having a thickness of 25-100 µm. Alternatively, the mercury cadmium telluride layer is grown by liquid phase epitaxy. The detectors have a p-type region 14 and an n-type region 16. Electrical contacts to the p-type regions 14 are made by backside diffusion or ion-implanted regions in the substrate 10. Electrical contacts to the n-type regions 16 are made by thin-film metallizations 20 which extend from regions 16 into valleys or vias 22 in the substrate. Source regions 24 of charge coupled devices are formed in the opposite surface 10b of the substrate. The photo signal from each detector is supplied to its corresponding source region 24. Interconnections between infrared detectors 12 and source regions 24 are provided by thin film metallizations 26 extending from the source regions 24 into valleys 28 which are immediately opposite valleys 22. The regions 30 between thin film metallizations 20 and 26 are either high-conductivity regions formed by ion-implantation, diffusion or metallized regions.

*

A detector structure which may be fabricated by planar processing and which permits fabrication of thin infrared detectors is shown in US-A-4197633 (Honeywell Inc., USA, 15.04.80).

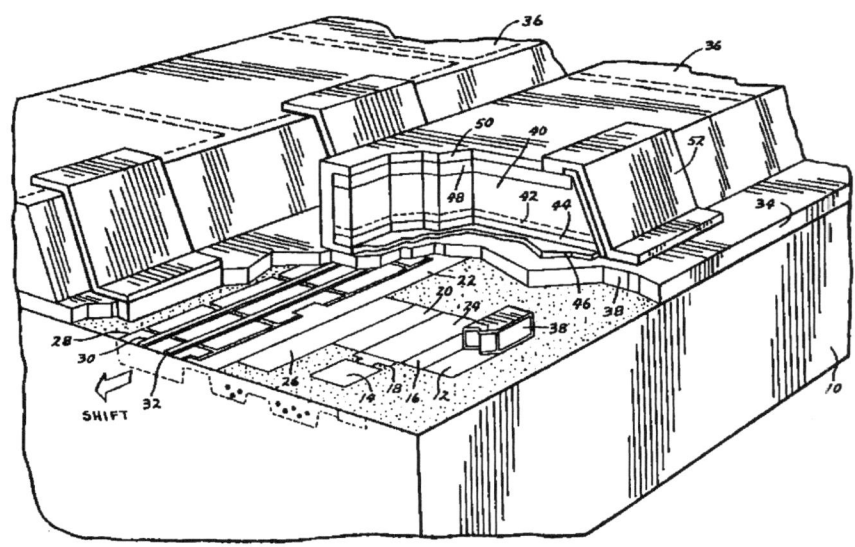

Fig. 2.5.9 (US-A-4197633 fig. 1)

A CCD signal processing circuitry is formed on a silicon substrate 10. The circuitry includes sources 12 and 14, gates 16, 18, 20, and 22, charge wells 24 and 26, shift registers 28, and clock lines 30 and 32. Photosignals from a detector of the detector array are received at source 12. The signals are transferred from source 12 by gates 16, 20, and 22 through charge wells 24 and 26 to shift register 28. Clock lines 30 and 32 cause charge to be advanced through shift register 28. An insulating epoxy layer 34 is formed on the substrate 10 in addition to a silicon dioxide passivation layer. Electrical connection between the individual detector elements in rows 36 and the CCD circuitry is provided by contact pads 38. These metal pads extend from sources 12 through the insulator layer 34. A p-type layer 42 is formed near the back surface of an HgCdTe body 40. A common electrical contact is provided by p^+-type layer 42 together with metal layer 44 and conductive epoxy layer 46. The rows of detector elements are separated mechanically by spray-etching and air abrasion, while columns of detector elements are separated electrically by junction ion-implantation. The epoxy layer 34 protects the CCD circuitry from damage during the air abrasion delineation procedure. A passivation layer 50 of ZnS covers the top and side surfaces of each row 36. Contact pads 38 are connected to regions 48 by thin film interconnects 52.

* *

An improved process for manufacturing the device disclosed in US-A-4197633, shown above, is taught in EP-A-0050512 (Honeywell Inc., USA, 28.04.82). Reference numbers below refer to the figure of US-A-4197633 above.

The low pressure air abrasion which was used to separate rows of detector elements turned out to be difficult to control and devices were frequently damaged during the process. To overcome this, the epoxy layer 34 is applied but prior to curing the same the detector wafer 40 is mounted onto the substrate 10. Pressure is applied and the assembly is heated. Substantially all of the adhesive layer that was previously formed between the wafer and contact pads 38 is eliminated.

*

A multicolour imager in which detectors having different wavelength sensitivities are stacked in a piggy-back fashion on top of one another and in which thin film interconnectors are used to connect the detectors with an integrated circuit, forming the detector substrate, is presented in US-A-4137625 (Honeywell Inc., USA, 06.02.79).

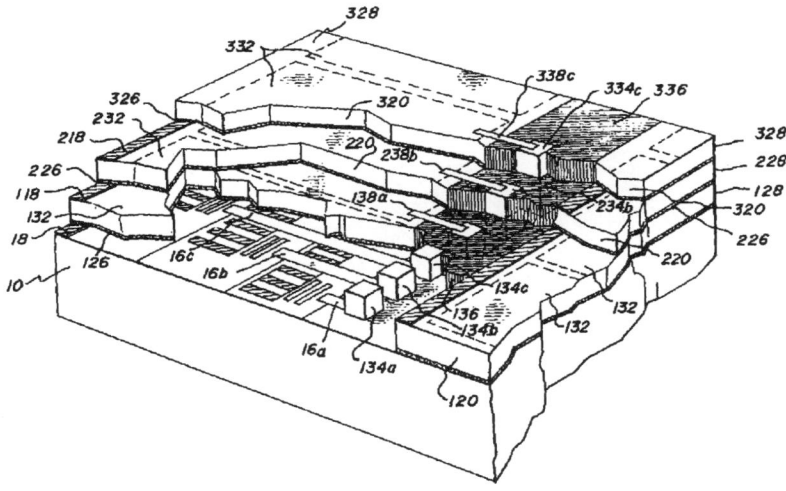

Fig. 2.5.10 (US-A-4137625 fig. 2)

A silicon integrated circuit substrate 10 includes three parallel CCD shift registers for each row of detectors. Thin film source contacts 16a, 16b and 16c are connected to source diffusion inputs of the shift registers. A first body 120 of p-type HgCdTe is bonded on top of the substrate by a non-conductive epoxy 126. The body is lapped and etched to a thickness of 10-12 microns before it is spray-etched to form rectangularly shaped rows. A low pressure air abrasion tool is used to erode away exposed portions of the epoxy layer to expose the source contacts. Next, n-type regions 132 are formed and first level of contact pads 134a, 134b and 134c are plated upon the source contacts so that they extend to the top surface of the rows. The regions between adjacent rows are filled with an inert material 136 such as a non-conductive epoxy. The epoxy is lapped to form a coplanar suface with the top suface of the rows. Thin film metal interconnects 138a are deposited to connect the n-type regions with the contact pads. Second and third bodies 220 and 320 of HgCdTe are formed and connected in a similar manner to the first body. The first body 120 has the lowest mole fraction of CdTe and therefore the longest wavelength response. Backside contacts 18, 118 and 218 are formed to provide a common contact for all detectors in a row.

*

A problem which arises when a read-out chip of for example silicon is attached to a detector chip of mercury cadmium telluride is the mechanical damage which may occur when the array is cooled to cryogenic temperatures for operation. The stress is due to a mismatch in the coefficients of thermal expansion between the two materials.

This problem is solved in FR-A-2484705 (Thomson-CSF, France, 18.12.81) by removing the detector material except for islands which form the detector elements.

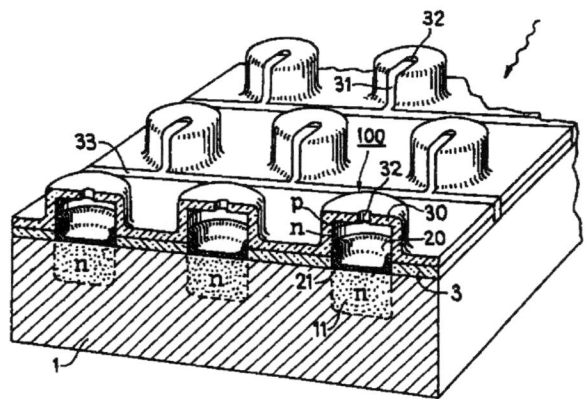

Fig. 2.5.11 (FR-A-2484705 fig. 1)

The detector array comprises a silicon read-out substrate 1 and detector elements 100. Input regions 11 are connected to corresponding detector elements by conductors 20 and 21. A common electrode 33 is formed on insulating layers 3 and 30 and connects to the detectors.

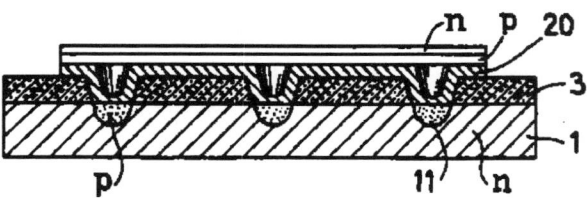

Fig. 2.5.12 (FR-A-2484705 fig. 8c)

A silicon substrate 1 is covered by an insulating layer 3 after that read-out circuitry including input regions 11 have been formed. Openings in the insulating layer are formed above the input regions. A p-type layer is formed at the top surface of an n-type detector substrate. An indium layer is formed on top of the p-type layer before the two substrates are joined with solder. The detector substrate is thinned to leave a thin n-type layer.

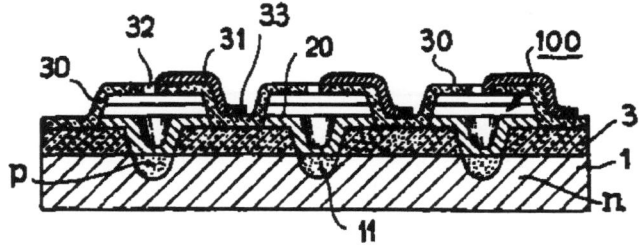

Fig. 2.5.13 (FR-A-2484705 fig. 8e)

Island-shaped individual detector elements are formed followed by insulating layer 30 and electrode 31.

*

The difference of thermal expansion coefficient between an HgCdTe detector material and a silicon read-out material is dealt with in FR-A-2494910 (Thomson-CSF, France, 28.05.82) where detector elements in the shape of islands are connected on one side to a silicon read-out chip and on the other side to a substrate having a similar thermal expansion coefficient as the silicon read-out chip.

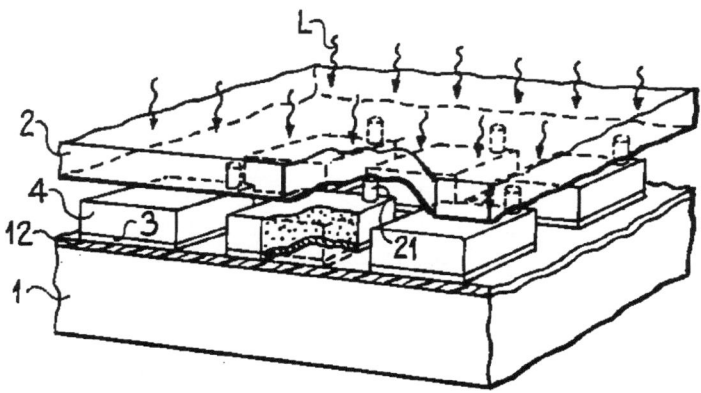

Fig. 2.5.14 (FR-A-2494910 fig. 1)

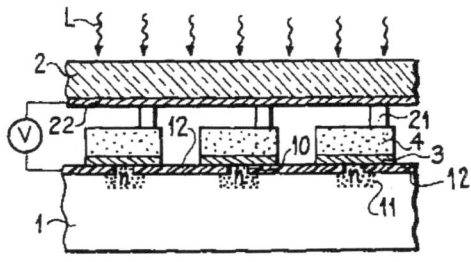

Fig. 2.5.15 (FR-A-2494910 fig. 2a)

Individual detector elements 4 of HgCdTe are formed into islands which are connected on a first side via indium connectors 3 to input regions 11 of a read-out chip 1. A second chip 2, which is transparent to infrared radiation L and which comprises common electrodes 22, is connected to an opposite side of the detector elements via indium connectors 21. It is important that the thermal expansion coefficient of the chip 1 and 2 are closely matched. Both chips may, for example, be made of silicon.

*

The problem of electrical and mechanical coupling is the basis of the invention in US-A-4311906 (Thomson-CSF, France, 19.01.82).

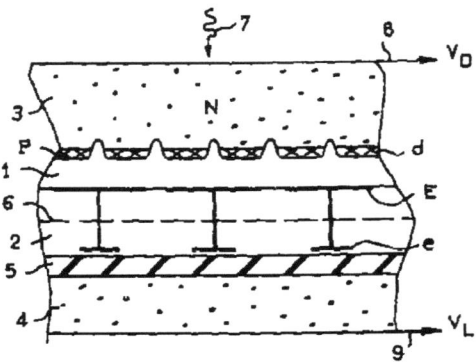

Fig. 2.5.16 (US-A-4311906 fig. 1)

Detectors d are formed as mesa-structure pn-junctions on a substrate 3. An insulation layer 1 is formed by a polymer film or by a vacuum layer. The radiation applied to the array may come from the side of the substrate opposite to the side which carries the detectors, indicated by the wavy arrow 7. In this case, in order to ensure practically complete absorption of the radiation in the vicinity of the detectors, the substrate of the detector is either given a small thickness or chosen so as to be lightly doped and to have a well-determined forbidden band width. It is also possible to employ a substrate consisting of two portions. The first portion is made of a semiconductor having a narrow forbidden-band width on which the junctions are formed, and the second portion, which covers the first portion, is made of a semiconductor having a larger forbidden-band width.

A semiconductor read-out device is formed on a silicon substrate 4. The substrate is covered by an oxide layer 5. A first array of metallic electrodes E collects electrical signals by capacitive coupling with the detectors d. A second array of metallic electrodes e which is separated from the first by an insultaing layer 2 and which is located at the interface between the insulating layer 2 and the oxide layer 5 is formed. Each electrode of the second array is connected electrically to one electrode of the first array which is located opposite and has larger dimensions. These two arrays of electrodes serve to reduce the surface area of the electrodes e in the read-out device, which is intended to consist of a large number of elements, while ensuring that the electrodes employed for collecting the signal E have a large surface area. A metallic screen 6, which is brought to a fixed potential, is inserted in the insulating layer 2. This screen serves to provide electrical insulation of the detection portion from the parasitic signals emitted by the read-out portion. The pitch of the electrodes E, corresponding to the spatial resolution in the picture pick-up system, is greater than the pitch of the detectors d. This configuration does not entail the need for any registration of the detectors d with respect to the first array of electrodes E. The operation of the array will therefore not be disturbed by any displacement of the array detectors.

*

An imager which uses Metal-Insulator-Semiconductor gate Static Induction Transistors (MISSIT), formed in silicon, for read-out and refresh and which has a photo-detecting layer of HgCdTe is presented in EP-A-0038697 (Semiconductor Research Foundation, Japan, 28.10.81).

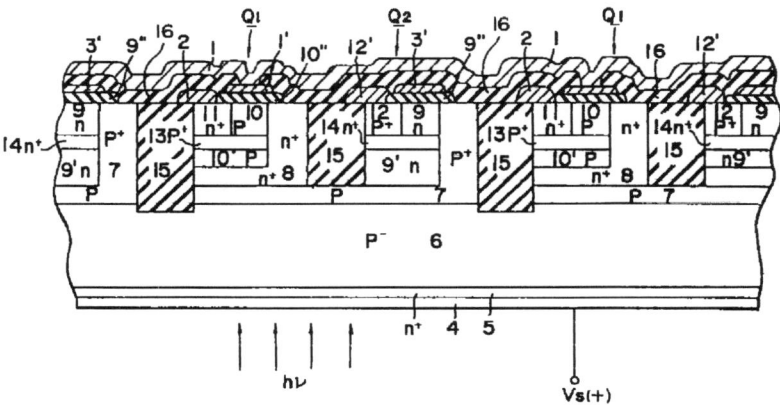

Fig. 2.5.17 (EP-A-0038697 fig. 1a)

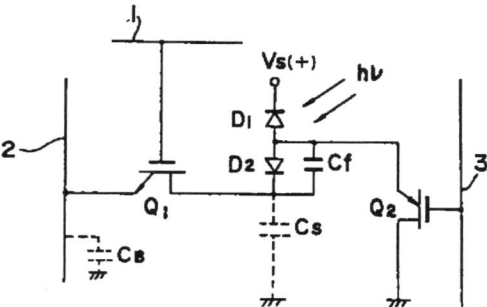

Fig. 2.5.18 (EP-A-0038697 fig. 1b)

The imager comprises a word line 1, which is connected to a gate electrode 1' of an n-channel MISSIT read-out transistor Q1, a bit line 2, which is connected to an n^+-type source region 11 of the transistor Q1, a gate electrode 3' of a p-channel MISSIT refresh transistor Q2 and a transparent electrode 4, which is biased by a power source voltage $V_S(+)$ and which is connected to an n^+-type region 5. An n^+-type region 8 serves as the drain of the transistor Q1 and a region from which electrons flow out in response to a light input and is charged positive, storing therein light information as a voltage. Holes, optically generated in a photo detecting region 6 are stored in a p region 7 which serves as a source region of the refresh transistor. A p region 10 and an n region 9, which are covered by insulating films 10" and 9", are channel regions of the transistors Q1 and Q2, respectively. A p^+-type region 13 and an n^+-type region 14 are buried regions provided for limiting the thickness of depletion layers in the channel regions so as to avoid voltage information stored in the n^+-type region 8 and the p region 7 varying under the influence of alpha rays or the like and to achieve a normally off characteristic with a short channel MOS gate structure. A storage capacitor C_S and an earth line are provided but are not illustrated in the sectional view of the imager. For detecting infrared light, HgCdTe is grown heterogeneously or deposited by a CVD method, sputtering or evaporation on the surface of the photo detecting portion.

The power source voltage $V_S(+)$ completely depletes the region 6. Optically generated electrons and holes are absorbed in the region 5 and stored in the p region 7, respectively. Accumulated holes charge the p region 7 positive. The potential barrier height for electrons in the n^+-type region 8 is lowered, permitting electrons to flow from the n^+-type region 8, across the p region 7 and towards the substrate. As a result, the n^+-type region 8, which is held floating relative to the earth line via the capacitor C_S, is depleted and its potential is increased. The potential is read out, non-destructively, to a signal output line (bit line) 2 by the read-out transistor Q1. It is shown that the potential on the signal output line is dependent on the junction capacitance between p region 7 and n^+-type region 8 but independent of the storage capacitance C_S and the capacitance of the signal output line. Furthermore, when the size of a detector element is decreased and a lens, which has a resolving power large enough to sufficiently collect the incident light, is provided in front of the optical input receiving surface the output voltage increases.

Several embodiments are disclosed in the patent application. In one of the embodiments the refresh transistor has been eliminated.

*

The imager of EP-A-0038697, shown above, has an complex structure and hence is difficult to produce. In EP-A-0042218 (Semiconductor Research Foundation, Japan, 23.12.81) a simplified structure is presented. In all the embodiments the refresh transistor present in most of the embodiments of EP-A-0038697 has been eliminated.

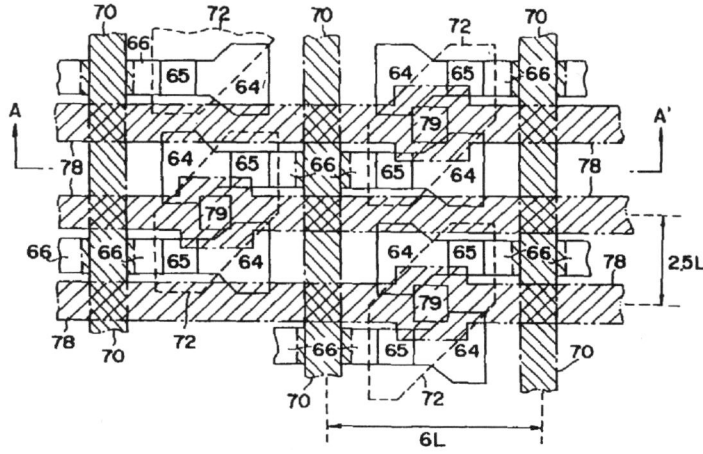

Fig. 2.5.19 (EP-A-0042218 fig. 5a)

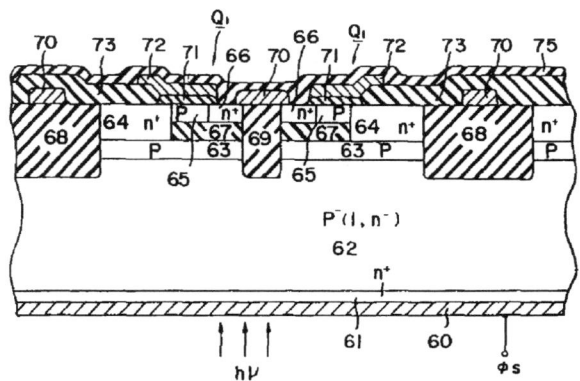

Fig. 2.5.20 (EP-A-0042218 fig. 5b)

A hook structure is formed by an n^+ 64, p 63, p^- 62 and n^+ 61 junction structure. A read-out MISSIT transistor Q1 is formed having a source, a channel and a drain formed by the n^+-type region 64, a p region 65 and an n^+-type region 66, respectively. The n^+-type region 66 is connected to a signal read-out line 75. A gate region 72 of the read-out transistor is connected via a contact hole 79 to a signal address line 78. An insulating layer 67 is formed between the p region 63 and the p channel region 65 and the n$^+$-type drain region 66. 68, 69, 73 and 75 denote insulating regions and layers. A region 71 serves as a gate insulating film of the read-out transistor. The signal address line 78 and the via 79 are common to two adjacent cells. By a pulse voltage Φ_S, connected to a transparent electrode 60, the substrate region 61 is supplied with the voltage $V_S(+)$ in the light integration period. In the refresh period the voltage Φ_S becomes 0 V or slightly negative, drawing out holes which are excess majority carriers stored as optical information in the p region 63.

By forming the n$^+$-type region 61, the p$^-$-type region 62 and the p layer 63 of HgCdTe, forming an insulating layer of ZnS on the p layer 63 at a predetermined position, growing thereon polysilicon by a CVD technique, leaving the polysilicon at a predetermined position and adding a desired impurity by ion-implantation, an infrared imager is formed. The polysilicon can be made reasonably well single crystallized through the use of a laser anneal technique.

*

An array which is compact and easy to manufacture is shown in JP-A-57031170 (Fujitsu Ltd, Japan, 19.02.82).

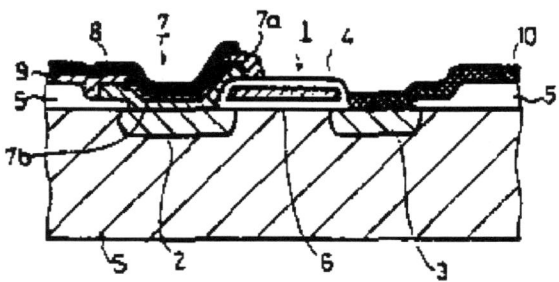

Fig. 2.5.21 (JP-A-57031170 fig. 1)

A MOSFET with a polysilicon gate electrode 4 is formed on a p-type silicon substrate. A thin HgCdTe film 7 is prepared on an n-type source 2. The film 7 is composed of an n-type layer 7a and a p-type layer 7b. An Ni film 8 is connected to layer 7b, the film being insulated from the silicon substrate by an oxide layer 5 and a ZnS layer 9.

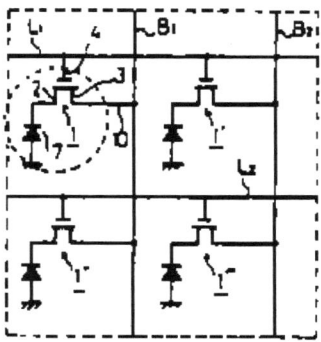

Fig. 2.5.22 (JP-A-57031170 fig. 2)

The structure described forms part of an area array having MOSFET switch read-out. A similar structure where a charge coupled device is used for read-out is also shown.

*

In GB-A-2113467 (Licentia Patent-Verwaltungs-GmbH, FRG, 03.08.83) a manufacturing process of an imager is disclosed using the step of converting a polycrystalline layer of HgCdTe to a monocrystalline layer.

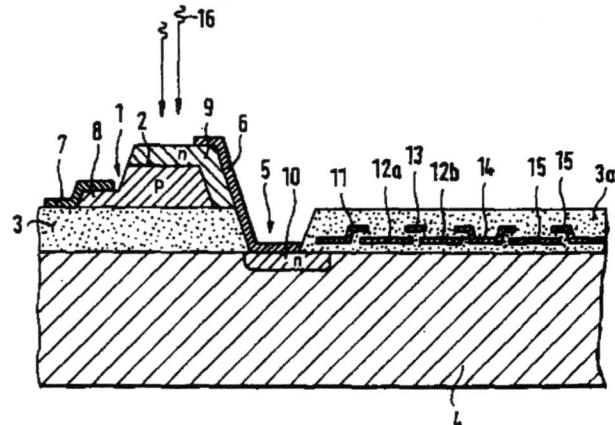

Fig. 2.5.23 (GB-A-2113467 fig. 1)

An individual mosaic region 1 comprising mercury cadmium telluride is located on an insulating layer 3 which covers a silicon semiconductor body 4. The individual mosaic regions 1 include a pn-junction 2 which separates an n-type region 9 from a p-type region 8. The p-type region is provided with a connecting contact 7. A connecting contact 6 is also applied to the n-type region, the contact being in electrically conductive connection with an input region 10 through openings 5 in the insulating layer. An insulating layer 3a and numerous gate structures 11-15 are arranged in the insulating layer performing background subtraction and CCD read-out.

In the manufacturing process, first the integrated semiconductor circuit is formed. The whole of the surface of the insulating layer 3 or 3a, respectively, is then covered by a layer comprising poly-crystalline HgCdTe material. This material is deposited chemically from the gas phase or vapour deposited in an ultra high vacuum. It is also possible to coat by means of cathode sputtering or to apply the layer in components. When applying the layer in components, a layer of CdTe is applied first and then reacted with HgTe vapour. It is also possible to apply an initial coating of HgTe and to let it react with CdTe vapour. Afterwards, the mosaic regions are etched out by the use of chemical or ion-etching. The polycrystalline mosaic regions are converted into monocrystalline regions by means of punctiform or continuous scanning radiation with a laser beam, without affecting the silicon semiconductor body.

*

A multi-colour imager formed by multi-level HgCdTe layers is disclosed in EP-A-0087842 (Philips Electronic and Associated Industries Limited, GB, 07.09.83).

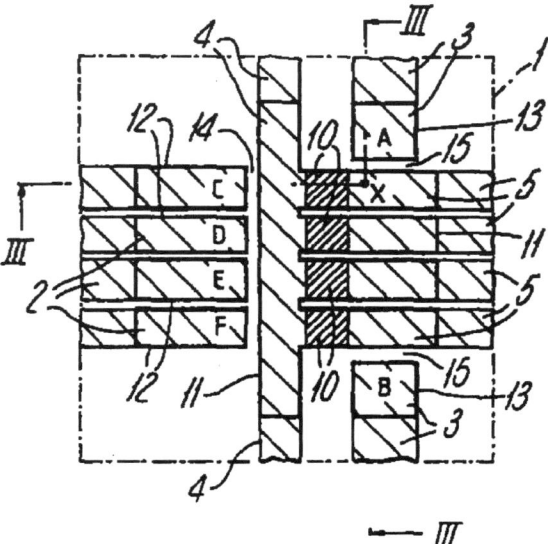

Fig. 2.5.24 (EP-A-0087842 fig. 1)

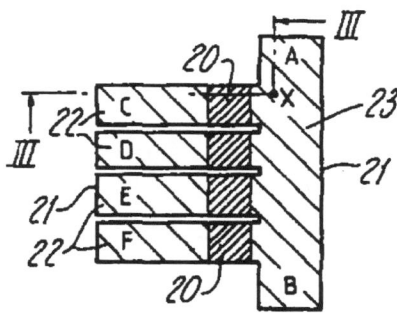

Fig. 2.5.25 (EP-A-0087842 fig. 2)

The next figure shows a cross-sectional view along the line III-III of the two previous figures.

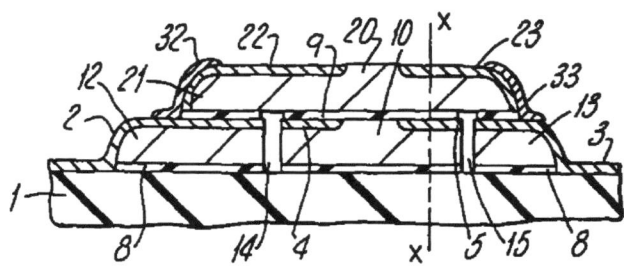

Fig. 2.5.26 (EP-A-0087842 fig. 3)

An upper detector element body 21 of $Hg_{0.73}Cd_{0.27}Te$ is mounted by an insulating epoxy adhesive 9 on lower detector element bodies 11, 12 and 13 of $Hg_{0.79}Cd_{0.21}Te$ which are mounted on a substrate 1 by an insulating epoxy adhesive 8. The substrate may be formed of sapphire, alumina or beryllia or it may be of silicon comprising signal-processing circuitry. Detector elements 10 and 20, present in the lower body and the upper body respectively, are positioned to overlap each other. Each of the detector elements 10 has electrical connections 4 and 5, and each of the detector elements 20 has electrical connections 2, 22, 32 and 3, 23, 33. Infrared radiation is incident at the top face of the upper body and the detector elements 10 sense radiation transmitted by the upper body. A third detector element body may be mounted on the upper detector element body so as to form a 3-colour detector. The detector element bodies are formed by ion-etching and have rounded edges. The detector elements are mesa-shaped as is described in EP-A-0007668, which is mentioned in chapter 2.1.

*

The imager of FR-A-2593642 (Societe Anonyme de Telecommunications, France, 31.07.87) comprises a silicon read-out substrate on which a detector substrate is glued. A general problem is addressed in forming reliable electrical connectors which are formed from a region on the read-out substrate, over the edge of the detector substrate, to a region on the detector substrate.

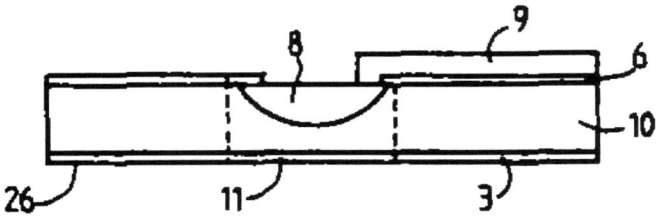

Fig. 2.5.27 (FR-A-2593642 fig. 3)

An n-type region 8 is formed in a p-type HgCdTe substrate. A thin strip of metal is formed on an insulating layer 6 by thermal evaporation or cathode sputtering. The strip is enlarged by immersing the substrate in a bath of electrolyte thereby forming a terminal 9. The substrate is thinned to form a layer 10 and a mask 11 is formed from a mask layer 26.

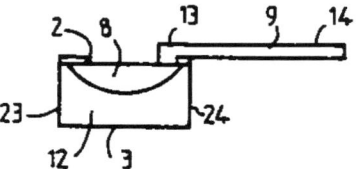

Fig. 2.5.28 (FR-A-2593642 fig. 4)

The layer 10 is removed by ion-abrasion or chemically etching except for a piece 12. A solution of pure bromine and alcohol may be used if the terminal 9 is made of gold.

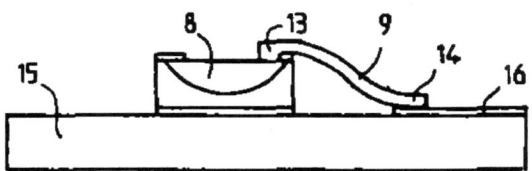

Fig. 2.5.29 (FR-A-2593642 fig. 6)

The piece 12, with the terminal 9, is attached to a silicon read-out substrate 15 by a glue. The glue may be conducting and, in that case, the glue forms a second electrical connection between the detector and the read-out substrate. If the glue is insulating, a second terminal, similar to terminal 9, is formed. Finally, the terminal 9 is bent down and connected to connection pad 16.

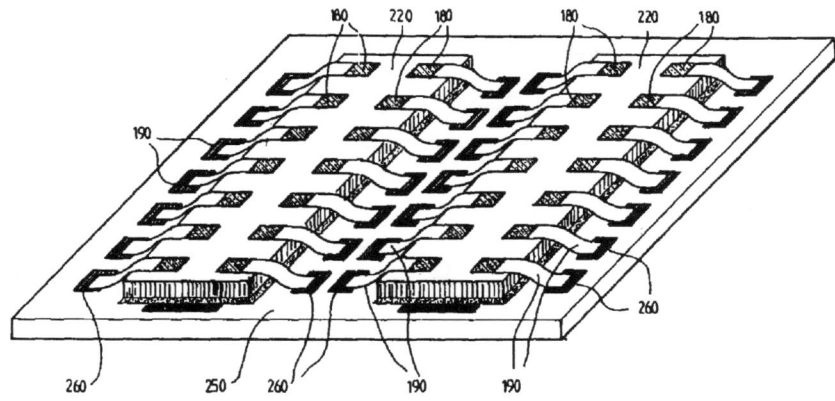

Fig. 2.5.30 (FR-A-2593642 fig. 8)

An area array is also shown comprising detector bars 220 in which detector elements 180 are formed together with terminals 190. The detector bars are glued to a read-out substrate 250 and the terminals 190 are bent and connected to connection pads 260.

* *

An imager with a similar process of manufacturing to that disclosed in FR-A-2593642, shown above, is presented in US-A-4532699 (Societe Anonyme de Telecommunications, France, 06.08.85).

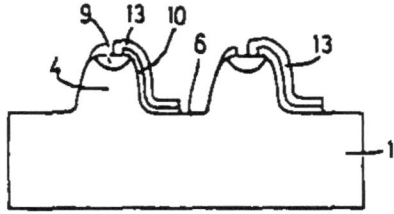

Fig. 2.5.31 (US-A-4532699 fig. 6)

Raised zones 4 and grooves 6 are formed on a substrate 1 of p-type HgCdTe. A light polishing is effected in order to soften the structure of the raised zones and to round their edges. Masking layer 10 of a dielectric material is deposited and n-type zones 9 are formed. Fine metal strips are deposited either by thermal evaporation or by cathode sputtering. The substrate is immersed in a bath of electrolyte to enlarge the strips whereby contact terminals 13 are formed. Thanks to the rounded edges of the raised zones, the contact terminals are deposited without risk of electrical discontinuity.

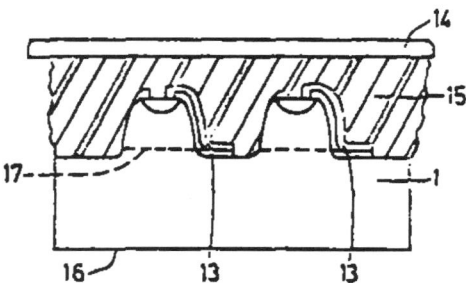

Fig. 2.5.32 (US-A-4532699 fig. 7)

The substrate 1 is attached by its active upper face onto a glass support plate 14 by means of wax 15. The plate ensures mechanical protection of the active face and holds and positions the substrate during the operations of thinning and isolation of the contact terminals. The substrate 1 is thinned from its rear face 16 by mechanical or mechano-chemical lapping or polishing. Further thinning is carried out up to the terminals (level 17) including the layer 10, by ionic abrasion or by a chemical process using for example a solution of pure bromine and alcohol.

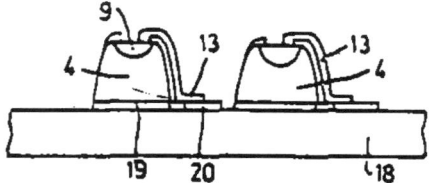

Fig. 2.5.33 (US-A-4532699 fig. 9)

Adhesion of the detector elements on to a silicon wafer 18, which is adapted to process the signals delivered by the detector elements, is carried out by a glue 19. The glue may be conducting to electrically connect the detector elements with the silicon wafer. If a non-conducting adhesive is used, the detector elements are connected with a terminal similar to terminal 13. The connection of projecting tabs of terminals 13 to the associated metallization zones 20 is effected by ultrasonic welding or by thermocompression using a welding tool adapted to the pattern of the detector elements.

*

Operation of photo-conductive detectors requires current bias. A steady bias current produces a standing DC output. The bias pedestal is difficult to compensate for accurately for each detector of an imager. An imager designed to be operated by a current bias which is cyclic in time and has a waveform and frequency to produce zero net average signal both in dark and in presence of radiation of uniform intensity is presented in GB-A-2128019 (The Secretary of State for Defence, GB, 18.04.84).

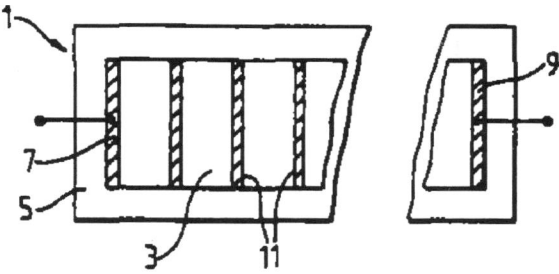

Fig. 2.5.34 (GB-A-2128019 fig. 1)

A detector 1 is comprised of a strip 3 of mercury cadmium telluride mounted upon a substrate 5. Bias contacts 7 and 9 are provided at each end of the strip and between these are a multiplicity of low-resistance read-out contacts 11. These contacts 11 are close-spaced and evenly distributed along the length of the strip. The spacing is at most comparable with the diffusion length of the minority carriers. The detector may be used as a position sensitive point source detector but if the signals from the read-out contacts 11 are spatially integrated some resemblance of an original scene is restored. In this case each read-out contact is connected to one of the differential inputs of a corresponding differential input integrating amplifier. These amplifiers are integrated in the substrate 5.

*

An imager comprising detectors formed in an HgCdTe layer which has been epitaxially grown on a semiconductor layer of a different material, the semiconductor layer comprising a charge transfer element, is disclosed in US-A-4553152 (Mitsubishi Denki Kabushiki Kaisha, Japan, 12.11.85).

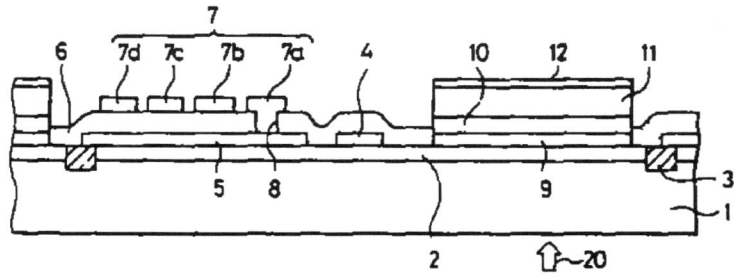

Fig. 2.5.35 (US-A-4553152 fig. 1)

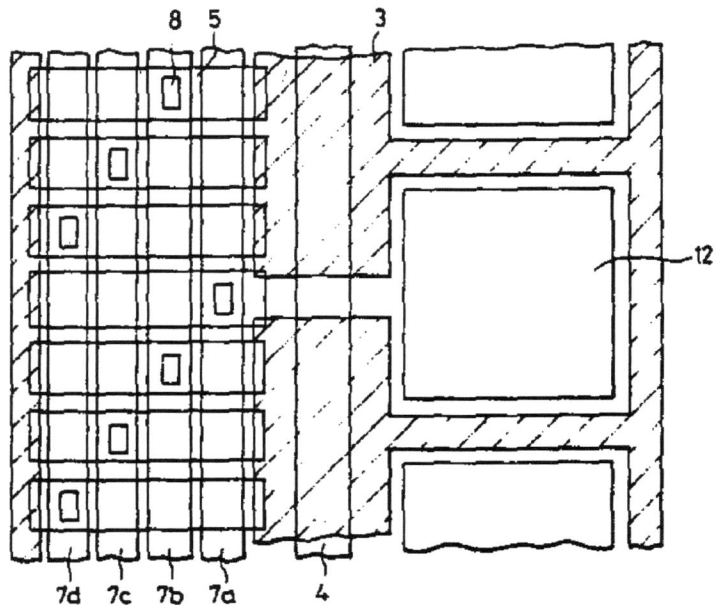

Fig. 2.5.36 (US-A-4553152 fig. 2)

An n-type gallium arsenide layer 2 is formed on a semi-insulating gallium arsenide substrate 1. A p-type gallium arsenide or semi-insulating gallium arsenide region 3 is formed in the n-type layer 2. A single elongated transfer gate 4 and a plurality of charge transfer gates 5 are formed on the surface of the layer 2. The gates are covered by insulating layer 6. On the insulating layer 6, a signal line portion 7 is formed. The signal line portion includes first to fourth signal lines 7a-7d. A plurality of contact holes 8 are formed in the insulating layer through which the signal lines are connected to the charge transfer gates. Also on the n-type layer 2 are a plurality of n-type cadmium telluride layers 9 on each of which an n-type mercury cadmium telluride layer 10 and a p-type mercury cadmium layer 11 are formed by the means of a molecular beam epitaxial process and a photolithographical process. Electrodes 12 of gold are formed on the p-type layer 11.

The transfer gates 4 and the charge transfer gates 5 may be formed as MIS structures or as Schottky barrier structures using aluminium or platinum. Furthermore, the gallium arsenide substrate may be substituted for a silicon substrate.

Radiation from a scene is received on the imager and electric charge is accumulated between the n-type layer 10 and the p-type layer 11. When the transfer gate 4 is opened, the charge is transferred to the n-type layer 2 beneath the charge transfer gate 5. Upon application of a signal on the signal portion 7, the charge is transferred to an output terminal.

*

In MIS-type detectors, radiation generated charge is stored in potential-wells in a semiconductor under MIS-gates. When MIS-gates are formed by depositing metal and patterning the metal by a photolithographic process, the device is exposed to potentially damaging mechanical, thermal and ultraviolet radiation stresses. A different approach to fabricate a MIS-type detector is presented in EP-A-0173074 (Texas Instruments Incorporated, USA, 05.03.86).

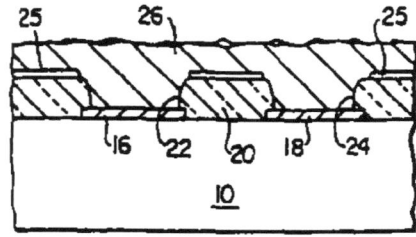

Fig. 2.5.37 (EP-A-0173074 fig. 1)

A substrate 10 of silicon comprising an array of switching elements is prepared. Contact pads such as 16 and 18 are connected to inputs of the switching elements. A protective nitride layer 20 is formed, with openings for the contact pads, on top of the substrate. A layer of photo-resist 25 is patterned on top of the nitride layer. A layer of indium alloy 26, fabricated from a combination of indium, bismuth, lead, cadmium and tin, is deposited on the substate so as to contact the contact pads. This layer is built up such that its upper surface is higher than the structures that constitute the topography of the substrate. Portions of the conductive layer overlying the photo-resist are removed by a lift off process.

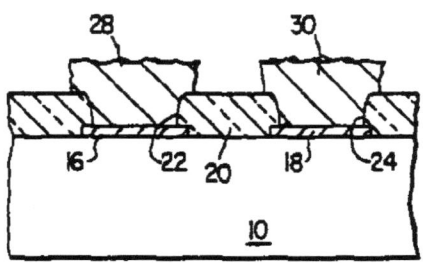

Fig. 2.5.38 (EP-A-0173074 fig. 2)

The remaining portions of the conductive layer form gates such as 28 and 30. The upper surface of the gates are irregular as a result of the use of an indium alloy. The height of the gates can be increased by placing the substrate in an oven at a temperature above the reflow temperature of the indium alloy.

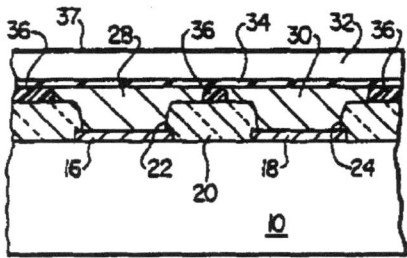

Fig. 2.5.39 (EP-A-0173074 fig. 3)

An $Hg_{0.8}Cd_{0.2}Te$ substrate 32 is formed and covered with a ZnS insulating layer 34. The two substrates 10 and 32 are pressed together to cause the gates such as 28 and 30 to deform and to be displaced laterally outward from the centers thereof. The gates will conform to the shape of the insulating layer. To maintain the substrates in this position a layer of low viscosity epoxy 36 is "wicked" into the interstices of the combined device and the epoxy is allowed to cure. The HgCdTe substrate is thinned and a layer of ZnS 37 is deposited to provide passivation of the substrate and also to serve as an anti-reflection coating.

*

An imager which has a detector substrate and a silicon read-out substrate attached on different sides of a sapphire substrate is presented in JP-A-61128564 (Fujitsu Ltd, Japan, 16.06.86).

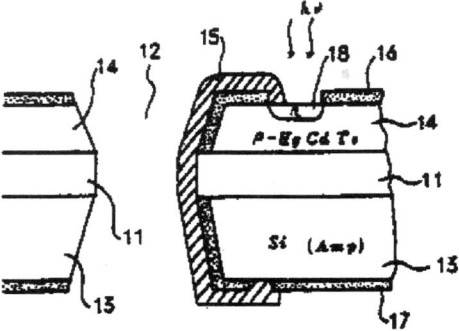

Fig. 2.5.40 (JP-A-61128564 fig. 2)

A silicon substrate 15 which comprises a read-out circuit is formed on one side of a sapphire substrate 11. A mercury cadmium telluride substrate 14 which comprises photodiodes 18 is formed on the other side of the sapphire substrate. The photodiodes are connected to corresponding read-out circuit by aluminium electrodes 15 via through-holes 12. The electrodes are formed by evaporating aluminium from both the detector side and the side of the silicon substrate side.

* *

An imager similar to the imager presented in JP-A-61128564 above is described in JP-A-63147366 (NEC Corp., Japan, 20.06.88) in which HgCdTe photodiodes are formed on one side of a CdTe substrate and a GaAs substrate comprising a read-out CCD is formed on the other side.

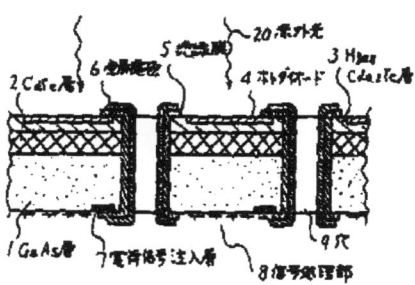

Fig. 2.5.41 (JP-A-63147366 fig. 1)

Infrared photodiodes 4 are formed by layers 3 and 4 of $Hg_{0.8}Cd_{0.2}Te$ on one side of a CdTe substrate 2. A GaAs layer 1 is formed on the other side of the substrate 2. Metallic electrodes 6 are formed to connect the photodiodes via through holes 9 with the input of a CCD 8 which is comprised in the GaAs substrate.

*

In JP-A-62022474 (Toshiba Corp., Japan, 30.01.87) vapor deposition of metal electrodes is used to connect an HgCdTe layer with a silicon read-out substrate. The metal electrodes form Schottky diodes or generate pn-junctions depending on the type of metal used.

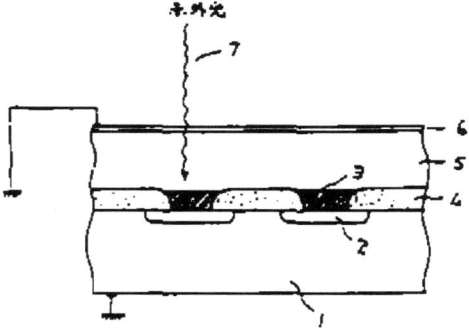

Fig. 2.5.42 (JP-A-62022474 fig. 1)

An CCD read-out circuit having p-type input regions 2 is formed in an n-type silicon substrate 1. A silicon oxide layer 4 is formed and openings corresponding to the input regions are provided. Metal electrodes 3 are formed by vapor deposition to connect to the input regions. A film 5 of HgCdTe is grown ontop of the structure before a transparent electrode film 6 is formed by vapor deposition. If the metal electrodes are formed of for example Al, Mo or Cu, Schottky junctions are formed between the electrodes and the HgCdTe film. Alternatively, if the metal electrodes are formed of an alloy containing indium, the indium is diffused into the HgCdTe film thereby forming pn-junctions.

*

An imager is presented in JP-A-62071270 (NEC Corp., Japan, 01.04.87) in which HgCdTe layers are grown directly on a read-out substrate by an epitaxial process.

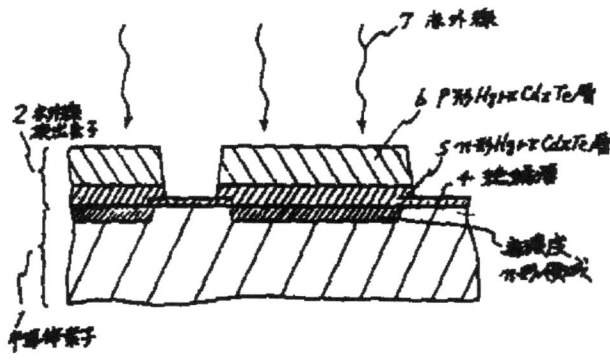

Fig. 2.5.43 (JP-A-62071270 fig. 1)

A read-out circuit having n-type input regions 3 is formed in a semiconductor substrate 1. An n-type HgCdTe layer 5 is epitaxially grown thereon, followed by a p-type HgCdTe layer 6. The n-type region 3 and the n-type HgCdTe layer 5 are made to have high doping concentrations.

*

Molecular Beam Epitaxy, MBE, is used to grow infrared detectors on a silicon read-out substrate in US-A-5003364 (Licentia Patent-Verwaltungs-GmbH, FRG, 26.03.91).

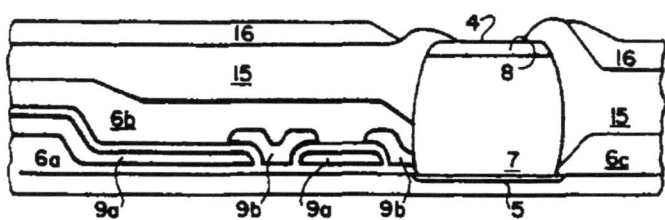

Fig. 2.5.44 (US-A-5003364 fig. 2)

Poly-silicon transfer gates 9a and 9b of a CCD are formed together with oxide layers 6a, 6b and 6c on a silicon substrate. A diffused region forms a detector contact 5. Detector layers 7 and 8 are grown to form a detector 4 using differential molecular beam epitaxy growth techniques. A passivation layer 15 and a metal layer 16 are applied to protect against incident radiation. The embodiment presented in the US patent comprises an i-type Ge layer 7 and a p-type Ge layer 8. HgCdTe or a Si/SiGe superlattice may be used as a detector 4. It is stated that by using molecular beam epitaxial growth techniques it is possible to grow semiconductor materials with different lattice constants on a silicon substrate.

*

In JP-A-63116459 (Mitsubishi Electric Corp., Japan, 20.05.88) an imager is made up of photodiodes formed in an HgCdTe substrate. The photodiodes are placed in columns and rows. The number of connections between the HgCdTe substrate and a silicon substrate, comprising a read-out device, is reduced by connecting the anodes of the photodiodes in each row to each other and connecting the cathodes of the photodiodes in each column to each other.

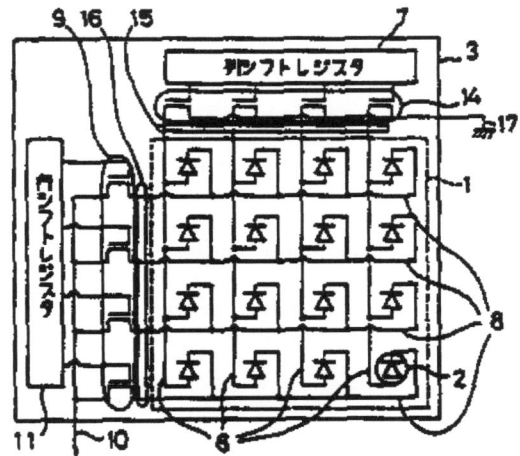

Fig. 2.5.45 (JP-A-63116459 fig. 1)

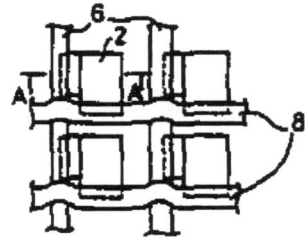

Fig. 2.5.46 (JP-A-63116459 fig. 2a)

An HgCdTe substrate 1 comprises photodiodes 2 placed in rows and columns. The anodes of all photodiodes belonging to the same row is connected to a row line 6 and the cathodes of all photodiodes belonging to the same column are connected to a column line 8. The row lines and column lines are connected to a read-out device formed in a silicon substrate 3.

*

In JP-A-63310165 (Nikon Corp., Japan, 19.12.88) infrared ray detector elements of HgCdTe are formed on a CdTe substrate in which a read-out CCD is integrated.

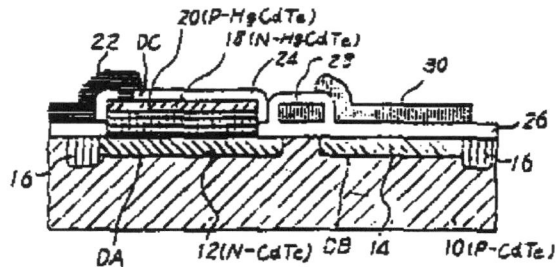

Fig. 2.5.47 (JP-A-63310165 fig. 1)

An n-type CdTe region 12 is formed on a surface of a p-type substrate 10 thereby forming an input diode DA. Transfer gates 30 are formed above an n-type CCD transfer region DB. A read-out gate 28 is formed at the same level as the transfer gates between the input diode and the transfer region. An n-type HgCdTe layer 19 is formed on the surface of the input diode 12 followed by a p-type HgCdTe layer 20 to constitute a detector element DC.

*

The lattice mismatch between silicon and mercury cadmium telluride makes it difficult to grow an epitaxial crystalline layer of mercury cadmium telluride on silicon. In EP-A-0343738 (Philips Electronic and Associated Industries Limited, GB, 29.11.89) an imager is formed by growing mercury cadmium telluride on a sapphire substrate at openings formed in a silicon layer which has been grown on the sapphire substrate (silicon-on-sapphire, SOS) at an earlier stage.

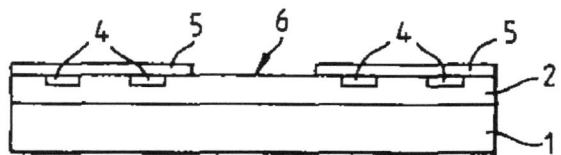

Fig. 2.5.48 (EP-A-0343738 fig. 2)

A silicon layer 2 comprising electronic circuits 4 is grown on a sapphire substrate 1. A first mask 5 is then applied except at a window 6.

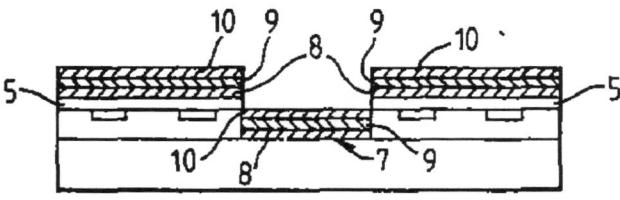

Fig. 2.5.49 (EP-A-0343738 fig. 4)

The silicon is etched away in the window 6 down to and exposing the sapphire surface 7. The surface 7 is cleaned and a buffer layer 8 of CdTe is grown. In the next step two layers 9 and 10 of HgCdTe are grown by metal-organic vapour phase epitaxy, MOVPE. The layers 9 and 10 form a pn-junction.

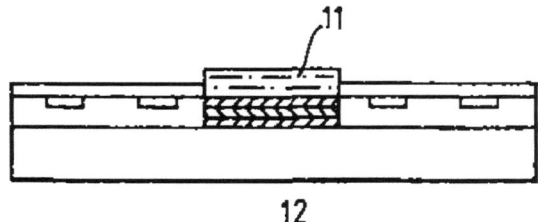

Fig. 2.5.50 (EP-A-0343738 fig. 6)

A second mask 11 is applied to the HgCdTe layer in the window area. Extraneous HgCdTe is then removed by etching and then the second mask is removed.

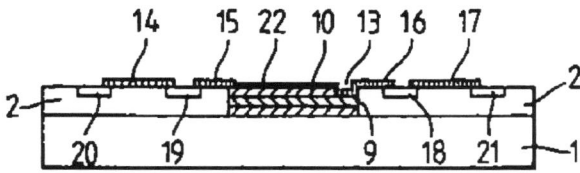

Fig. 2.5.51 (EP-A-0343738 fig. 10)

In a further mask and etch process, a contact opening 13 is made down to the lower layer 9. The first mask is now removed by etching. Conductive metal connections 14 to 17 are then evaporated onto the device through a mask.

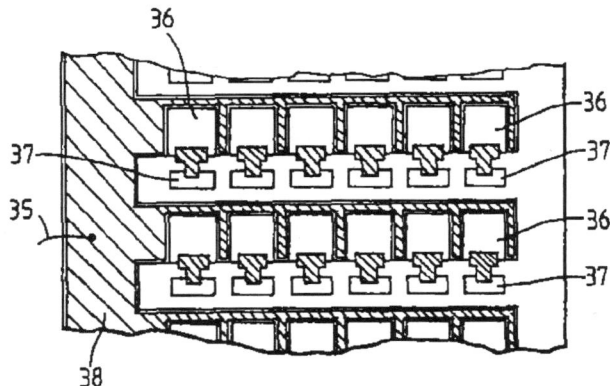

Fig. 2.5.52 (EP-A-0343738 fig. 13)

A two-dimensional detector array is shown above, in which the technique described above is used to form detector elements 36 each associated with one electronic circuit 37. A metallization 38 provides a common connection 35. In this case the HgCdTe layer 9 and the CdTe buffer layer 8 are not removed at the region corresponding to the metallization 38.

*

GaAs, CdTe and HgCdTe are deposited epitaxially on a silicon substrate in US-A-4910154 (Ford Aerospace Corporation, USA, 20.03.90). If islands of individual detector elements are formed, the deposited layers will have a combined thickness which makes it difficult to interconnect the detector elements with corresponding silicon circuitry due to the steep sides of the detector elements. It is therefore proposed to grow the GaAs, CdTe and HgCdTe layers in recesses formed in the silicon substrate, and interconnect with a planar technology.

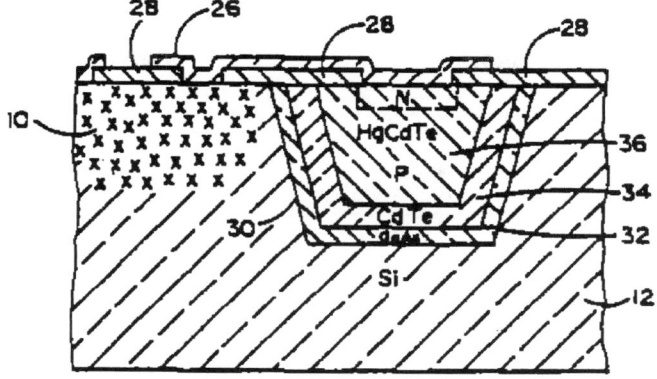

Fig. 2.5.53 (US-A-4910154 fig. 3)

Recesses 30 with steeply inclined sides are formed in a silicon substrate 12 by plasma or chemical etching techniques. The recess formation exposes crystal facets of the silicon substrate on which a particularly beneficial epitaxial growth of GaAs can take place. Next, an epitaxial layer of GaAs is grown through evaporation in an ultra-high vacuum by molecular beam epitaxy, MBE, or metalorganic chemical vapor deposition, MOCVD. A thin monocrystalline layer 32 is formed on the floor of the recess but a polycrystalline layer, if any, is formed on the silicon dioxide and/or silicon nitride layer which covers the substrate except for the recesses. The polycrystalline phase is removed by etching. A thin CdTe layer 34 is formed in the recesses in the same way as the GaAs layer. A thick HgCdTe layer is grown to fill the recesses by liquid phase epitaxy or a vapor phase epitaxy process such as MBE, MOCVD or close-spaced vapor epitaxy. Two HgCdTe layers may be grown to form a heterojunction. Finally, n-type regions are implanted to form diodes and metallic conductor interconnection contacts are vapor-deposited.

*

Manufacturing of an imager, such as the imager presented in US-A-4197633 above, comprises the steps of bonding a mercury cadmium telluride body via a first surface by an adhesive layer to a substrate on which a conductor pattern is present, removing portions of the body thoughout its thickness and providing electrical connections from the substrate to regions of the body. A problem is that the conductor pattern is exposed in the removal step. Furthermore, pn-junctions formed in the body adjacent to a second surface, opposite to the first surface, can be degraded by heat treatments. In GB-A-2231199 (Philips Electronic and Associated Industries Limited, GB, 07.11.90) the body is adhered to the substrate via the second surface thereby protecting the pn-junctions against material degradation during heat treatments.

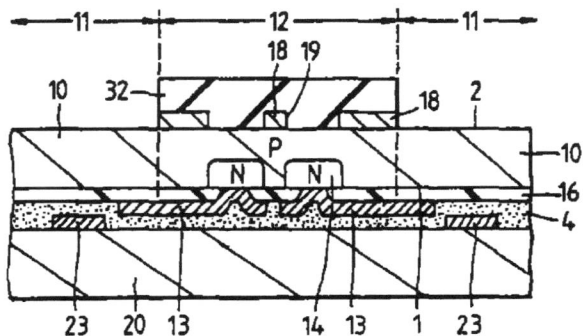

Fig. 2.5.54 (GB-A-2231199 fig. 3)

The manufacturing method comprises the steps of, forming n-type regions 14 in a p-type mercury cadmium telluride body 10, forming electrode leads 13 on an passivation layer of ZnS or cadmium telluride, bonding the body 10 by an insulating adhesive layer 4 to a silicon substrate 20 which comprises a processing circuit and contact pads 23, forming an infrared transmission array 19 which comprises windows in an infrared mask layer 18, removing portions 11 by spray-etching of the body throughout its thickness and removing exposed adhesive 4 from contact pads 23 by plasma etching. The body 10 may be formed by growing a mercury cadmium telluride layer on a base of cadmium telluride. The n-type regions are either formed by ion-milling through the passivation layer and continuing slightly into the mercury cadmium telluride surface or by localised mercury in-diffusion or by donor dopant implantation.

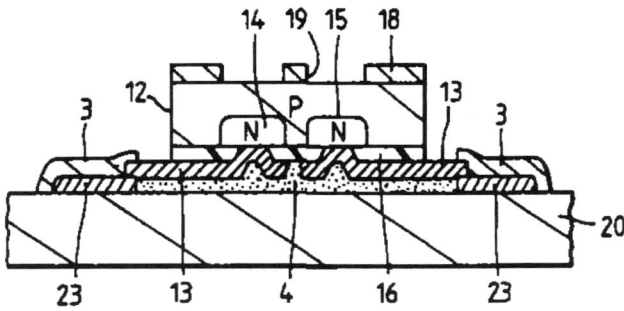

Fig. 2.5.55 (GB-A-2231199 fig. 5)

The next step is to provide electrical connections 3 between the electrode leads 13 and the contact pads 23. Because the active regions are present, adjacent to the passivated lower surface during the heat treatments, for example when the adhesive layer 4 is cured, they do not suffer significantly from degradation of the material properties of the mercury cadmium telluride which occur most easily and rapidly at the top surface and exposed sides of the body

10 due to out-diffusion of mercury. Note that the protruding parts of the electrode leads 13 are supported throughout their length first by the body 10 and after the bonding by the substrate 20. An imager having two rows of detector elements are shown above. However, the number of rows may be different. Below, an imager having four rows is shown.

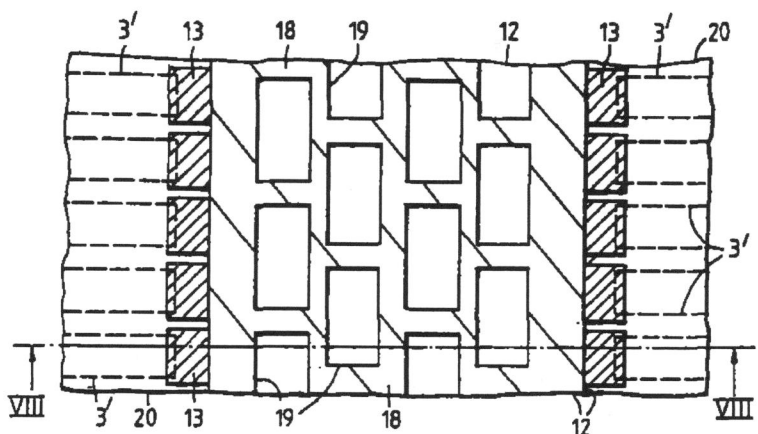

Fig. 2.5.56 (GB-A-2231199 fig. 7)

Several embodiments of the invention are disclosed. For example, an imager having an array of lenses corresponding to the array of detector elements and an imager provided with a cold-shield are shown. Another embodiment is shown in chapter 2.1.

*

When electrical connections from a detector array to processor circuits are formed by an array of small diameter vias or electrical conduits through a wafer substrate, a problem is that sub-micron vias can be formed by either etching or laser drilling only as long as they do not extend more than a few microns in length. The imager of US-A-5315147 (Grumman Aerospace Corporation, USA, 24.05.94) comprises an integrated circuit wafer having a signal processing circuitry formed on a first surface thereof and detector elements formed upon a surface opposite to the first surface. The wafer has an array of waffle-like hollows, each hollow has a dense array of small diameter vias which extend through the wafer. Conductive conduits are formed through the hollows and vias to connect the detectors to associated signal processing circuitry.

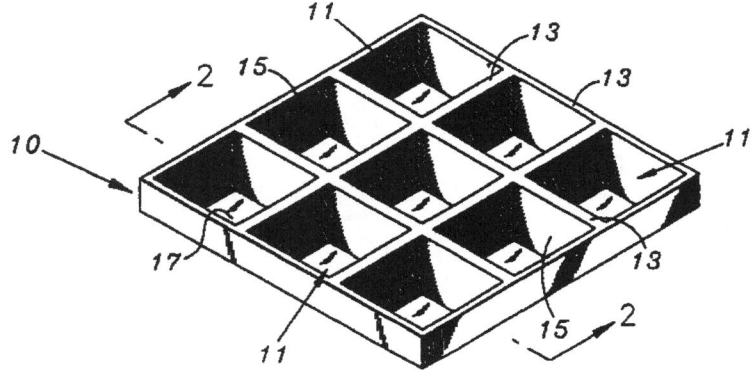

Fig. 2.5.57 (US-A-5315147 fig. 1)

A (100) silicon wafer 10 is anisotropically etched to form hollows 11 having sloped sides 15. A plurality of via holes are etched or laser drilled through the floor 17 of each hollow. Next, the wafer is oxidized to coat its surfaces with a first insulator of SiO_2. A refractory conductor is deposited within the vias to form conductive conduits. A perspective view of a single hollow is shown below.

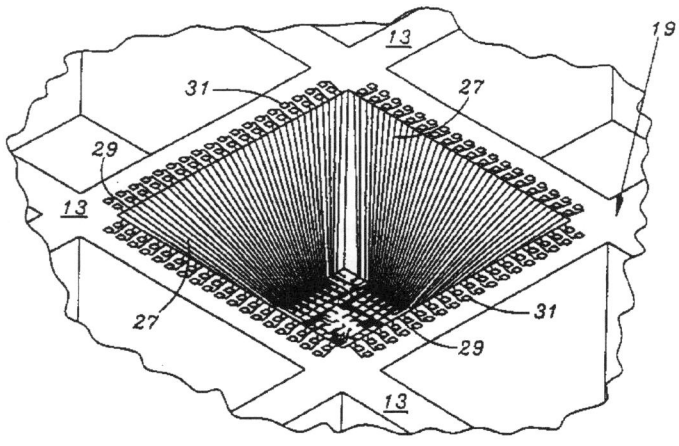

Fig. 2.5.58 (US-A-5315147 fig. 7)

Traces 27, 29 are formed upon the floor, the beveled sides and the ridges 13 of each hollow. Next, a portion of the surface 19 is coated with a second insulator except for the termination pads 31 of the traces 29. The hollows are filled with polysilicon and the surface is coated with a third insulating layer. Again the pads are not coated. A single crystalline layer is grown across the third insulator. The opposite side of the wafer is shown below.

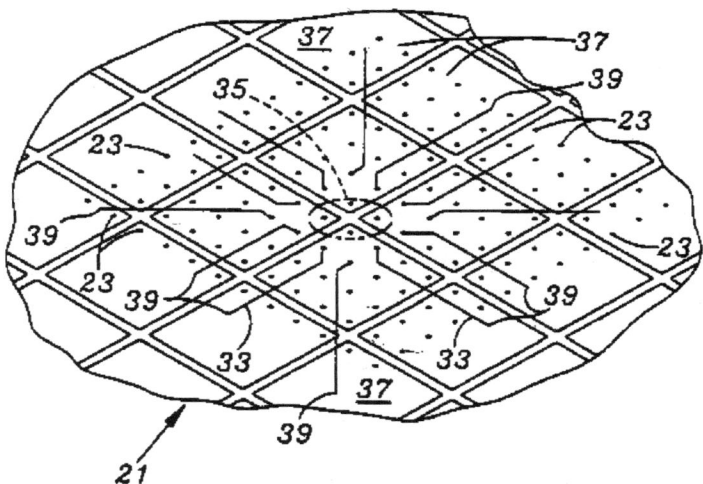

Fig. 2.5.59 (US-A-5315147 fig. 9)

Conductive conduits or traces 33 are formed which contact the conductive conduits within the vias 23 at one end and which will contact the individual detector element electrodes 37 at their opposite end. The traces are insulated by coating the surface with a fourth insulator, except for the ends of the traces 33 which are to be connected to the detector element electrodes. Next, a signal processing circuitry is formed on the opposite side in the graphotaxially grown layer of single crystalline silicon. A separate graphotaxially grown layer of silicon can be grown over the signal processing circuitry to form the substrate for a second layer of signal processing circuitry. Finally, detector element electrodes 37 are formed and the detector elements are formed by growing HgCdTe graphotaxially from a window to the silicon wafer or epitaxially over an intermediate layer of gallium arsenide, which has been graphotaxially grown on the silicon wafer.

*

The imager disclosed in WO-A-9202959 (Minnesota Mining and Manufacturing Company, USA, 20.02.92) comprises thin film transistors deposited on a substrate, a planarization layer deposited on the transistors followed by a mercury cadmium telluride layer and a top electrode layer.

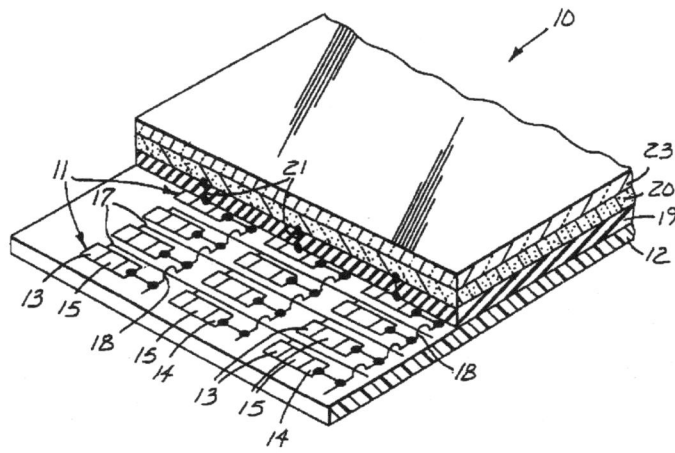

Fig. 2.5.60 (WO-A-9202959 fig. 1)

A plurality of thin film field effect transistors 11 are deposited onto a substrate 12. Each of the transistors has a source electrode 13, a drain electrode 14 and a gate electrode 15. Source lines 17 link the source electrodes in each row of the transistors and drain lines 18 link the drain electrodes in each column of the transistors. The source lines and drain lines are electrically isolated by a planarization layer 19. A mercury cadmium telluride layer 20 is deposited onto the planarization layer followed by a top electrode layer 23. The gate electrodes are connected with the mercury cadmium telluride layer by connectors 21. A cross-section of the imager is shown below.

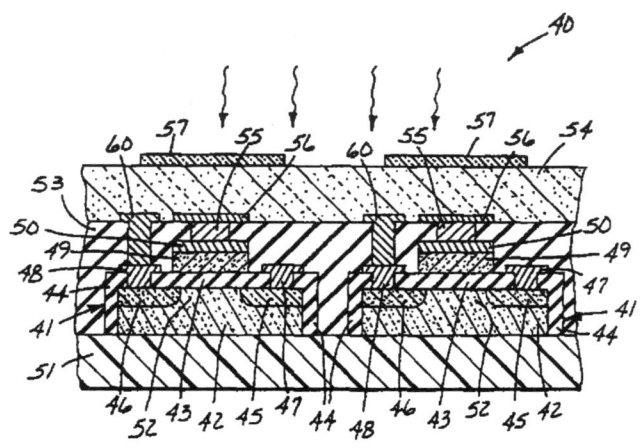

Fig. 2.5.61 (WO-A-9202959 fig. 4)

Here, 41 indicates the thin film transistors, 51 the substrate, 43 a dielectric layer, 49 polysilicon gates, 50 gate electrodes, 55 contact plugs, 56 bottom electrodes, 53 the planarization layer, 54 the mercury cadmium telluride layer and 57 the top electrode layer. The planarization layer is formed from silicon oxide, silicon nitride, silicon oxide nitride or from a polyimide. The planarization layer may be formed as a double or triple layer.

*

Each detector element of the imager disclosed in JP-A-5315580 (Fujitsu Ltd, Japan, 26.11.93) comprises a photodiode formed from HgCdTe layers which have been grown in a recess in a silicon substrate having a surface orientation selection layer and a buffer layer formed in between.

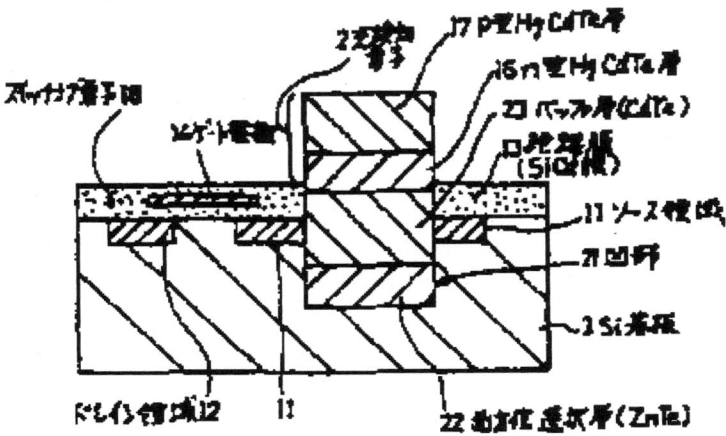

Fig. 2.5.62 (JP-A-5315580 fig. 1a)

A recess 21 is formed in a silicon substrate 3 for each detector element. A MOS read-out transistor comprising a gate 14 formed on an insulating layer 13, a drain region 12 and a ring-shaped source region 11 is formed in the substrate adjacent the recess. A surface orientation selection layer 22 composed of ZnTe layers or GaAs and ZnTe layers which are laminated and a buffer layer 23 of n-type CdTe are provided in the recess. An n-type HgCdTe layer 16 and a p-type HgCdTe layer 17 are laminated on the buffer layer to form a photodetecting element 2.

*

In JP-A-6021419 (Fujitsu Ltd, Japan, 28.01.94) HgCdTe layers are formed on a silicon substrate with an high-resistance compound semiconductor in between.

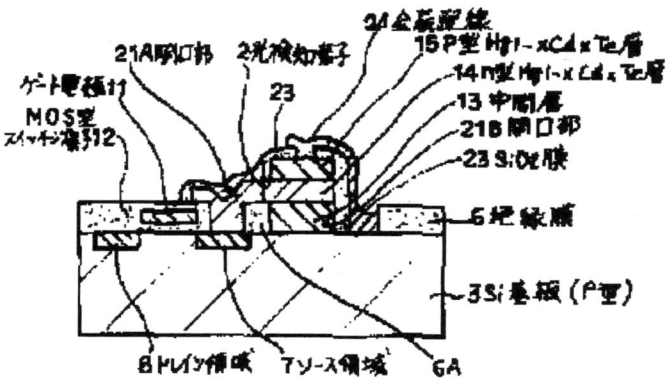

Fig. 2.5.63 (JP-A-6021419 fig. 1a)

A high-resistance compound semiconductor film 13 is formed on a silicon substrate 3 at openings in an insulating film 6. A detector element 2 is composed of an n-type and a p-type HgCdTe layer 14 and 15 formed on the film 13. The detector element is connected to an electrode 24 and to a MOS transistor in the silicon substrate by a connector 21A.

*

The imager of JP-A-6089991 (Toshiba Corp., Japan, 29.03.94) comprises HgCdTe detector regions which have been grown on a silicon substrate. A method to clean the silicon surface before the HgCdTe regions are grown thereon is disclosed.

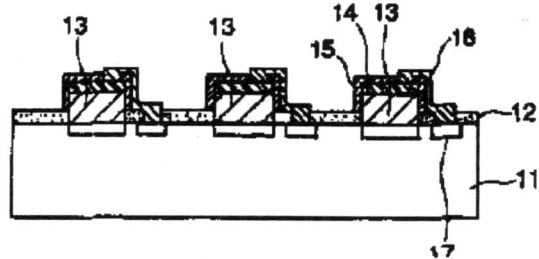

Fig. 2.5.64 (JP-A-6089991 fig. 1)

A charge transfer device is formed in a silicon substrate 11. The substrate is treated with a hydrofluoric acid solution to remove the surface oxide film. Cleaning is performed by using ultrapure water having a remaining dissolved oxygen concentration not exceeding 30ppb.

Thereafter, HgCdTe detector regions 13 are grown on the silicon substrate and individual detector elements are connected to the charge transfer device.

*

The detector elements of the imager presented in JP-A-6326342 (NEC Corp., Japan, 25.11.94) are separated from each other by providing amorphous material only at regions between the detector elements and removing this amorphous material in a wet etching process.

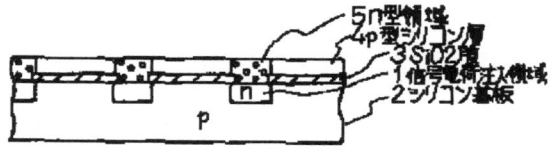

Fig. 2.5.65 (JP-A-6326342 fig.1a)

N-type regions 1 are formed in a p-type substrate 2. N-type regions 5 are formed on the n-type regions 1 and a p-type layer 4 is formed on a SiO_2 layer at regions outside the n-type regions 5.

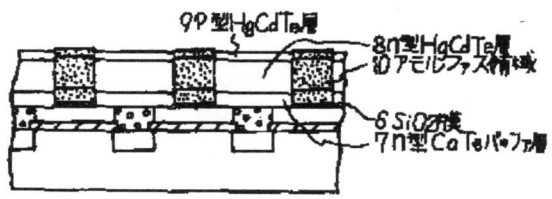

Fig. 2.5.66 (JP-A-6326342 fig.1b)

A SiO_2 film 6 is formed on the p-type regions 5. CdTe buffer layer 7, n-type HgCdTe layer 8 and p-type HgCdTe layer 9 are formed using MBE.

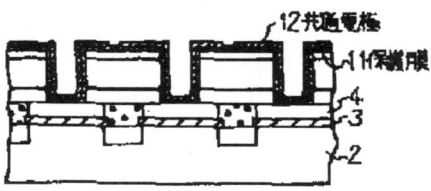

Fig. 2.5.67 (JP-A-6326342 fig. 1d)

Wet etching is carried out to remove amorphous regions 10 which have been formed on the SiO_2 film 6 there by providing separate detector elements. The array is also provided with a protective film 11 and a common electrode 12.

*

Connections made by Through Hole Technologies

A different approach to the problem of connecting terminals of individual detectors formed in an HgCdTe layer with corresponding terminals of a read-out device integrated in a semiconductor substrate is to provide connections via holes formed through the thickness of the HgCdTe layer. One particular technique is known as the loophole technique. The characteristic features of this technique are that an HgCdTe substrate is adhered to a semiconductor substrate which comprises a read-out circuit and that holes are drilled through the HgCdTe substrate where the pn-junctions are formed and the detector terminals are connected to input terminals of the read-out circuit by filling the apertures with a metallization layer. The technique gives rise to a robust device which is rather easy to manufacture and which may have a small size of the individual detector elements.

An imager and a method of manufacturing an imager using the loophole technique described above is presented in EP-A-0061803 (Philips Electronic and Associated Industries Limited, GB, 06.10.82).

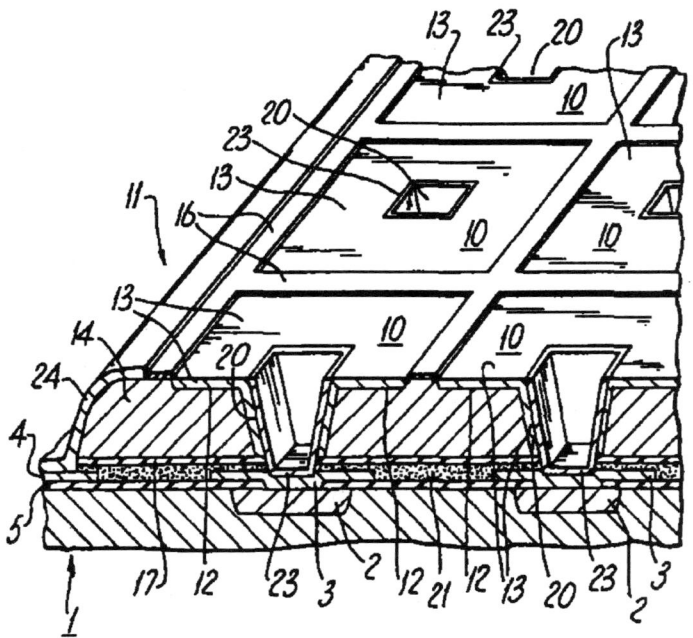

Fig. 2.5.68 (EP-A-0061803 fig. 1)

The imager comprises a silicon substrate 1 in which a read-out circuit such as a CCD is integrated. The read-out circuit has input regions 2 and metal electrodes 3. An array of photovoltaic detector elements 10 are formed in a p-type mercury cadmium telluride body 11 which is mounted by an insulating adhesive 21 on the substrate. Each detector element comprises an n-type region 13 which forms a pn-junction 12 with an adjacent part 14 of the body. Apertures 20 corresponding to each detector element extend through the thickness of the body and also through the adhesive layer to reach the metal electrodes. The n-type regions extend through the thickness of the body at the side-walls of the apertures and are electrically connected to the metal electrodes of the substrate by a metallization layer 23 formed in the apertures. The p-type part 14 of the body is connected to the metal electrodes 4 of the substrate by metallization 24. This metallization may contact the part 14 all around the outer side-walls of the body 11 or it may contact at apertures in the body. A passivation layer 17 and a grid of passivation of an anodic oxide or ZnS are formed on the body.

The apertures 20 are formed by etching with an ion beam of for example argon ions. The dose, energy and mass of the bombarding ions can be chosen such that a sufficient excess concentration of mercury is produced from the etched-away parts of the body 11 as to act as a diffusion source for converting the p-type body parts adjacent the side-walls of the resulting apertures into n-type side-wall regions which form part of the regions 13. The method of type-conversion by ion bombardment is further described in EP-A-0062367 (Philips Electronic and Associated, GB, 13.10.82) which is presented in chapter 2.2. An imager manufactured by two photolithographic steps may be formed if the regions 13 are formed entirely by the n-type side-wall regions. In this case a first photolithographic step is used to provide a masking layer on the body surface having windows located above parts of the metal electrodes below the body. The apertures are formed using localized ion-etching through the thickness of the body at the windows. The metallization layer is provided in the apertures by depositing metal on the masking layer and on exposed parts of the metal electrodes at the apertures and then removing the masking layer to leave the metallization layer in the apertures. The second photolithographic step is used to determine the area of the peripheral portion of the body at which electrical connection is made to the p-type part of the body.

**

In GB-A-2208256 (Philips Electronic and Associated Industries Limited, GB, 15.03.89) detector elements are provided in an hexagonal arrangement whereby each detector in the array is surrounded by six neighbouring detector elements. Compared to a square X-Y matrix arrangement, in which each detector is surrounded by eight neighbouring detector elements, the detector elements can have increased detectivity and lower leakage currents. Furthermore, the array can have a higher spatial resolution for a given detecting area and a higher probability of detecting small objects of less than one element subtense.

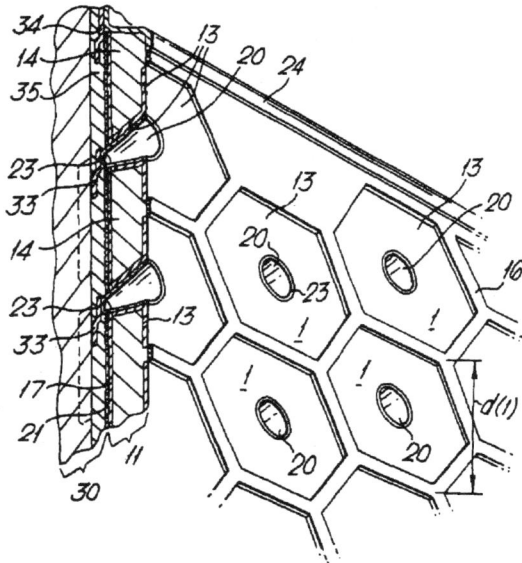

Fig. 2.5.69 (GB-A-2208256 fig. 3)

The hexagonal array of detector elements 1 are formed in a single body 11 of cadmium mercury telluride, which is present on a surface of a circuit substrate 30. The substrate comprises circuitry such as a silicon CCD or a MOSFET switch array for reading out signals produced by the detector elements. At the surface of the substrate there is a conductor pattern 33 which extends through an insulating layer 35 to provide contact areas in an hexagonal arrangement to which n-type regions 13 of the detector elements are connected by metallization 23. The metallization is present in apertures 20 which are present at the centre of the detector elements. The p-type main body part 14 is connected to a substrate conductor 34 by edge metallization 24. Passivation layers 16 and 17 are present on the surfaces of the detector element array body, and the back surface of the body is secured to the substrate by an electrically insulating adhesive layer 21.

Several advantages are achieved with this structure. The ratio of the detecting area to the perimeter around the edge of the detecting area, a/L is greater for the hexagonal diode elements compared to square diode elements in a corresponding X-Y arrangement. Higher detectivity is achieved since the detectivity of photovoltaic diodes increases with increasing ratio a/L. Furthermore, detector elements of hexagonal shape have corners which are less sharp than the square corners. This decreases the leakage-currents since sharp corners can produce high electric field gradients which increase the leakage-currents. The spatial resolution of a radiation image is determined principally by the pitch of the detector elements along different orientations of the array. The hexagonal array has a significantly smaller average pitch compared to a corresponding X-Y array. The greater symmetry of the hexagonal arrangement permits detection systems to use simpler shape correlators for identifying objects when a object image extends on only a few elements in the array.

* *

It is conventional to cool mercury cadmium telluride photodiodes to about 80 K for 8-14 μm waveband detection material and to about 200 K for 3-5 μm waveband detection material. By designing the photodiodes so as to incorporate a minority-carrier extraction region and to inhibit minority-carrier injection, the minority-carrier concentration in the photosensitive region can be depressed to produce low noise and high sensitivity characteristics similar to those as are obtained by cooling. The resulting photodiodes for 3-5 μm waveband detection may be operated at ambient temperature, about 295 K, and those for the 8-14 μm waveband at about 190 K. An imager comprising photodiodes designed in this manner is presented in GB-A-2239555 (Philips Electronic and Associated Industries Limited, GB, 03.07.91).

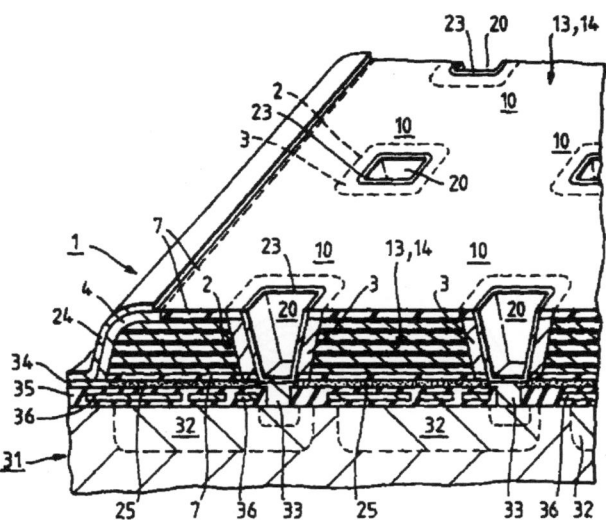

Fig. 2.5.70 (GB-A-2239555 fig. 1)

Photodiodes 10 are formed in an HgCdTe body 1 on a circuit substrate 31. Each photodiode comprises a vertical region 3 which extends through the thickness of the body to form a diode junction 2. The diode junction is connected to an electric conductor pattern 33 by a metallization layer 23. The body comprises a plurality of alternating first and second HgCdTe layers 13 and 14 having junctions 12 therebetween. The composition of HgCdTe of layers 13 and 14 may be different thereby forming heterojunctions or the layers 13 may be formed of cadmium telluride so that a compositional superlattice structure is formed. The junctions 12 serve to extract minority charge carriers from the layers 13 and 14 during operation to reduce the minority carrier concentration. The layers 13 are connected in parallel to the vertical regions 3 to provide an output signal from each photodiode. The layers 13 have sufficiently low doping levels to be depleted throughout their thickness and thereby isolate electrically the vertical regions 3 of individual photodiodes from each other. The layers 14 have sufficiently high doping levels to be undepleted over a part of their thickness. These parts provide parallel conductive paths from all the photodiodes to the common electrical connection 4, 24, 34.

Electron-hole pairs generated in the body are separated by the pn-junctions 12 by extraction of electrons from the p-layer 14 into the depleted layers 13 where they drift to the n-regions 3 and by extraction of holes from the depleted layers 13 into the p conductive paths of the layers 14 where they drift to the common connection 4, 24, 34. A close spacing of the junctions 12 gives a high efficiency of minority-carrier extraction from the layers 13 and 14.

The device may be formed by depositing alternating layers of cadmium telluride and mercury telluride by vapour phase deposition techniques and interdiffuse the layers, either during growth or subsequently, so as to form a mercury cadmium telluride layer. Reference is made to GB-A-2146663 (The Secretary of State for Defence, GB, 24.04.85) and GB-A-2203757 (Philips Electronic and Associated Industries Limited, GB, 26.10.88).

* *

An anodized film is used in JP-A-2272765 (NEC Corp., Japan, 07.11.90) to suppress cross-talk between detector elements.

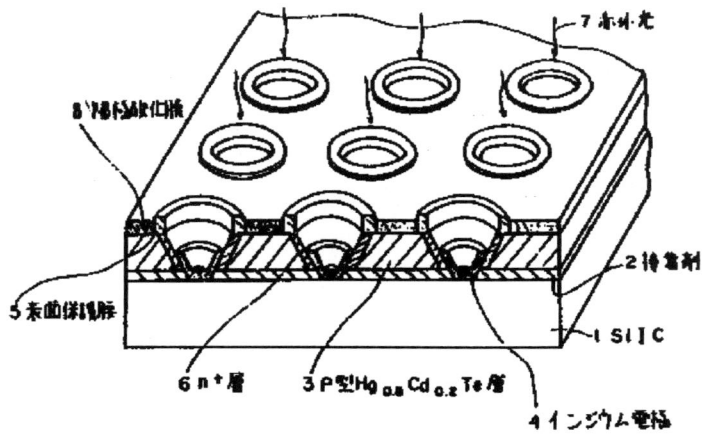

Fig. 2.5.71 (JP-A-2272765 fig. 1)

A p-type HgCdTe layer 3 is attached to a silicon read-out substrate 1. N-type regions 6 are formed on the side-walls of holes which go through the HgCdTe layer. A protective film 5 is formed in the vicinity of the regions 6. An anodized film 8 is formed on the layer 3 and an inverted layer (n-type) will be formed underneath the film. Electron-hole pairs generated between the detector elements will be absorbed by the inverted layer thereby suppressing cross-talk between the detector elements.

* *

Cross-talk may also be reduced by forming trenches between individual detectors. In JP-A-4267562 (NEC Corp., Japan, 24.09.92) and JP-A-4269869 (NEC Corp., Japan, 25.09.92) gate electrodes are also provided in the trenches.

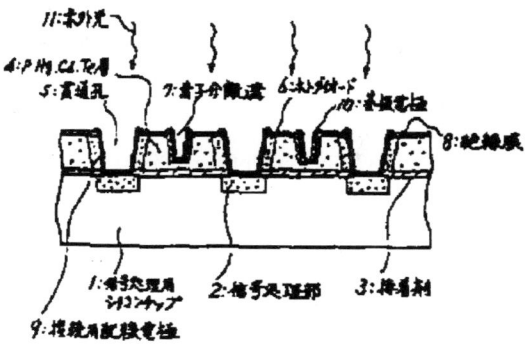

Fig. 2.5.72 (JP-A-4269869 fig. 1)

Photodiodes 6 are formed in a p-type HgCdTe layer 4. Grooves 7 are formed between the photodiodes and gate electrodes 10 are placed on an insulating layer 8 in the grooves. A potential is applied to the gate electrodes to create accumulation layers which will guide generated electrons to the photodiodes. This potential also stabilizes the potential of the photodiodes.

* *

Two or more HgCdTe layers having different bandgaps are formed as different levels on a read-out substrate in EP-A-0475525 (Philips Electronic and Associated Industries Limited, GB, 18.03.92) to provide different wavelength response. The individual detectors are connected by the same method as described in EP-A-0061803 above.

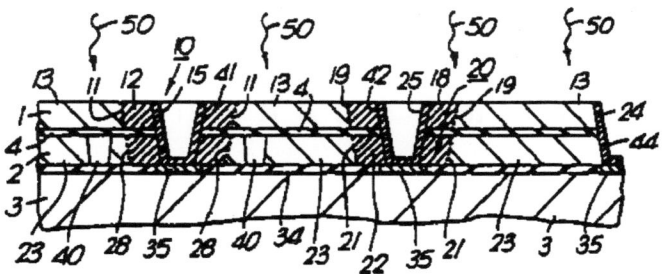

Fig. 2.5.73 (EP-A-0475525 fig. 2)

A thin p-type body 2 of $Hg_{0.79}Cd_{0.21}Te$ is bonded to a silicon integrated circuit 3 by an intermediate film of epoxy adhesive. The upper and lower surfaces of the body 2 are passivated by a surface layer of cadmium telluride or ZnS. Gaps 40 are etched in the body 2 to separate the islands 28 from the remainder of the body. Then, a thin p-type body 1 of $Hg_{0.7}Cd_{0.3}Te$ is bonded to the upper surface of the body 2 by an intermediate film or epoxy adhesive.

Apertures and n-type regions 12, 28 and 18, 22 are formed, followed by metallization 15 and 25. A peripheral side-wall 44 is contacted by a metallization 24.

Although exposed to the incident infrared radiation 50, the separate island 28 has only one connection 15 and so does not contribute to the output signal of the shorter-wavelength photodiode 10. The further region 18 in the shorter-wavelength response material 1 forms a further pn-junction 19 which is connected in parallel with the longer-wavelength photodiode 20 by connections 24 and 25. However, because of the difference in bandgap, the photocurrent generated at this parallel junction 19 by the incident radiation is insignificant compared with that generated at the photodiode junction 21.

* *

The shorter-wavelength photodiode 10 and the longer-wavelength photodiode 20 of the imager presented above in EP-A-0475525 are placed side-by-side and not on top of each other. The signals from the photodiodes will therefore not refer to the same source from a scene projected on the imager. This problem is solved in EP-A-0481552 (Philips Electronics UK Limited, GB, 22.04.92) by the use of infrared lenses placed on the imager.

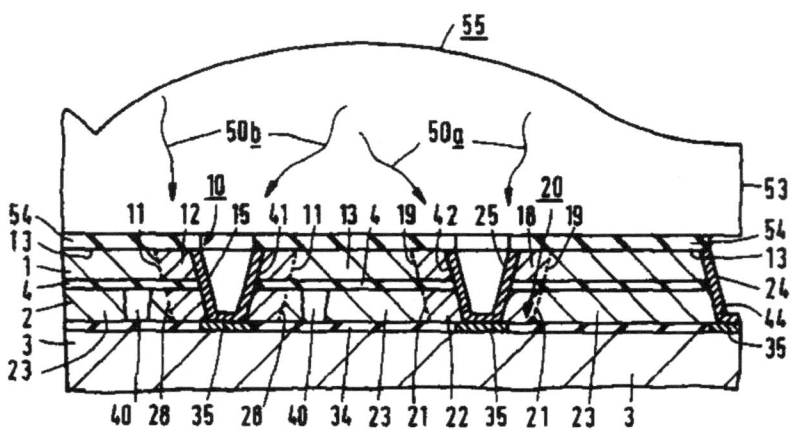

Fig. 2.5.74 (EP-A-0481552 fig. 4)

Lenses 55, made of cadmium telluride, germanium or silicon, are bonded to the imager by a thin film 54 of epoxy adhesive. Different wavelengths to be detected are directed at different angles towards the lenses so that each lens 55 concentrates the longer-wavelength 50a to the longer-wavelength photodiode 20 while also concentrating the shorter-wavelength 50b onto the shorter-wavelength photodiode 10. Lenses may also be used to adjust for uneven centre-to-centre spacing of individual detector elements when a plurality of discrete imaging arrays are placed side-by-side to form a large composite imaging array device. See for example EP-A-0510267 (N.V. Philips' Gloeilampenfabrieken, The Netherlands, 28.10.92).

* *

EP-A-0561615 (Philips Electronics UK Limited, GB, 22.09.93) teaches how to form an electrical connection on a side-wall, and only on the side-wall, of a step structure. This technique is used to form a compact two-wavelength imager.

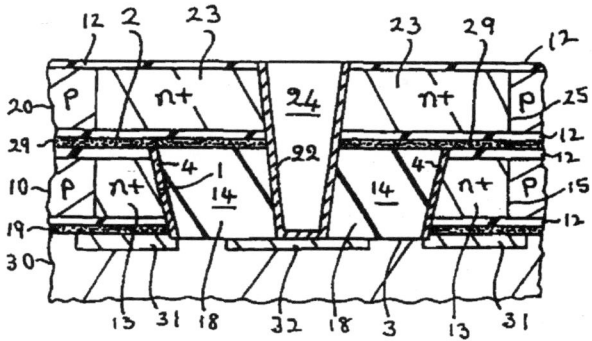

Fig. 2.5.75 (EP-A-0561615 fig. 4)

A p-type HgCdTe body portion 10 of a first bandgap is secured by an epoxy adhesive layer 19 on a silicon integrated circuit 30 with contact pads 31 and 32. Side-walls 1 are formed by ion-milling apertures 14 through the thickness of the body portion 10. At the same time annular n-type regions 13 are formed thereby creating photodiodes with pn-junctions 15. Gold or chrome are deposited over the structure and electrical connections 4 on the side-walls, connecting pads 31, are formed by a directional etching treatment, for example by ion-milling. The apertures 14 are filled with insulating material 18, for example cadmium telluride or a glass, before a p-type HgCdTe body portion 20 of a second bandgap is mounted on the body portion 10 by an intermediate epoxy adhesive film 29. Apertures 24 are etched through the body portion 20 and the insulating material 18. N-type regions 23 are formed creating photodiodes with pn-junctions 25. Conductive layers 22 connect the regions 23 with pads 32.

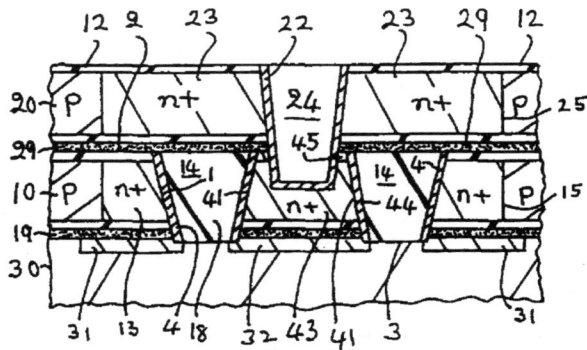

Fig. 2.5.76 (EP-A-0561615 fig. 5)

Alternatively, n-type pillar portions 43 are formed when apertures 14 are formed. Side-wall connectors 41 connect to pads 32 and n-type regions 23 are connected to side-wall connectors 41 by the conductive layers 22 and 45.

*

A detector involving an ambipolar drift of radiation-generated free minority carriers using the loophole technique is disclosed in GB-A-2207802 (Philips Electronic and Associated Industries Limited, GB, 08.02.89). (Ambiplar drift field imagers are presented in Chapter 1.2.)

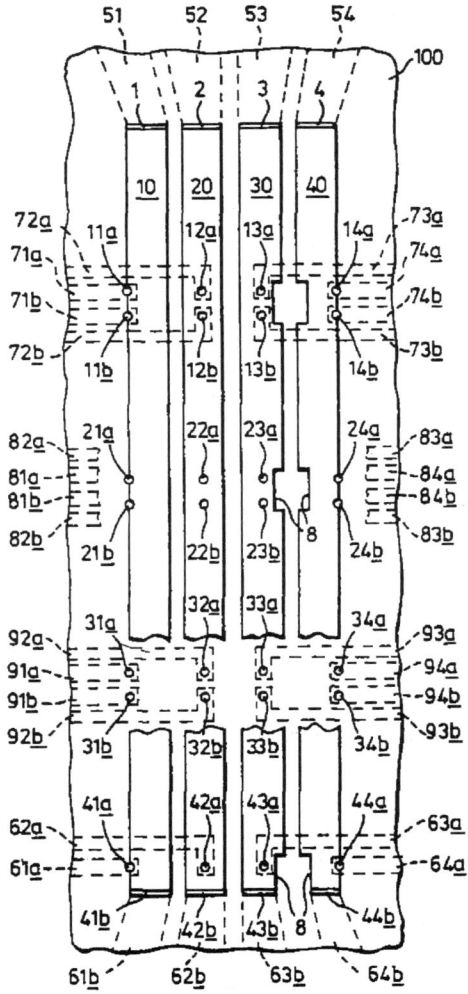

Fig. 2.5.77 (GB-A-2207802 fig. 2)

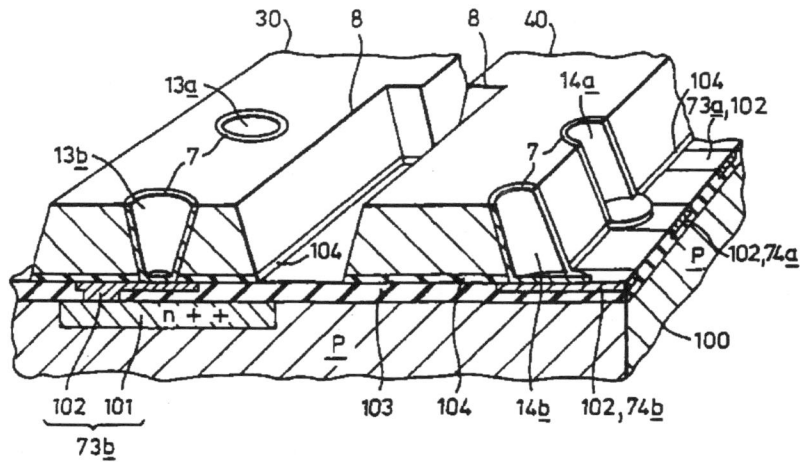

Fig. 2.5.78 (GB-A-2207802 fig. 3)

Biasing-electrode means 1-4 and 41b-44b are spaced so as to cause a bias current predominantly of majority charge-carriers to flow along each semiconductor strip 10, 20, 30, 40 of n-type mercury cadmium telluride, the bias current supporting an ambipolar drift of charge carriers in the opposite direction. Metallized vias provide read-out means in the drift path of each strip. The strips are secured to a silicon substrate 100 by a layer of electrically-insulating epoxy adhesive. Conductor pattern 61-64, 71-74, 81-84, 91-94 provides electrical connections to each read-out means. The substrate comprises the read-out means and signal processing circuitry. Each read-out means is formed at vias (reference a and b) which extend through the thickness of the strips to the conductor pattern. These vias are formed with small transverse dimensions by ion-etching. Metallization 7 is deposited over substantially the whole side-wall of each via without extending on the upper surface of the strips. As a result, the read-out means and their connections do not interrupt the ambipolar drift path to any great extent. In the figures, different read-out via configurations across the width of the different strips 10, 20, 30, 40 are illustrated.

*

In US-A-4720738 (Texas Instruments Incorporated, USA, 19.01.88) an HgCdTe layer is attached to a read-out substrate. Detector elements are connected to aluminium buses which are connected to bonding pads on the substrate via conductors that occupy holes through the HgCdTe layer.

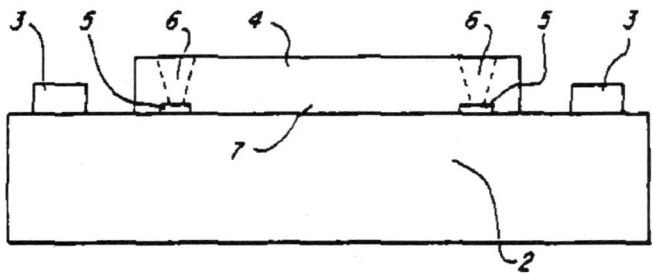

Fig. 2.5.79 (US-A-4720738 fig. 1)

A silicon chip 2 supports a mercury cadmium telluride detector array 4. The detector array is bonded to the silicon chip by a low viscosity epoxy 7 before it is lapped and polished to a final thickness. Areas 6 located at the edge borders of the detector array are etched into the mercury cadmium telluride to expose openings to bonding areas 5 on the silicon processor chip. These openings can be made with ion-milling or with a 1% bromium methanol solution static etch. Bonding areas 3 are formed to provide for external bonding. A cross-sectional area of a single via area is shown below.

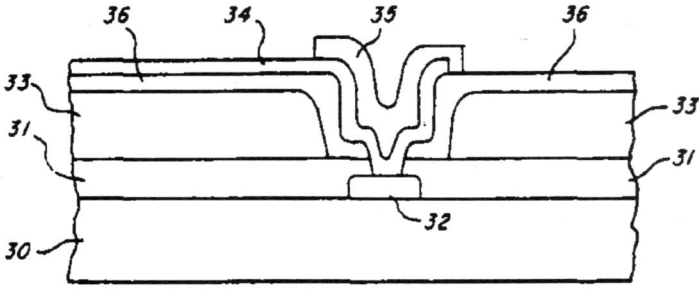

Fig. 2.5.80 (US-A-4720738 fig. 5)

In this figure the silicon chip, the epoxy layer, the bonding area, and the mercury cadmium telluride layer are labeled 30, 31, 32 and 33, respectively. A detector element is connected to an aluminium bus line 34 which is insulated from the mercury cadmium telluride layer by a ZnS layer 36. A layer of indium 35 is located over the aluminium bus connection at the bonding pad region.

*

An imager connecting pn-junction detectors or MIS detectors to connection pads of a silicon read-out chip via through holes is presented in EP-A-0137988 (Texas Instruments Incorporated, USA, 24.04.85). The through holes are created by first ion-milling small holes and then enlarge the same by spray etching.

A MIS detector consists of an insulated gate provided over an expanse of semiconductor. The gate is charged up to create a depletion well in the semiconductor beneath the gate, and then floated. The well then collects carriers from photo-generated electron-hole pairs. At the end of the collection cycle the voltage on the MIS gate is sensed. The gate voltage is then controlled to collapse the well and stored carriers are recombined. A new depletion well is created to begin a new detection cycle.

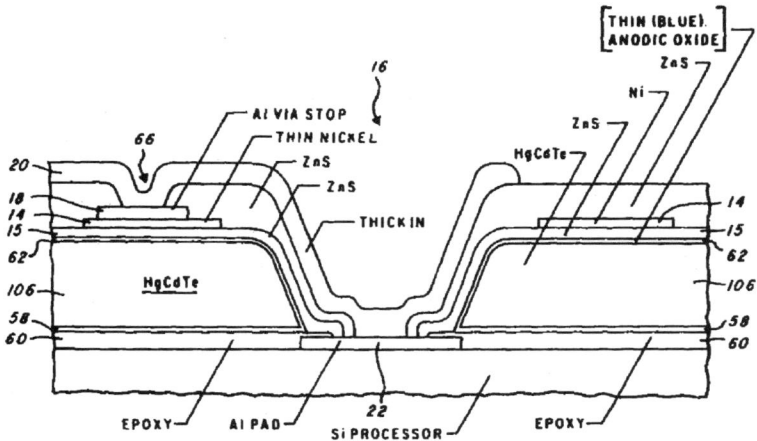

Fig. 2.5.81 (EP-A-0137988 fig. 1)

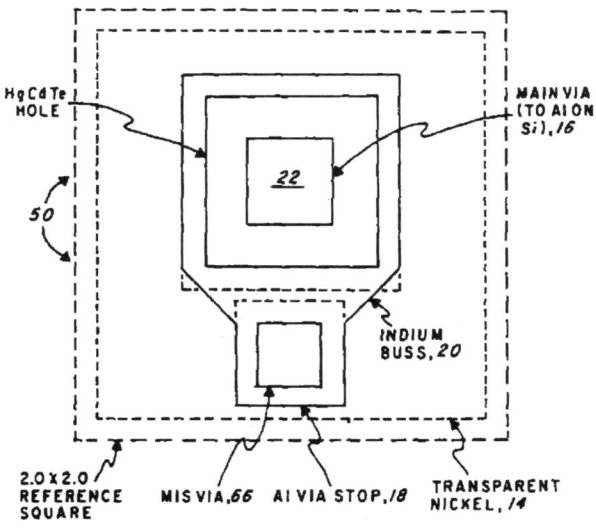

Fig. 2.5.82 (EP-A-0137988 fig. 2)

In a first embodiment an HgCdTe bar 106 is lapped to have parallel surfaces and one side is polished smooth and anodically oxidized 58 before it is mounted on a silicon substrate by a very low-viscosity and low surface tension glue 60. The silicon is then polished down to 12 microns. A patterning step patterns a via 16 through the HgCdTe. This is a crucial step - the via must not be etched too narrow or too wide. Furthermore, the walls of the via must be very smooth to provide good step coverage. For example, simple ion-milling may leave rough walls. Instead, ion-milling is used to make a small via hole and then spray etching is used to enlarge the via. This technique of making vias is claimed in EP-A-0288792 (Texas Instruments Incorporated, USA, 02.11.88). An anodic oxide layer 62 is formed followed by a ZnS layer 15. Nb_2O_5, Ta_2O_5, TiO_2 or TiO_2/Al_2O_3 composites may be used instead of ZnS. A thin, 10nm, storage gate 14 of nickel or chromium is formed and connected to an indium interconnect 20 having a aluminium via stop 18 formed in between. The interconnect makes contact to a silicon processor via bond pad 22. The silicon processor may comprise averaging capacitors as claimed in EP-A-0137704 (Texas Instruments Incorporated, USA, 17.04.85).

Due to the anodic oxidation process a substantial amount of positive trapped charge remains in the oxide. If the HgCdTe bar is of n-type this charge will form channel stops around the MIS storage gates. On the other hand if the HgCdTe bar is of p-type the space charge of electrons which is found all over the surface of the bar will immediately flow into the depletion wells underneath the storage gates, swamping the desired signal provided by photoelectrons. In this case one additional masking step is used to deposit a field plate over the whole surface of the HgCdTe except for the storage gates.

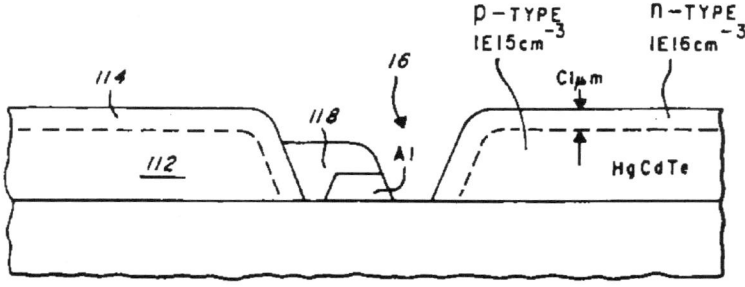

Fig. 2.5.83 (EP-A-0137988 fig. 16)

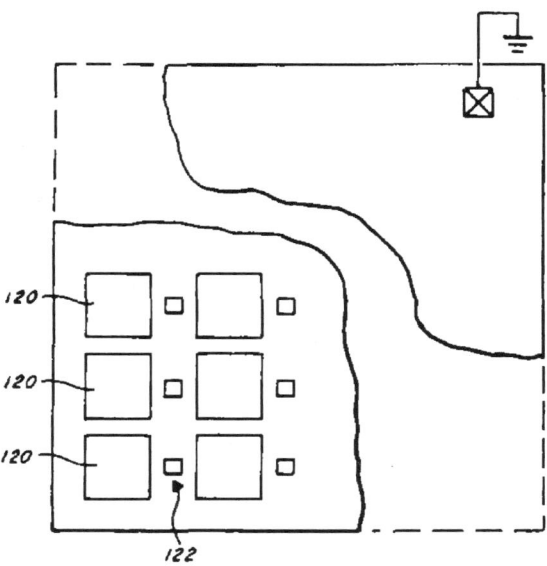

Fig. 2.5.84 (EP-A-0137988 fig. 17)

A second embodiment using pn-junction detectors is shown above. In this case a p-type HgCdTe bar 112 is attached to a read-out silicon substrate and vias 16 are formed as described above. An n-type layer 114 is formed by ion-implantation. Almost any species of ion can be used since the implant damage tends to produce n-type material in HgCdTe. Since this is done after the via hole has been patterned, the walls of the via hole are also implanted, so that formation of an ohmic contact from the n-type layer to the contact pad 22 on the silicon can be performed by simply depositing a small amount of metal 118 at the bottom of the via hole 16. Individual detectors are separated by forming mesas 120 by etching away HgCdTe. An alternative way of connecting the n-type regions 114 to the contact pads 22 on the silicon is through a via which is not located within the active device mesa 120, but is adjacent to it, as shown by the vias 122 above.

* *

The storage gate of EP-A-0137988, presented above, is formed of nickel or chromium. There is a tradeoff between transmittance and electrical conductivity of the gate. This problem is further analysed in EP-A-0416299 (Texas Instruments Incorporated, USA, 13.03.91). It is proposed to use bismuth (Bi), antimony (Sb) or titanium oxynitride (TiN_xO_y).

*

A method of passivation of HgCdTe by anodic selenidization is disclosed in US-A-4726885 (Texas Instruments Incorporated, USA, 23.02.88).

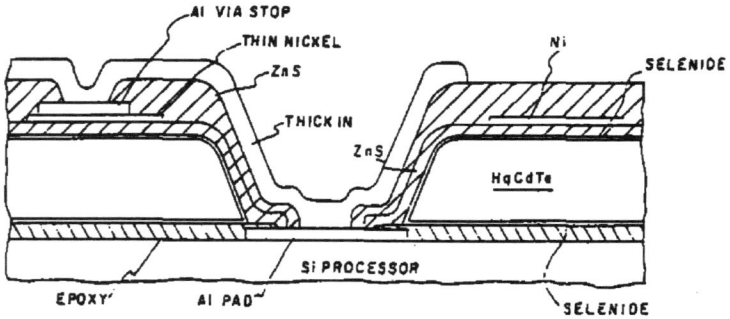

Fig. 2.5.85 (US-A-4726885 fig. 1B)

An HgCdTe slice which has been cleaned and polished is anodically selenidized. The slice is then attached with epoxy to a silicon processor chip and MIS detectors are formed similar to the MIS detectors presented in EP-A-0137988 above.

* *

A process for forming a ZnS insulative layer on an HgCdTe material characterized by the absence of an intermediate native oxide layer is taught in EP-A-0366886 (Texas Instruments Incorporated, USA, 09.05.90).

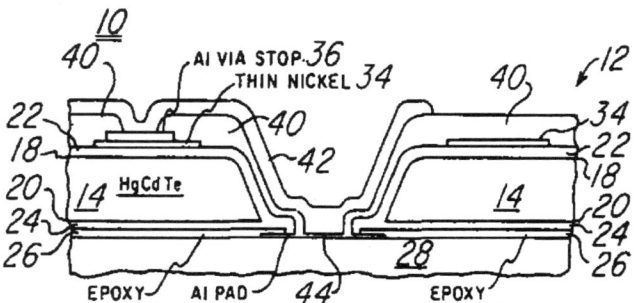

Fig. 2.5.86 (EP-A-0366886 fig. 2)

In the process proposed, the surface of an HgCdTe material is treated to remove and/or prevent formation of surface contaminants including oxides. A layer of passivating material is formed in direct contact with the semiconductor surface, the direct contact allowing the passivating material to tie up dangling semiconductor bonds. Furthermore, it is shown how the process is applied to form the imager described in the first embodiment of EP-A-0137988 presented above. Passivation layers 22 and 24 are formed according to this method.

**

As shown in EP-A-0366886 above, a sulphide passivation layer may be formed without an intermediate oxide layer. If this technique is used to form the insulation layer of the MIS detector array shown above in EP-A-0137988, channel stops can not be provided by trapped positive charge in the passivation layer. An approach to solve this problem is to etch away the sulphide passivation layer in channel stop areas and grow anodic oxide to regain the fixed positve charge, but this has the problem of extra wet processing steps. In EP-A-0416320 (Texas Instruments Incorporated, USA, 13.03.91) n-type channel stops are provided by lattice damage implants using boron or by ion-milling using argon ions.

**

During the operation of a MIS detector of the kind presented in EP-A-0137988 above, a voltage is applied to the gate to form a depletion region in the HgCdTe. Photons penetrate the gate and create electron-hole pairs. The holes accumulate in the HgCdTe at the interface with the gate insulator and form an inversion layer which reduces the size of the depletion region and lowers the absolute value of the gate potential. An increase in the magnitude of the gate voltage implies a larger depletion region so that more holes can be collected. However, too large a gate voltage leads to breakdown in the HgCdTe. In US-A-5144138 (Texas Instruments Incorporated, USA, 01.09.92) the photocapacitors include a heterojunction. This structure allows increased potential well capacity, reduced dark current and detection of two colours. Other features of this document are presented in chapter 1.5.

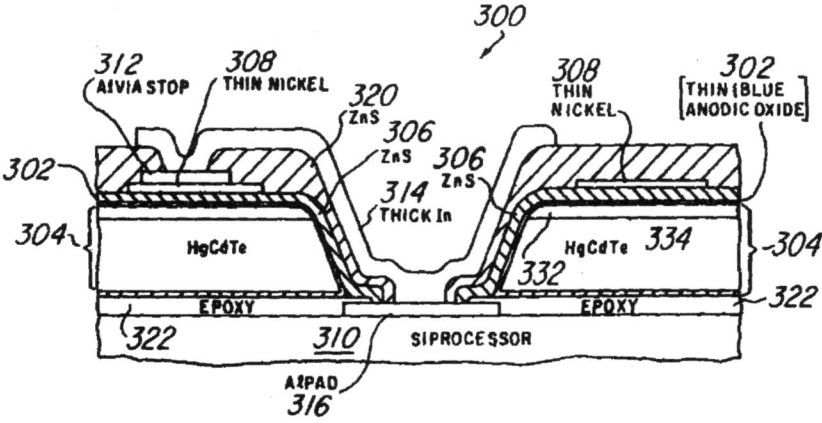

Fig. 2.5.87 (US-A-5144138 fig. 3)

The MIS detector comprises a gate 308, a gate dielectric 306 of ZnS, an anodic oxide passivation 303, a layer 332 of $Hg_{0.73}Cd_{0.27}Te$ and a layer 334 of $Hg_{0.8}Cd_{0.2}Te$.

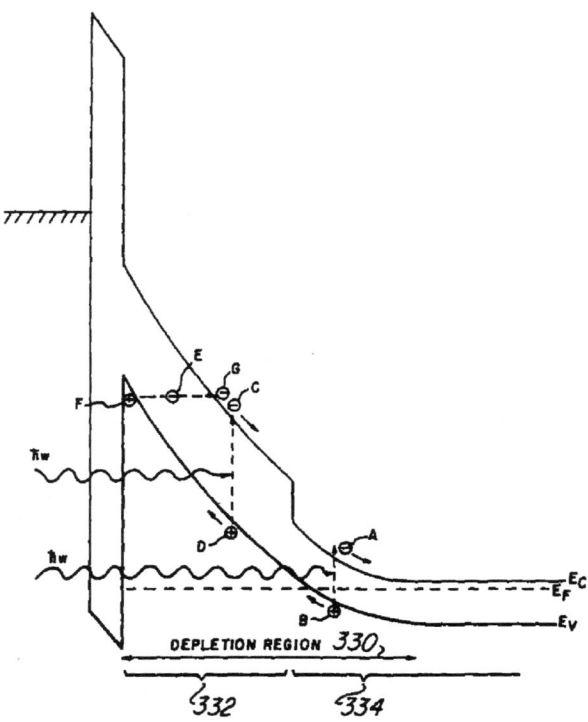

Fig. 2.5.88 (US-A-5144138 fig. 4b)

A band diagram for the detector along a direction from the gate through the gate dielectric and the oxide into the HgCdTe with a first bias applied to the gate is shown above. This first bias creates a depletion region 330 that extends through layer 332 into layer 334. An analogous band diagram when a second bias which is smaller than the first bias is applied to the gate is shown below. In this case the depletion region 331 is confined to the wide bandgap layer 332.

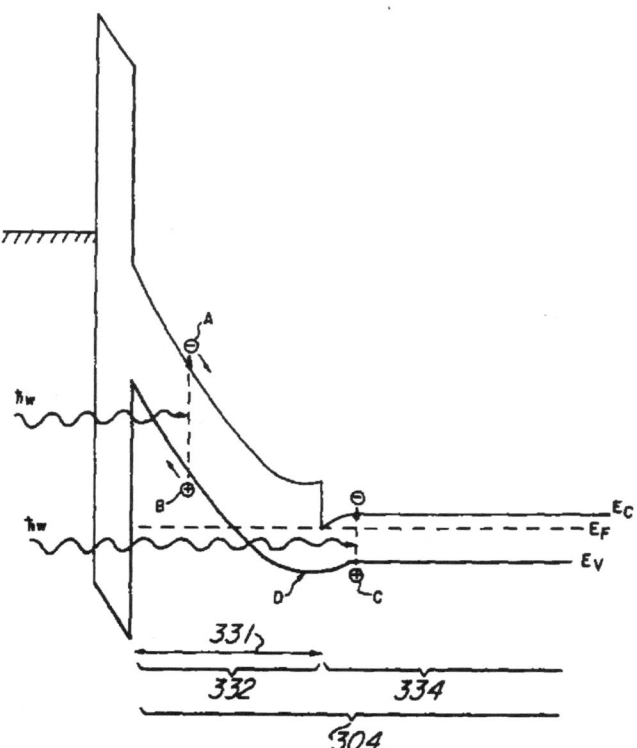

Fig. 2.5.89 (US-A-5144138 fig. 4c)

When the first larger gate bias is applied 8-12 µm wavelength photons are absorbed in the narrow bandgap layer 334 after having passed through the wide bandgap layer 332. Photons with 3-5 µm wavelength are also detected by absorption in the wide bandgap layer 332. This is indicated by A-B and C-D respectively in the first bandgap diagram above. The detector can also detect 3-5 µm wavelength photons while excluding detection of 8-12 µm wavelength photons. The second lower gate bias is applied and 3-5 µm wavelength photons are absorbed in the wide bandgap layer 332. Photons with 8-12 µm wavelength are absorbed in the narrow bandgap layer 334. This is indicated by A-B and C respectively in the second bandgap diagram above. The holes generated from the latter photons are however precluded form collection into the inversion layer by a barrier in the valence band due to the heterojunction. The barrier is indicated by D in the second bandgap diagram above. Two colour operation is therefore achieved by using two different gate biases.

Also shown in the first bandgap diagram above is a dark current mechanism of an electron E tunneling from the valence band into an unoccupied midgap state leaving behind a hole F which is collected and then tunneling into the conduction band, electron G, leaving behind an unoccupied midbandgap state. Compared to a detector comprising no wide bandgap layer 332, the tunneling distance, E-F-G is greater, the tunneling barrier is higher and so the dark current is smaller when the same gate voltage is used. It is also shown that the potential well capacity is greater.

An alternative to the heterostructure is also shown. In this case, a compositionally graded structure is formed where the composition x of $Hg_{1-x}Cd_xTe$ varies with the distance from the passivation layer. Methods of manufacturing such structures are discussed.

Other aspects of the invention are shown in chapter 1.5.

*

Each pixel of the imager presented in GB-A-2197984 (The Marconi Company Limited, GB, 02.06.88) comprises a homojunction formed in parallel with a heterojunction.

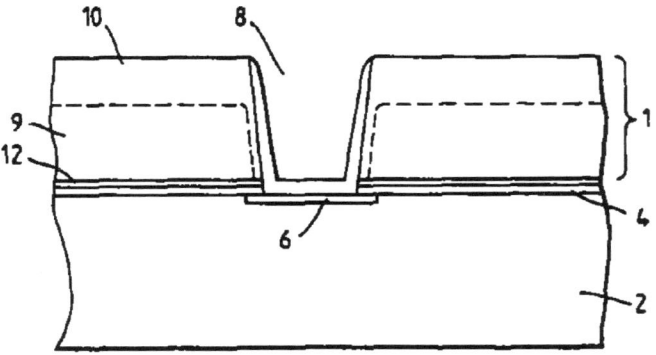

Fig. 2.5.90 (GB-A-2197984 fig. 2)

A pn-junction layer 1 of HgCdTe, which has a passivation layer 12, is bonded to a silicon multiplexer chip 2 by a bonding layer 4. The lower layer 9 of the heterojunction 1 is p-type and the upper layer 10 is n-type. Holes 8 are ion-etched right through the semiconductor layer to contact pads 6. The regions of the layer 9 forming the walls of the holes are converted to n-type by the ion beam. A homojunction is thus formed in parallel with the heterojunction. The compositions are selected so that the heterojunctions determine the infrared response and noise characteristics and the homojunctions act as noiseless isolation devices.

*

In the invention disclosed in JP-A-1061055 (NEC Corp., Japan, 08.03.89) a read-out chip is attached to an HgCdTe layer and connection is made via holes through the silicon chip.

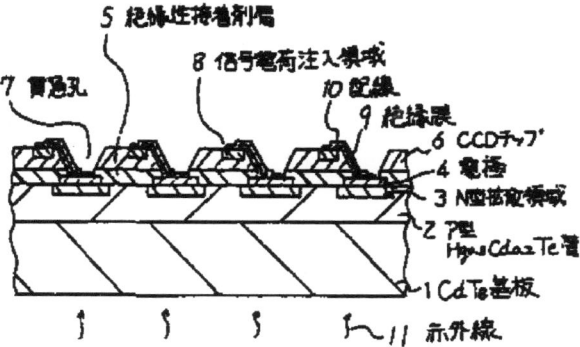

Fig. 2.5.91 (JP-A-1061055 fig. 1)

A p-type HgCdTe layer 2 is grown on a semi-insulating CdTe substrate 1. N-type regions 3 are formed followed by electrodes 4. A CCD read-out chip 6 is attached to the HgCdTe layer by an insulating adhesive layer 5, and through holes are formed in the read-out chip to expose the electrodes. Implanted input regions 8 of the CCD are connected to corresponding electrodes by wirings 10.

*

A problem encountered when aluminium is used as gate metal for a MIS structure is the significant etching of the aluminium by the bromine solution which is used to form vias or contacts in overlaying ZnS insulator films. To solve this problem, it is proposed in EP-A-0407062 (Texas Instruments Incorporated, USA, 09.01.91) to use refractory metals as the metallization layers of an infrared detector.

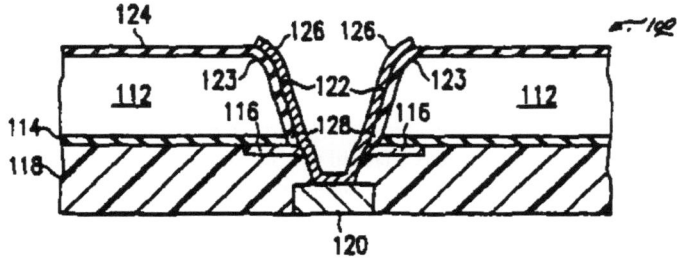

Fig. 2.5.92 (EP-A-0407062 fig. 9)

A ZnS insulating layer 114 is deposited on a sensing layer 112 of HgCdTe. A refractory metal, such as tantalum, molybdenum, tungsten, titanium or refractory metal alloys such as titanium/tungsten, is sputter-deposited onto the insulating layer. The structure is connected to a signal processor 120 utilizing an epoxy 118. A via 122 is etched through the sensing layer, the insulating layer, the gate and the epoxy. Rounded edges 123 are formed in the sensing layer by etching the via with a bromine solution. The bromine solution does not significantly reduce the thickness of the refractory metal gate. The refractory metal gate forms an oxide layer over itself which protects the metal from the bromine solution. After an insulator 124 has been deposited, a layer of metal 126 is formed within the via to electrically connect the gate to the processor 120. The fill factor of this structure is increased by forming vertically integrated MIS in the pixel cells.

*

A structure having low amount of surface leakage currents and a high quantum efficiency is presented in JP-A-5343729 (NEC Corp., Japan, 24.12.93). The HgCdTe detecting layer of the presented structure is overlayed on both sides by HgCdTe layers having a larger bandgap than the bandgap of the detecting layer.

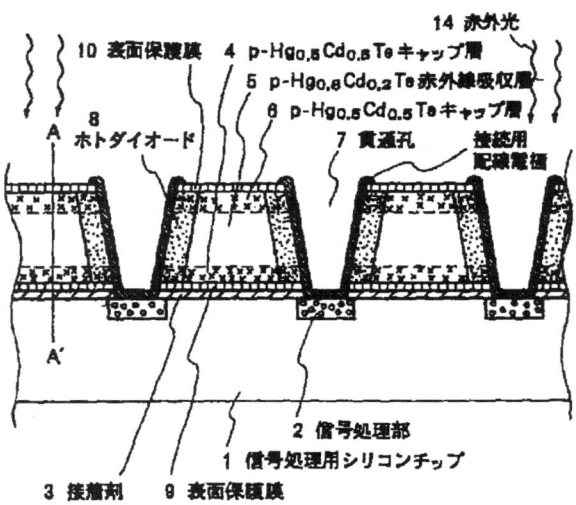

Fig. 2.5.93 (JP-A-5343729 fig. 1a)

Two layers of p-type $Hg_{0.5}Cd_{0.5}Te$ 4 and 6 are formed on each side of a detecting layer 5 made of p-type $Hg_{0.8}Cd_{0.2}Te$. The structure is attached to a silicon chip 1. Electrical contact between n-type regions of the detector layer and a signal processor 2 formed in the silicon chip is made via through holes. The two layers 4 and 6 prevent electrons which have been generated in the detecting layer diffusing to the surface regions of the HgCdTe material where the density of crystal defects is high.

*

The electrical connection between detector elements and an input pad of a read-out circuit is improved in JP-A-6204448 (NEC Corp, Japan, 22.07.94) by providing an indium film on the detector substrate corresponding to the input pad prior to forming a through hole.

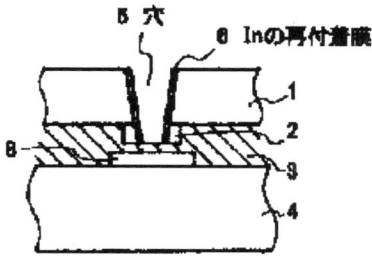

Fig. 2.5.94 (JP-A-6204448 fig. 1b)

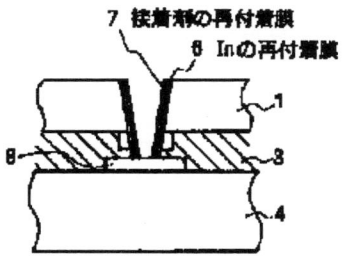

Fig. 2.5.95 (JP-A-6204448 fig. 1c)

An indium film 2 is formed on an HgCdTe substrate 1 at regions which corresponds to input pads 8 of a silicon read-out chip. The HgCdTe substrate is bonded to the read-out chip by a bonding agent 3. Next, ion-milling is used to provide a hole through the HgCdTe substrate and an indium coating 6 and a redeposited film 7 of the bonding agent is formed.

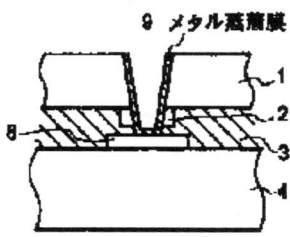

Fig. 2.5.96 (JP-A-6204448 fig. 1e)

The structure is dipped in hydrochloric acid to remove the indium coating and the redeposited film before a metal film 9 is formed. The structure including the indium film 2 reduces the risk of possible breakage of the metal film 9 due to the fact that the bonding agent 3 being removed to a larger extent than the HgCdTe substrate during the ion-milling process.

*

In US-A-5318666 (Texas Instruments Incorporated, USA, 07.06.94) photodiodes and vias are formed simultaneously by employing a dry reactive etching process. Portions of an HgCdTe body adjacent to the vias are thereby type converted.

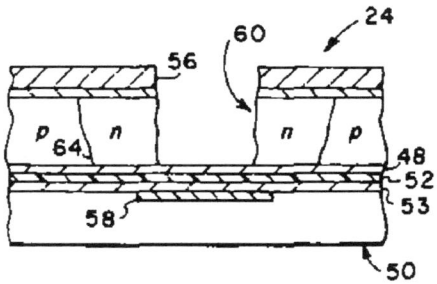

Fig. 2.5.97 (US-A-5318666 fig. 3)

Passivation layers of ZnS or cadmium telluride are depostited on the upper and lower surfaces of a vacancy doped p-type HgCdTe body. The body is mounted on a read-out chip by means of an epoxy adhesive 52. A mask of photoresist material including openings 56 is applied over the upper passivation layer. Each opening is aligned in register with a conductive electrical input pad 58 of the read-out device 50. Vias 60 and n-type regions 64 around the vias are formed by a dry reactive etching process.

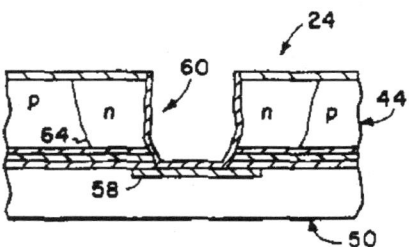

Fig. 2.5.98 (US-A-5318666 fig. 6)

The vias are extended down through layers 48, 52 and 53 by using a plasma dry etching process. A wet etching process is then used to smooth the surfaces of the vias. An electrically conductive layer of indium is formed over the photoresist masking layer and within the vias in electrical contact with the region 64 and the input pad 58. The photoresist masking layer and the portions of the conductive layer extending above the upper passivation layer are removed by a lift-off process.

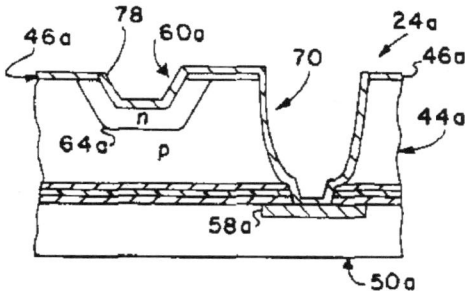

Fig. 2.5.99 (US-A-5318666 fig. 13)

In a second embodiment, the vias 60a are not formed completely through the p-type body 44a, but instead are formed as cavities extending only partially within the body. Vias 70 are formed and n-type regions 64a are connected to input pads 58a by conductive layers formed in the vias.

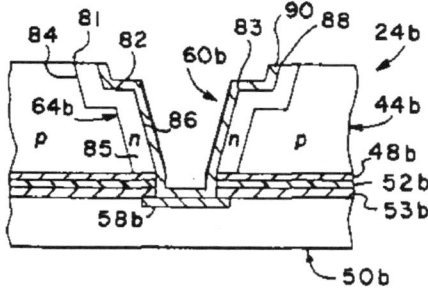

Fig. 2.5.100 (US-A-5318666 fig. 14)

In a third embodiment, vias 60b are formed by a two-phase reactive etching procedure. The body 44b is initially masked to define openings above input pads 58b. The structure is then placed in a reactive etching chamber and reactively etched only for a period sufficient to form lateral via portions 82 and n-doped regions 84. The first mask is removed and a second mask with openings is applied to the structure. The openings in the second mask have a smaller diameter than the openings in the first mask. The structure is again placed in the reactive etching chamber and reactively etched to form vertical via portions 86. The structure is then completed as described in the first embodiment.

*

A metal interconnect fabrication process is disclosed in US-A-5384267 (Texas Instruments Incorporated, USA, 24.01.95). A metal layer and a photoresist layer are formed on an array of HgCdTe detectors. The photoresist layer is patterned to form a positive mask and the metal interconnect is formed by using a dry etching technique.

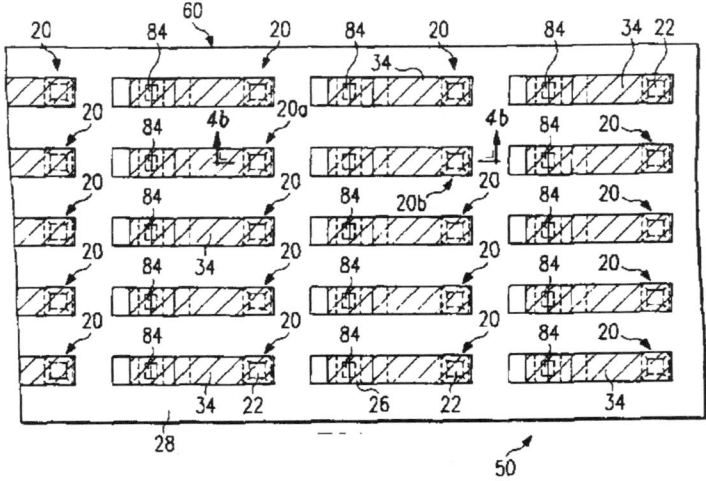

Fig. 2.5.101 (US-A-5384267 fig. 4a)

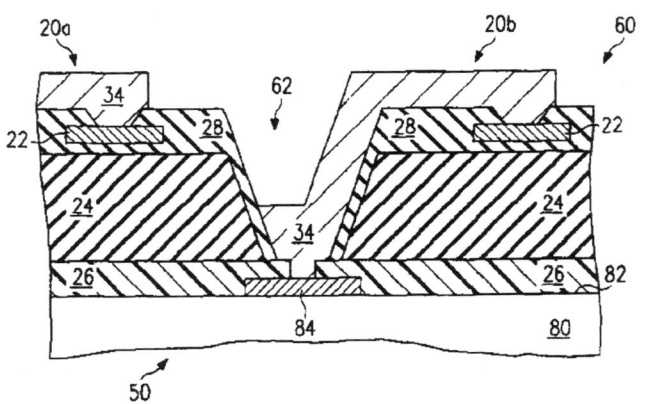

Fig. 2.5.102 (US-A-5384267 fig. 4b)

A number of HgCdTe bars 24 are mounted on surface 82 of a silicon processor 80 by the use of an epoxy or adhesive compound. Each detector element 20 comprises a storage gate 22 which is associated with a via 62 and a contact pad 84. An insulating layer 28 of ZnS is formed on the HgCdTe bars. A metal layer of indium, indium/lead/indium, aluminium, aluminium/titanium-tungsten, titanium, tungsten, titanium-tungsten alloys, tantalum, molybdenum or alloys such as titanium silicon or titanium nitride is formed over the array.

Next, a photoresist material is applied to the structure and the photoresist material is patterned to form a positve mask. The metal layer not protected by the mask is removed by dry etching thereby forming metal interconnects 34. By forming the metal layer on the array prior to applying the photoresist reduces potential surface contamination and increases the integrity of the metal bonds with the respective storage gates 22 and contact pads 84.

A method to form indium bumps for hybrid arrays using dry etching of an indium layer is also disclosed.

*

Patent Number Index

EP-A-0007667 06.02.80 92
Applicant: Philips Electronic and Associated Industries Limited, GB
Inventor: Withers R B
Priority: GB780031750 31.07.78
Family:
CA-A-1150806 26.07.83
GB-A-2027986 27.02.80
JP-A-55048981 08.04.80
JP-A-59188178 25.10.84
US-A-4301591 24.11.81
US-A-4435462 06.03.84

EP-A-0007668 06.02.80 93
Applicant: Philips Electronic and Associated Industries Limited, GB
Inventors: Baker Ian Martin, Withers Richard Breton
Priority: GB780031751 31.07.78
Family:
CA-A-1143461 22.03.83
GB-A-2027556 20.02.80
JP-A-55048952 08.04.80
US-A-4310583 12.01.82

EP-A-0007669 06.02.80 93
Applicant: Philips Electronic and Associated Industries Limited, GB
Inventors: Jenner Michael David, Blackman Maurice Victor
Priority: GB780031749 31.07.78
Family:
CA-A-1143460 22.03.83
GB-A-2027985 27.02.80
JP-A-55022896 18.02.80
JP-A-63107074 12.05.88
US-A-4321615 23.03.82

EP-A-0024970 11.03.81 202; 254
Applicant: Thomson-CSF, France
Inventors: Reboul Jean Philippe, Villard Michel
Priority: FR790021903 31.08.79
Family:
FR-A-2464563 06.03.81
JP-A-56036174 09.04.81

EP-A-0038697 28.10.81 44; 345
Applicant: Semiconductor Research Foundation, Japan
Inventors: Nishizawa Jun-ichi, Ohmi Tadahiro, Tamamushi Takashige
Priority: JP800054001 22.04.80
JP800060316 06.05.80
Family:
JP-A-56150878 21.11.81
JP-A-56157073 04.12.81

EP-A-0042218 23.12.81 45; 347
Applicant: Semiconductor Research Foundation, Japan
Inventors: Jun-ichi Nishizawa, Tadahiro Ohmi, Takashige Tamamushi
Priority: JP800069257 24.05.80
Family:
JP-A-56165473 19.12.81
US-A-4454526 12.06.84

EP-A-0044610 04.08.82 57
Applicant: Texas Instruments Incorporated, USA
Inventors: Chapman Richard A, Kinch Michael A, Hynecek Jaroslav
Priority: US800159991 16.06.80
Family:
JP-A-57030468 18.02.82
US-A-4327291 27.04.82

EP-A-0050512 28.04.82 340
Applicant: Honeywell Inc., USA
Inventor: Young Miriam F
Priority: US800198319 20.10.80
Family:
CA-A-1176763 23.10.84
US-A-4351101 28.09.82

EP-A-0061801 06.10.82 29
Applicant: N.V. Philips' Gloeilampenfabrieken, The Netherlands
Inventors: Blackburn Anthony, Readhead John Barry
Priority: GB810009818 30.03.81
Family: GB-A-2095899 06.10.82

JP-A-57170676 20.10.82
US-A-4467201   21.08.84

EP-A-0061802   06.10.82   30
Applicant: Philips Electronic and Associated
           Industries, GB
Inventors: Blackburn Anthony, Readhead
           John Barry
Priority: GB810009819   30.03.81
Family: GB-A-2095900 06.10.82
           JP-A-57173279 25.10.82
           US-A-4482807 13.11.84

EP-A-0061803   06.10.82   374
Applicant: Philips Electronic and Associated
           Industries Limited, GB
Inventor: Baker Ian Martin
Priority: GB810009778   27.03.81
Family: GB-A-2095905 06.10.82
           JP-A-57171226 21.10.82
           US-A-4521798 04.06.85
           US-A-4559695 24.12.85

EP-A-0062367   13.10.82   147; 375
Applicant: Philips Electronic and Associated
           Industries Limited, GB
Inventor: Wotherspoon John Thomas Mclean
Priority: GB810009774   27.03.81
Family: GB-A-2095898 06.10.82
           JP-A-57170577 20.10.82
           US-A-4411732 25.10.83

EP-A-0064918   17.11.82   148
Applicant: Thomson-CSF, France
Inventors: Rebondy Jacques, Villard Michel
Priority: FR810009430   12.05.81
Family: FR-A-2506079 19.11.82
           JP-A-57197857 04.12.82

EP-A-0066020   08.12.82   15
Applicant: Texas Instruments Incorporated,
           USA
Inventor: Chapman Richard A
Priority: EP810302462   03.06.81

EP-A-0068652   05.01.83   149
Applicant: The Secretary of State for Defence
           in Her Britannic Majesty's
           Government of The United
           Kingdom of Great Britain and
           Northern Ireland Whitehall, GB
Inventors: Dean Anthony Brian, Farrow
           Robin Frederick Charles,
           Migliorato Piero, White Anthony
           Michael, Williams Gerald Martin
Priority: GB810019440   24.06.81
Family: DE-A-3278553 30.06.88
           GB-A-2100927 06.01.83
           JP-A-58002078 07.01.83
           US-A-4494133 15.01.85

EP-A-0087842   07.09.83   350
Applicant: Philips Electronic and Associated,
           GB
Inventor: Readhead John Barry
Priority: GB820006290   03.03.82
Family: GB-A-2116363 21.09.83
           JP-A-58161834 26.09.83
           US-A-4555720 26.11.85
           US-A-4625389 02.12.86

EP-A-0094973   30.11.83   47
Applicant: Semiconductor Research
           Foundation, Japan
Inventor: Nishizawa Jun-ichi
Priority: JP810192417   30.11.81
Family: JP-A-58093386 03.06.83
           US-A-4608587 26.08.86
           WO-A-8302037 09.06.83

EP-A-0094974   30.11.83   48
Applicant: Semiconductor Research
           Foundation, Japan
Inventor: Nishizawa Jun-ichi
Priority: JP810194286   01.12.81
Family: JP-A-58095877 07.06.83
           US-A-4641167 03.02.87
           WO-A-8302038 09.06.83

EP-A-0137704   17.04.85   386
Applicant: Texas Instruments Incorporated,
           USA
Inventors: Tew Claude E, Coale Cecil R Jr
Priority: US830528304   31.08.83
Family: JP-A-60155932 16.08.85

EP-A-0137988   24.04.85   384
Applicant: Texas Instruments Incorporated,
           USA
Inventors: Tew Claude E, Lewis Adam J,
           Schulte Eric F
Priority: US830528206   31.08.83
           US830528207   31.08.83
           US830528317   31.08.83
Family: EP-A-0288792 02.11.88
           JP-A-60094764 27.05.85
           US-A-4447291 08.05.84
           US-A-4684812 04.08.87
           US-A-4686373 11.08.87
           US-A-4779004 18.10.88

EP-A-0151311   14.08.85   36
Applicant: Honeywell Inc., USA
Inventors: Mortensen Craig A, McCullough
           John B
Priority: US840569012   09.01.84
Family: AU-B-565155   03.09.87
           AU-D-3737985D 18.07.85
           CA-A-1207893 15.07.86
           JP-A-60158324 19.08.85
           US-A-4531059 23.07.85

EP-A-0167305	**08.01.86**	159		**EP-A-0293216**	**30.11.88**	115
Applicant:	The Secretary of State for Defence in Her Britannic Majesty's Government of the United Kingdom of Great Britain and Northern Ireland Whitehall, GB			Applicant:	The Marconi Company Limited, GB	
				Inventor:	Singh Surendra Pratap	
				Priority:	GB870012450	27.05.87
				Family:	GB-A-2205442	07.12.88
Inventors:	Elliott Charles Thomas, Ashley Timothy			**EP-A-0317083**	**24.05.89**	320
Priority:	GB840017303	06.07.84		Applicant:	Grumman Aerospace Corporation, USA	
Family:	DE-D-3587973D	23.02.95		Inventor:	Solomon Allen L	
	JP-A-61026269	05.02.86		Priority:	US870122178	18.11.87
	US-A-5016073	14.05.91		Family:	CA-A-1296814	03.03.92
EP-A-0171801	**19.02.86**	160			DE-A-3879629	29.04.93
Applicant:	Honeywell Inc., USA				JP-A-1168041	03.07.89
Inventors:	Marciniec John W, Cochran Bruce C, Hartley Martha A				US-A-4794092	27.12.88
Priority:	US840641615	17.08.84		**EP-A-0317084**	**24.05.89**	320
Family:	JP-A-61084062	28.04.86		Applicant:	Grumman Aerospace Corporation, USA	
EP-A-0173074	**05.03.86**	357		Inventor:	Solomon Allen L	
Applicant:	Texas Instruments Incorporated, USA			Priority:	US870122177	18.11.87
				Family:	DE-A-3879109	15.04.93
Inventors:	Schulte Eric F, Lewis Adam J				JP-A-1168040	03.07.89
Priority:	US840646659	31.08.84			US-A-4784970	15.11.88
Family:	JP-A-61070755	11.04.86		**EP-A-0343738**	**29.11.89**	362
	US-A-4616403	14.10.86		Applicant:	Philips Electronic and Associated Industries Limited, GB	
	US-A-4729003	01.03.88				
EP-A-0188241	**23.07.86**	36		Inventor:	Ard Christopher Kyle	
Applicant:	Honeywell Inc., USA			Priority:	GB880012591	27.05.88
Inventors:	Reine Marion B, Kusner Ronald R			Family:	GB-A-2219132	29.11.89
Priority:	US850690908	14.01.85			JP-A-2032568	02.02.90
Family:	AU-B-571360	14.04.88		**EP-A-0366886**	**09.05.90**	388
	AU-D-5071485D	17.07.86		Applicant:	Texas Instruments Incorporated, USA	
	CA-A-1254293	16.05.89				
	JP-A-61164127	24.07.86		Inventor:	Strong Roger Lynn	
	US-A-4628203	09.12.86		Priority:	US880251726	30.09.88
EP-A-0281026	**07.09.88**	322		**EP-A-0405865**	**02.01.91**	299
Applicant:	Honeywell Inc., USA			Applicant:	Hughes Aircraft Company, USA	
Inventors:	Gurnee Mark N, White William J			Inventors:	Hu Willliam C, Longerich Ernest P, D'Agostino Saverio A	
Priority:	US870020618	02.03.87				
Family:	JP-A-63308362	15.12.88		Priority:	US890373972	30.06.89
	US-A-4807000	21.02.89		Family:	CA-A-2017743	31.12.90
EP-A-0288792	**02.11.88**	386			JP-A-3044955	26.02.91
Applicant:	Texas Instruments Incorporated, USA				US-A-5092036	03.03.92
Inventor:	Schulte Eric F			**EP-A-0406696**	**09.01.91**	262
Priority:	US830528206	31.08.83		Applicant:	Santa Barbara Research Center, USA	
	US830528207	31.08.83				
	US830528317	31.08.83		Inventors:	Rosbeck Joseph P, Cockrum Charles A	
Family:	EP-A-0137988	24.04.85				
	US-A-4447291	08.05.84		Priority:	US890375229	03.07.89
	US-A-4684812	04.08.87		Family:	DE-D-69005048D	20.01.94
	US-A-4686373	11.08.87			DE-T-69005048T	07.07.94
	US-A-4779004	18.10.88			IL-A-94636	07.10.94
					JP-A-3046279	27.02.91
					US-A-4961098	02.10.90

EP-A-0407062   09.01.91   393
Applicant: Texas Instruments Incorporated, USA
Inventors: York Rudy L, Luttmer Joseph D, Wan Chang F, Orent Thomas N, Hutchins Larry, Simmons Art
Priority: US890373951   29.06.89
Family: JP-A-3054857   08.03.91
           US-A-5132761   21.07.92
           US-A-5188970   23.02.93

EP-A-0416299   13.03.91   387
Applicant: Texas Instruments Incorporated, USA
Inventors: Wan Chang-Feng, Sulzbach Frank C
Priority: US890400603   30.08.89
Family: EP-A-0637085   01.02.95
           US-A-4968886   06.11.90

EP-A-0416320   13.03.91   389
Applicant: Texas Instruments Incorporated, USA
Inventors: Luttmer Joseph D, Kinch Michael A
Priority: US890403536   06.09.89

EP-A-0445545   11.09.91   235
Applicant: Santa Barbara Research Center, USA
Inventors: Huang Chao, Norton Paul R
Priority: US900490011   07.03.90
Family: JP-A-4217363   07.08.92
           US-A-5049962   17.09.91

EP-A-0475525   18.03.92   379
Applicant: Philips Electronic and Associated Industries Limited, GB
Inventors: Baker Ian Martin, Dunn William Andrew Ernest
Priority: GB900019897   12.09.90
Family: GB-A-2247985   18.03.92
           IL-A-99412   30.05.94
           JP-A-4234170   21.08.92
           US-A-5185648   09.02.93

EP-A-0481552   22.04.92   380
Applicant: Philips Electronics UK Limited, GB
Inventor: Baker Ian Martin
Priority: GB900022464   17.10.90
Family: GB-A-2248964   22.04.92
           IL-A-99730   07.10.94
           JP-A-4260368   16.09.92
           US-A-5239179   24.08.93

EP-A-0485115   13.05.92   239; 300
Applicant: Cincinnati Electronics Corporation, USA
Inventors: Timlin Harold A, Martin Charles J

Priority: US900609678   06.11.90
Family: CA-A-2054934   07.05.92
           IL-A-99856   29.12.94
           JP-A-4290265   14.10.92
           US-A-5227656   13.07.93
           US-A-5304500   19.04.94

EP-A-0497326   05.08.92   186; 242
Applicant: Fujitsu Limited, Japan
Inventors: Arinaga Kenji, Kajihara Nobuyuki, Sudo Gen, Fujiwara Koji, Hikida Soichiro, Ito Yuichiro
Priority: JP910012058   01.02.91
Family: JP-A-4246860   02.09.92
           US-A-5196692   23.03.93

EP-A-0510267   28.10.92   380
Applicant: N.V. Philips' Gloeilampenfabrieken, The Netherlands
Inventor: Horne Russell John
Priority: EP910303666   24.04.91

EP-A-0543537   26.05.93   189
Applicant: Fujitsu Ltd, Japan
Inventors: Awamoto Kenji, Ito Yuichiro
Priority: JP910304862   20.11.91
Family: JP-A-5145055   11.06.93
           US-A-5311006   10.05.94

EP-A-0561615   22.09.93   381
Applicant: Philips Electronics UK Limited, GB
Inventors: Matthews Brian Edward, Jenner Michael David
Priority: GB920005933   18.03.92
Family: GB-A-2265253   22.09.93
           JP-A-6021423   28.01.94
           US-A-5367166   22.11.94

FR-A-2484705   18.12.81   342
Applicant: Thomson-CSF, France
Inventor: Reboul Jean-Philippe, Lenoble Claude
Priority: FR770039610   29.12.77

FR-A-2494910   28.05.82   343
Applicant: Thomson-CSF, France
Inventors: Munier Bernard, Reboul Jean-Philippe
Priority: FR790013189   23.05.79

FR-A-2526227   04.11.83   155; 203
Applicant: Thomson-CSF, France
Inventors: Arques Marc, Berger Jean-Luc
Priority: FR820007425   29.04.82

FR-A-2593642   31.07.87   351
Applicant: Societe Anonyme de Telecommunications, France

Inventor: Gauthier André
Priority: FR820009263 27.05.82

GB-A-1337968 21.11.73 336
Applicant: Selenia Industrie Elettroniche Associate S.P.A., Italy
Inventor:
Priority: GB700057369 02.12.70

GB-A-1488258 12.10.77 23
Applicant: Secretary of State for Defence, GB
Inventor: Elliott Charles Thomas
Priority: GB740051498 27.11.74
Family: DE-A-2553378 12.08.76
FR-A-2293063 25.06.76
JP-A-51088276 02.08.76
US-A-3995159 30.11.76

GB-A-1559473 16.01.80 90
Applicant: Mullard Limited, GB
Inventor: Blackman Maurice Victor, Jenner Michael David
Priority: GB750030800 23.07.75
Family: CA-A-1059646 31.07.79

GB-A-1559474 16.01.80 91
Applicant: Mullard Limited, GB
Inventor: Blackman Maurice Victor, Jenner Michael David
Priority: GB750030806 23.07.75
Family: CA-A-1075375 08.04.80

GB-A-1568958 11.06.80 92
Applicant: Mullard Limited, GB
Inventor: Blackman Maurice Victor, Jenner Michael David
Priority: GB750030799 22.10.76

GB-A-1597581 09.09.81 144
Applicant: Selenia Industrie Elettroniche Associate S.p.A., Italy
Inventor:
Priority: IT760052606 14.12.76
Family: US-A-4312115 26.01.82

GB-A-1600599 21.10.81 96
Applicant: Philips Electronic and Associated Industries Limited, GB
Inventor: Blackman Maurice Victor, Jenner Michael David
Priority: GB770016402 31.05.78

GB-A-1605321 19.07.89 26
Applicant: Philips Electronic and Associated Industries Limited, GB
Inventor: Charlton David Eric, Elliott Charles Thomas
Priority: GB780012694 31.03.78
Family: DE-C-3125292 21.12.89
FR-A-2633730 05.01.90

US-A-4931648 05.06.90

GB-A-2007909 23.05.79 23
Applicant: The Secretary of State for Defence, GB
Inventor: Day Derek Joseph, Elliott Charles Thomas
Priority: GB770045995 04.11.77
GB780041968 25.10.78
Family: DE-A-2847778 10.05.79
FR-A-2408193 01.06.79
JP-A-54074691 14.06.79
US-A-4231052 28.10.80

GB-A-2019649 31.10.79 27
Applicant: The Secretary of State for Defence, GB
Inventor: Elliott Charles Thomas
Priority: GB780016279 25.04.78
GB790012154 06.04.79
Family: DE-A-2916770 08.11.79
FR-A-2424679 23.11.79
JP-A-54148424 20.11.79
US-A-4258254 24.03.81

GB-A-2094548 15.09.82 33
Applicant: The Secretary of State for Defence, GB
Inventor: Elliott Charles Thomas
Priority: GB810007605 11.03.81
Family: FR-A-2501942 17.09.82
JP-A-57160158 02.10.82

GB-A-2095906 06.10.82 30
Applicant: The Secretary of State for Defence, GB
Inventor: White Anthony Michael, Elliot Charles Thomas
Priority: GB810009928 30.03.81
GB820007250 12.03.82
Family: DE-A-3211769 16.12.82
FR-A-2502846 01.10.82
JP-A-57175224 28.10.82
US-A-4926228 15.05.90

GB-A-2113467 03.08.83 349
Applicant: Licentia Patent-Verwaltungs-GmbH, FRG
Inventors: Maier Dr Horst, Schulz Dr Max
Priority: DE823200853 14.01.82
Family: DE-A-3200853 21.07.83
FR-A-2519803 18.07.83
US-A-4661168 28.04.87

GB-A-2128019 18.04.84 354
Applicant: The Secretary of State for Defence, GB
Inventors: Elliott Charles Thomas, White Anthony Michael
Priority: GB820027181 23.09.82

GB-A-2132017   27.06.84   205
Applicant: The Secretary of State for Defence, GB
Inventors: Williams Gerald Martin, Gordon Neil Thomson, Dean Anthony Brian, Young Ian Malcolm
Priority: GB820035841   16.12.82
           GB830033116   12.12.83

GB-A-2146663   24.04.85   378
Applicant: The Secretary of State for Defence in Her Britannic Majesty's Government of The United Kingdom of Great Britain and Northern Ireland Whitehall, GB
Inventors: Irvine Stuart James Curzon, Mullin John Brian, Tunnicliffe Jean
Priority: GB830024531   13.09.83
Family: CA-A-1229290   17.11.87
           DE-A-3466898   26.11.87
           EP-A-0135344   27.03.85
           JP-A-60077431 02.05.85
           US-A-4566918   28.01.86

GB-A-2197984   02.06.88   392
Applicant: The Marconi Company Limited, GB
Inventors: Knowles Peter, Jenkin Graham Thomas
Priority: GB860027886   21.11.86

GB-A-2201834   07.09.88   28
Applicant: Philips Electronic and Associated Industries Limited, GB
Inventor: Dyson Clive Malcolm
Priority: GB790018368   25.05.79
Family: AU-B-570397   02.06.88
           AU-B-596782   17.05.90
           CA-A-1251548   21.03.89
           DE-A-3019481   12.01.89
           FR-A-2625369   30.06.89
           GB-A-2199986   20.07.88
           NL-A-8002890   01.08.88
           SE-A-8003777   06.08.88
           SE-A-8903287   06.10.89
           SE-B-464736   03.06.91
           SE-B-464900   24.06.91
           US-A-4801802   31.01.89

GB-A-2203757   26.10.88   378
Applicant: Philips Electronic and Associated Industries Limited, GB
Inventors: Easton Brian Colin, Whiffin Peter Arthur Charles
Priority: GB870009185   16.04.87
Family: EP-A-0288108   26.10.88
           JP-A-63283135 21.11.88
           US-A-4874634   17.10.89

GB-A-2207801   08.02.89   31
Applicant: The Secretary of State for Defence, GB
Inventors: Dean Anthony Brian, Dennis Peter Neil John, Elliott Charles Thomas
Priority: GB790026455   30.07.79
Family: DE-A-3028915   03.05.89
           FR-A-2625398   30.06.89
           NL-A-8004275   02.01.89
           NL-B-190883   02.05.94
           US-A-4883962   28.11.89
           US-A-5321290   14.06.94

GB-A-2207802   08.02.89   33; 382
Applicant: Philips Electronic and Associated Industries Limited, GB
Inventor: Wotherspoon John Thomas Mclean
Priority: GB820024643   27.08.82
Family: DE-A-3330673   10.08.89
           FR-A-2626109   21.07.89
           US-A-4859851   22.08.89

GB-A-2208256   15.03.89   375
Applicant: Philips Electronic and Associated Industries Limited, GB
Inventor: Jackson John
Priority: GB830010279   15.04.83

GB-A-2229036   12.09.90   210
Applicant: Mitsubishi Denki Kabushiki Kaisha, Japan
Inventor: Hisa Yoshihiro
Priority: JP890058763   10.03.89
Family: FR-A-2644293   14.09.90
           JP-A-2237154   19.09.90
           US-A-5075748   24.12.91
           US-A-5156980   20.10.92

GB-A-2231199   07.11.90   117; 365
Applicant: Philips Electronic and Associated Industries Limited, GB
Inventors: Jenner Michael David, Blackman Maurice Victor, Crimes Graham Joseph
Priority: GB890008219   12.04.89

GB-A-2238165   22.05.91   175
Applicant: Mitsubishi Denki Kabushiki Kaisha, Japan
Inventor: Hisa Yoshihiro
Priority: JP890285221   31.10.89
Family: FR-A-2653937   03.05.91
           JP-A-3145763   20.06.91
           US-A-5115295   19.05.92

GB-A-2239555   03.07.91   377
Applicant: Philips Electronic and Associated Industries Limited, GB
Inventor: Baker Ian Martin
Priority: GB890004611   01.03.89

GB-A-2241377   28.08.91   108
Applicant: Philips Electronics and Associated
            Industries Limited, GB
Inventors: Charlton David Eric, Jenner
            Michael David, King Raymond
            Edward John
Priority: GB870012132   22.05.87

GB-A-2241605   04.09.91   108
Applicant: Philips Electronic and Associated
            Industries Limited, GB
Inventor: Jenner Michael David
Priority: GB850022539   11.09.85

GB-A-2246662   05.02.92   181
Applicant: Mitsubishi Denki Kabushiki
            Kaisha, Japan
Inventor: Yoshida Yasuaki
Priority: JP900201768   30.07.90
Family: FR-A-2665302   31.01.92
            JP-A-4085961   18.03.92

GB-A-2246907   12.02.92   264
Applicant: Mitsubishi Denki Kabushiki
            Kaisha, Japan
Inventor: Takiguchi Tohru
Priority: JP900211159   07.08.90
Family: FR-A-2665800   14.02.92
            JP-A-4092481   25.03.92
            US-A-5187378   16.02.93

GB-A-2260218   07.04.93   35
Applicant: The Secretary of State for Defence,
            GB
Inventors: Brewitt Taylor Colin Raymond,
            Elliot Charles Thomas, Rees Huw
            David, White Anthony Michael
Priority: GB830027208   11.10.83
Family: DE-A-3437334   08.07.93
            FR-A-2689685   08.10.93
            NL-A-8403092   01.04.93
            US-A-5248884   28.09.93

GB-A-2261323   12.05.93   81
Applicant: Mitsubishi Denki Kabushiki
            Kaisha, Japan
Inventor: Hisa Yoshihiro
Priority: JP910321317   06.11.91
Family: FR-A-2683391   07.05.93
            JP-A-5267695   15.10.93
            US-A-5410168   25.04.95

GB-A-2261767   26.05.93   191
Applicant: Mitsubishi Denki Kabushiki
            Kaisha, Japan
Inventor: Takami Akihiro
Priority: JP910334323   20.11.91
Family: JP-A-5145093   11.06.93
            US-A-5229321   20.07.93

GB-A-2274739   03.08.94   182
Applicant: Mitsubishi Denki Kabushiki
            Kaisha, Japan
Inventor: Yoshida Yasuaki
Priority: JP930013608   29.01.93
Family: FR-A-2701164   05.08.94
            JP-A-6232234   19.08.94
            US-A-5371352   06.12.94

GB-A-2284930   21.06.95   252
Applicant: Mitsubishi Denki Kabushiki
            Kaisha
Inventors: Mitsui Kotaro, Kawazu Zempei,
            Mizuguchi Kazuo, Ochi Seiji,
            Ohkura Yuji, Hayafuji Norio,
            Kizuki Hirotaka, Tsugami Mari,
            Takami Akihiro
Priority: JP930147162   18.06.93
Family: JP-A-7014996   17.01.95

JP-A-55102280 05.08.80   13
Applicant: Fujitsu Ltd, Japan
Inventors: Takigawa Hiroshi, Doi Shiyouji,
            Imai Souichi
Priority: JP790010213   30.01.79

JP-A-55150279 22.11.80   289
Applicant: Fujitsu Ltd, Japan
Inventors: Imai Souichi, Takigawa Hiroshi,
            Doi Shiyouji
Priority: JP790057723   10.05.79

JP-A-57010986 20.01.82   59
Applicant: Fujitsu Ltd, Japan
Inventor: Miyamoto Yoshihiro
Priority: JP800086865   25.06.80

JP-A-57024580 09.02.82   147
Applicant: Fujitsu Ltd, Japan
Inventors: Itou Makoto, Imai Souichi
Priority: JP800099170   18.07.80

JP-A-57031170 19.02.82   348
Applicant: Fujitsu Ltd, Japan
Inventors: Tanigawa Kunihiro, Takigawa
            Hiroshi
Priority: JP800105934   31.07.80

JP-A-57062563 15.04.82   13
Applicant: Fujitsu Ltd, Japan
Inventors: Takigawa Hiroshi, Ueda Tomoshi,
            Maekawa Tooru
Priority: JP800137885   30.09.80

JP-A-57073984 08.05.82   276
Applicant: Fujitsu Ltd, Japan
Inventors: Yoshikawa Mitsuo, Hamashima
            Shigeki, Takigawa Hiroshi, Itou
            Michiharu, Ueda Tomoshi
Priority: JP800150929   27.10.80

JP-A-57091557   07.06.82   98
Applicant: Fujitsu Ltd, Japan
Inventors: Itou Makoto, Imai Souichi
Priority: JP800168258   28.11.80

JP-A-57169278   18.10.82   14
Applicant: Mitsubishi Denki KK, Japan
Inventors: Nagahama Kouki, Nishitani Kazuo, Ishii Jiyun
Priority: JP810054645   10.04.81

JP-A-58164261   29.09.83   154
Applicant: Tokyo Shibaura Denki KK, Japan
Inventor: Nagasaka Hiroo
Priority: JP820046245   25.03.82

JP-A-58171848   08.10.83   16
Applicant: Fujitsu KK, Japan
Inventors: Takigawa Hiroshi, Yoshikawa Mitsuo, Ueda Tomoshi
Priority: JP820054578   31.03.82

JP-A-59047759   17.03.84   62
Applicant: Fujitsu KK, Japan
Inventors: Maekawa Tooru, Tanigawa Kunihiro
Priority: JP820158408   10.09.82

JP-A-59061080   07.04.84   63
Applicant: Fujitsu K.K., Japan
Inventor: Sakai Hiroshi
Priority: JP820171781   29.09.82

JP-A-59112652   29.06.84   279
Applicant: Fujitsu KK, Japan
Inventors: Nomura Shiyouji, Tanigawa Kunihiro
Priority: JP820222498   17.12.82

JP-A-60028266   13.02.85   103
Applicant: Fujitsu KK, Japan
Inventors: Itou Makoto, Gotou Jiyunjirou, Koseto Masaru, Rokushiya Kiyoshi
Priority: JP830138063   27.07.83

JP-A-60074568   26.04.85   63
Applicant: Fujitsu KK, Japan
Inventor: Miyamoto Yoshihiro
Priority: JP830180582   30.09.83

JP-A-60102530   06.06.85   159
Applicant: Yokokawa Hokushin Denki KK, Japan
Inventor: Takeuchi Youji
Priority: JP830210065   09.11.83

JP-A-60103666   07.06.85   64
Applicant: Fujitsu KK, Japan
Inventor: Miyamoto Yoshihiro
Priority: JP830211235   10.11.83

JP-A-60140869   25.07.85   17
Applicant: Fujitsu KK, Japan
Inventors: Hikita Souichirou, Kajiwara Nobuyuki
Priority: JP830250072 28.12.83

JP-A-60180162   13.09.85   64
Applicant: Fujitsu KK, Japan
Inventors: Ueda Tomoshi, Takigawa Hiroshi, Yoshikawa Mitsuo, Itou Michiharu, Saitou Tetsuo
Priority: JP840037135   27.02.84

JP-A-60182766   18.09.85   75
Applicant: Fujitsu KK, Japan
Inventors: Ueda Tomoshi, Takigawa Hiroshi, Yoshikawa Mitsuo
Priority: JP840037925   29.02.84

JP-A-60193375   01.10.85   65
Applicant: Fujitsu KK, Japan
Inventors: Ueda Tomoshi, Takigawa Hiroshi, Yoshikawa Mitsuo, Itou Michiharu, Maruyama Kenji
Priority: JP840050500   15.03.84

JP-A-61059771   27.03.86   290
Applicant: Fujitsu Ltd, Japan
Inventors: Yamamoto Toshiro, Watanabe Shuji, Ito Yuichiro
Priority: JP840180359   31.08.84

JP-A-61067958   08.04.86   66
Applicant: Fujitsu Ltd, Japan
Inventors: Yoshikawa Mitsuo, Ito Michiharu, Maruyama Kenji
Priority: JP840190809   12.09.84

JP-A-61128564   16.06.86   358
Applicant: Fujitsu Ltd, Japan
Inventor: Maruyama Kenji
Priority: JP840250758   28.11.84

JP-A-61147118   04.07.86   162
Applicant: Fujitsu Ltd, Japan
Inventors: Koseto Masaru, Goto Junjiro, Rokushiya Kiyoshi, Nomura Shoji
Priority: JP840270005   20.12.84

JP-A-61198787   03.09.86   205
Applicant: Fujitsu Ltd, Japan
Inventors: Maekawa Toru, Takigawa Hiroshi
Priority: JP850039098   28.02.85

JP-A-61214462   24.09.86   163
Applicant: NEC Corp., Japan
Inventor: Maejima Yukihiko
Priority: JP850053478   19.03.85

JP-A-61220459 30.09.86   107
Applicant: Fujitsu Ltd, Japan
Inventors: Hamashima Shigeki, Ito Makoto,
           Yamamoto Kosaku, Kawachi
           Tetsuya
Priority: JP850062443   27.03.85

JP-A-61222161 02.10.86   164; 206
Applicant: Fujitsu Ltd, Japan
Inventors: Nomura Shoji, Goto Junjiro,
           Rokushiya Kiyoshi, Koseto
           Masaru, Shibata Shusaku
Priority: JP850064506   27.03.85

JP-A-61251167 08.11.86   165
Applicant: Fujitsu Ltd, Japan
Inventors: Hamashima Shigeki, Maekawa
           Toru
Priority: JP850092506   30.04.85

JP-A-61268075 27.11.86   255
Applicant: Fujitsu Ltd, Japan
Inventor: Maekawa Toru
Priority: JP850109429   23.05.85

JP-A-62011265 20.01.87   165; 280
Applicant: Toshiba Corp., Japan
Inventor: Naruse Yujiro
Priority: JP850149241   09.07.85

JP-A-62013085 21.01.87   166; 281
Applicant: Fujitsu Ltd, Japan
Inventors: Sakai Hiroshi, Tanigawa Kunihiro
Priority: JP850152880   11.07.85

JP-A-62022474 30.01.87   359
Applicant: Toshiba Corp., Japan
Inventor: Naruse Yujiro
Priority: JP850161193   23.07.85

JP-A-62036858 17.02.87   207
Applicant: Fujitsu Ltd, Japan
Inventor: Maekawa Toru
Priority: JP850176438   10.08.85

JP-A-62071270 01.04.87   360
Applicant: NEC Corp., Japan
Inventor: Yamagata Toshio
Priority: JP850211538   24.09.85

JP-A-62090986 25.04.87   256
Applicant: NEC Corp., Japan
Inventor: Maejima Yukihiko
Priority: JP850231609   17.10.85

JP-A-62104163 14.05.87   208
Applicant: Fujitsu Ltd, Japan
Inventors: Kawachi Tetsuya, Maekawa Toru,
           Ito Makoto, Ueda Tomoshi
Priority: JP850244428   31.10.85

JP-A-62165973 22.07.87   257
Applicant: Fujitsu Ltd, Japan
Inventor: Maekawa Toru
Priority: JP860008409   17.01.86

JP-A-62224982 02.10.87   258
Applicant: Fujitsu Ltd, Japan
Inventors: Miyamoto Yoshihiro, Hikita
           Soichiro
Priority: JP860070030   27.03.86

JP-A-62224983 02.10.87   259
Applicant: Fujitsu Ltd, Japan
Inventors: Miyamoto Yoshihiro, Hikita
           Soichiro
Priority: JP860070029   27.03.86

JP-A-62272564 26.11.87   281
Applicant: Fujitsu Ltd, Japan
Inventors: Hikita Soichiro, Tanigawa
           Kunihiro
Priority: JP860116414   20.05.86

JP-A-63005560 11.01.88   18
Applicant: NEC Corp., Japan
Inventor: Maejima Yukihiko
Priority: JP860150385   25.06.86

JP-A-63043366 24.02.88   209
Applicant: Fujitsu Ltd, Japan
Inventors: Hikita Soichiro, Maekawa Toru
Priority: JP860187508   08.08.86

JP-A-63046765 27.02.88   19
Applicant: NEC Corp., Japan
Inventor: Maejima Yukihiko
Priority: JP860191385   15.08.86

JP-A-63070454 30.03.88   112
Applicant: Fujitsu Ltd, Japan
Inventor: Ito Makoto
Priority: JP860215274   11.09.86

JP-A-63116459 20.05.88   361
Applicant: Mitsubishi Electric Corp., Japan
Inventor: Hisa Yoshihiro
Priority: JP860262385   04.11.86

JP-A-63133580 06.06.88   210
Applicant: Fujitsu Ltd, Japan
Inventors: Hikita Soichiro, Miyamoto
           Yoshihiro
Priority: JP860281341   25.11.86

JP-A-63147366 20.06.88   359
Applicant: NEC Corp., Japan
Inventor: Maejima Yukihiko
Priority: JP860295346   10.12.86

JP-A-63148677 21.06.88 114
Applicant: Fujitsu Ltd, Japan
Inventor: Hamashima Shigeki
Priority: JP860296108 11.12.86

JP-A-63150976 23.06.88 167
Applicant: Fujitsu Ltd, Japan
Inventor: Sudo Hajime
Priority: JP860297382 12.12.86

JP-A-63170960 14.07.88 211
Applicant: Fujitsu Ltd, Japan
Inventors: Yamamoto Kosaku, Ito Michiharu,
           Hirota Koji, Hamashima Shigeki,
           Ueda Tomoshi
Priority: JP870002808 08.01.87

JP-A-63170961 14.07.88 282
Applicant: Fujitsu Ltd, Japan
Inventor: Sudo Hajime
Priority: JP870002805 08.01.87

JP-A-63229751 26.09.88 212
Applicant: Fujitsu Ltd, Japan
Inventors: Yamamoto Kosaku, Ito Michiharu,
           Hirota Koji
Priority: JP870062441 19.03.87

JP-A-63260171 27.10.88 168
Applicant: Fujitsu Ltd, Japan
Inventors: Maruyama Kenji, Miyamoto
           Yoshihiro
Priority: JP870095540 17.04.87

JP-A-63268271 04.11.88 291
Applicant: Fujitsu Ltd, Japan
Inventor: Sudo Hajime
Priority: JP870102867 24.04.87

JP-A-63273365 10.11.88 212
Applicant: Mitsubishi Electric Corp., Japan
Inventor: Okata Ryoji
Priority: JP870108033 30.04.87

JP-A-63281460 17.11.88 213; 292
Applicant: Fujitsu Ltd, Japan
Inventors: Hikita Soichiro, Ueda Tomoshi,
           Miyamoto Yoshihiro
Priority: JP870117780 13.05.87

JP-A-63296272 02.12.88 214; 293
Applicant: Fujitsu Ltd, Japan
Inventors: Hikita Soichiro, Ueda Tomoshi,
           Miyamoto Yoshihiro
Priority: JP870132955 27.05.87

JP-A-63300559 07.12.88 283
Applicant: NEC Corp., Japan
Inventor: Maejima Yukihiko
Priority: JP870137282 29.05.87

JP-A-63310165 19.12.88 362
Applicant: Nikon Corp., Japan
Inventor: Udagawa Kenji
Priority: JP870145322 12.06.87

JP-A-1050557 27.02.89 170
Applicant: Fujitsu Ltd, Japan
Inventors: Kawachi Tetsuya, Maekawa Toru,
           Hamashima Shigeki, Ueda
           Tomoshi, Tanaka Masahiro
Priority: JP870208512 21.08.87

JP-A-1050560 27.02.89 283
Applicant: Fujitsu Ltd, Japan
Inventors: Hikita Soichiro, Sudo Hajime,
           Tanigawa Kunihiro
Priority: JP870208513 21.08.87

JP-A-1061055 08.03.89 393
Applicant: NEC Corp., Japan
Inventor: Maejima Yukihiko
Priority: JP870218825 31.08.87

JP-A-1061056 08.03.89 295
Applicant: NEC Corp., Japan
Inventor: Maejima Yukihiko
Priority: JP870219461 01.09.87

JP-A-1201971 14.08.89 220
Applicant: Fujtisu Ltd, Japan
Inventors: Hikita Soichiro, Miyamoto
           Yoshihiro
Priority: JP880025908 05.02.88

JP-A-1205476 17.08.89 220
Applicant: Fujitsu Ltd, Japan
Inventor: Yoshikawa Mitsuo
Priority: JP880029253 10.02.88

JP-A-1218062 31.08.89 222
Applicant: Fujitsu Ltd, Japan
Inventors: Wakayama Hiroyuki, Ito Yuichiro
Priority: JP880044801 26.02.88

JP-A-1227472 11.09.89 285
Applicant: Fujitsu Ltd, Japan
Inventors: Wakayama Hiroyuki, Ito Yuichiro
Priority: JP880054524 07.03.88

JP-A-1228180 12.09.89 221
Applicant: Fujitsu Ltd, Japan
Inventor: Yoshikawa Mitsuo
Priority: JP880055799 08.03.88

JP-A-1233777 19.09.89 222
Applicant: Fujitsu Ltd, Japan
Inventors: Ueda Toshiyuki, Ito Michiharu,
           Maekawa Toru, Hikita Soichiro,
           Miyamoto Yoshihiro
Priority: JP880060778 14.03.88

Patent Number Index

JP-A-2009180   12.01.90   173
Applicant: Mitsubishi Electric Corp., Japan
Inventor: Yoshida Yasuaki
Priority: JP880160273   28.06.88

JP-A-2086177   27.03.90   225
Applicant: Fujitsu Ltd, Japan
Inventors: Miyamoto Yoshihiro, Hikita Soichiro
Priority: JP880238331   22.09.88

JP-A-2155269   16.06.90   226
Applicant: Fujitsu Ltd, Japan
Inventors: Maekawa Toru, Tanaka Masahiro, Ueda Toshiyuki, Hashimoto Arihiro
Priority: JP880310779   07.12.88

JP-A-2213174   24.08.90   227
Applicant: Mitsubishi Electric Corp., Japan
Inventor: Takiguchi Toru
Priority: JP890034186   13.02.89

JP-A-2214159   27.08.90   215; 293
Applicant: Mitsubishi Electric Corp., Japan
Inventor: Hisa Yoshihiro
Priority: JP890034381   14.02.89

JP-A-2248077   03.10.90   227
Applicant: NEC Corp., Japan
Inventor: Maejima Yukihiko
Priority: JP890067660   22.03.89

JP-A-2272765   07.11.90   378
Applicant: NEC Corp., Japan
Inventor: Maejima Yukihiko
Priority: JP890092913   14.04.89

JP-A-2272766   07.11.90   78
Applicant: Fujitsu Ltd, Japan
Inventors: Takigawa Hiroshi, Miyamoto Yoshihiro
Priority: JP890095013   13.04.89

JP-A-2303160   17.12.90   228
Applicant: Fujitsu Ltd, Japan
Inventors: Wakayama Hiroyuki, Sudo Hajime
Priority: JP890126441   18.05.89

JP-A-3077373   02.04.91   229
Applicant: Fujitsu Ltd, Japan
Inventors: Yamamoto Kosaku, Kawachi Tetsuya, Saito Tetsuo, Yamamoto Tamotsu, Maekawa Toru
Priority: JP890213933   19.08.89

JP-A-3085762   10.04.91   230
Applicant: Fujitsu Ltd, Japan
Inventor: Arinaga Kenji
Priority: JP890224037   29.08.89

JP-A-3104278   01.05.91   113
Applicant: Mitsubishi Electric Corp, Japan
Inventor: Komine Yoshiharu
Priority: JP890243000   19.09.89

JP-A-3108371   08.05.91   231
Applicant: Fujitsu Ltd, Japan
Inventors: Ueda Toshiyuki, Saito Tetsuo, Tanaka Masahiro, Yamamoto Tamotsu
Priority: JP890247317   21.09.89

JP-A-3133181   06.06.91   232
Applicant: Fujitsu Ltd, Japan
Inventors: Ueda Toshiyuki, Tanaka Masahiro, Maekawa Toru
Priority: JP890272309   18.10.89

JP-A-3175682   30.07.91   230
Applicant: Mitsubishi Electric Corp., Japan
Inventors: Yasumura Kenji, Takada Yutaka, Fukita Muneyoshi
Priority: JP890316030   04.12.89

JP-A-3196568   28.08.91   221
Applicant: Mitsubishi Electric Corp., Japan
Inventor: Takiguchi Toru
Priority: JP890337861   25.12.89

JP-A-3219670   27.09.91   232
Applicant: Fujitsu Ltd, Japan
Inventors: Kawachi Tetsuya, Maekawa Toru, Yamamoto Kosaku, Saito Tetsuo, Sugiyama Iwao
Priority: JP900015312   24.01.90

JP-A-3241774   28.10.91   233
Applicant: Fujitsu Ltd, Japan
Inventors: Ozaki Kazuo, Yamamoto Kosaku, Ueda Toshiyuki, Saito Tetsuo, Maekawa Toru
Priority: JP900039233   19.02.90

JP-A-3268463   29.11.91   235
Applicant: Fujitsu Ltd, Japan
Inventors: Sudo Hajime, Hikita Soichiro, Saito Tetsuo
Priority: JP900068881   19.03.90

JP-A-3270269   02.12.91   236
Applicant: Fujitsu Ltd, Japan
Inventors: Daiku Hiroshi, Hikita Soichiro, Sudo Hajime, Saito Tetsuo
Priority: JP900071441   20.03.90

JP-A-4133363   07.05.92   183
Applicant: Mitsubishi Electric Corp., Japan
Inventor: Komine Yoshiharu
Priority: JP900258463   25.09.90

JP-A-4199877   21.07.92   301
Applicant: Mitsubishi Electric Corp., Japan
Inventor: Komine Yoshiharu
Priority: JP900336131   29.11.90

JP-A-4253344   09.09.92   184
Applicant: Fujitsu Ltd, Japan
Inventor: Goto Junijiro
Priority: JP910026648   29.01.91

JP-A-4267562   24.09.92   378
Applicant: NEC Corp., Japan
Inventor: Ajisawa Akira
Priority: JP910050705   22.02.91

JP-A-4269869   25.09.92   378
Applicant: NEC Corp., Japan
Inventor: Ajisawa Akira
Priority: JP910053928   26.02.91

JP-A-4288882   13.10.92   243
Applicant: Mitsubishi Electric Corp., Japan
Inventor: Takiguchi Toru
Priority: JP910040130   06.03.91

JP-A-4293240   16.10.92   288
Applicant: Mitsubishi Electric Corp., Japan
Inventor: Yasunaga Masatoshi
Priority: JP910083399   20.03.91

JP-A-4313267   05.11.92   244
Applicant: Mitsubishi Electric Corp., Japan
Inventor: Fukita Muneyoshi, Yasumura
           Kenji
Priority: JP910070857   03.04.91

JP-A-4318970   10.11.92   245
Applicant: Mitsubishi Electric Corp., Japan
Inventor: Endo Yutaro
Priority: JP910085162   17.04.91

JP-A-4318979   10.11.92   265
Applicant: NEC Corp., Japan
Inventor: Ajisawa Akira
Priority: JP910085323   17.04.91

JP-A-4337676   25.11.92   246
Applicant: Mitsubishi Electric Corp., Japan
Inventor: Endo Yutaro
Priority: JP910109960   15.05.91

JP-A-4337677   25.11.92   247
Applicant: Mitsubishi Electric Corp., Japan
Inventor: Endo Yutaro
Priority: JP910109961   15.05.91

JP-A-5055620   05.03.93   247
Applicant: Mitsubishi Electric Corp., Japan
Inventor: Kawatsu Yoshihei
Priority: JP910215007   27.08.91

JP-A-5129580   25.05.93   249
Applicant: Fujitsu Ltd, Japan
Inventor: Sugiyama Iwao
Priority: JP910286938   01.11.91

JP-A-5175476   13.07.93   248
Applicant: Fujitsu Ltd, Japan
Inventors: Ozaki Kazuo, Yamamoto Kosaku,
           Yamamoto Tamotsu
Priority: JP910344282   26.12.91

JP-A-5226626   03.09.93   249
Applicant: Fujitsu Ltd, Japan
Inventors: Ueda Toshiyuki, Miyamoto
           Yoshihiro, Yamamoto Tamotsu
Priority: JP920023682   10.02.92

JP-A-5315580   26.11.93   371
Applicant: Fujitsu Ltd, Japan
Inventor: Sugiyama Iwao
Priority: JP920113333   06.05.92

JP-A-5343727   24.12.93   250
Applicant: NEC Corp., Japan
Inventor: Oikawa Ryuichi
Priority: JP920145181   05.06.92

JP-A-5343729   24.12.93   394
Applicant: NEC Corp., Japan
Inventor: Ajisawa Akira
Priority: JP920153167   12.06.92

JP-A-6013642   21.01.94   194
Applicant: Fujitsu Ltd, Japan
Inventors: Kajiwara Nobuyuki, Sudo Hajime,
           Arinaga Kenji, Fujiwara Koji,
           Nakamura Hiroko
Priority: JP920166474   25.06.92

JP-A-6021419   28.01.94   372
Applicant: Fujitsu Ltd, Japan
Inventors: Nishino Hiroshi, Sugiyama Iwao
Priority: JP920174610   02.07.92

JP-A-6089991   29.03.94   372
Applicant: Toshiba Corp., Japan
Inventors: Shigenaka Keitaro, Sugiura Risa
Priority: JP920240663   09.09.92

JP-A-6125108   06.05.94   250
Applicant: Fujitsu Ltd, Japan
Inventors: Sudo Hajime
Priority: JP920276083   14.10.92

JP-A-6163865   10.06.94   304
Applicant: Fujitsu Ltd, Japan
Inventors: Awamoto Kenji, Sakachi Yoichiro,
           Doi Shoji
Priority: JP920316671   26.11.92

Patent Number Index

JP-A-6163969   10.06.94   251; 267
Applicant: Fujitsu Ltd, Japan
Inventors: Ueda Toshiyuki, Hikita Soichiro, Watanabe Yoshio, Ozaki Kazuo, Tanaka Masahiro
Priority: JP920309124   19.11.92

JP-A-6204448   22.07.94   394
Applicant: NEC Corp., Japan
Inventor: Miyamoto Keiji
Priority: JP920348172   28.12.92

JP-A-6204449   22.07.94   252
Applicant: Fujitsu Ltd, Japan
Inventors: Fujiwara Koji, Sudo Hajime, Nakamura Hiroko, Arinaga Kenji, Kajiwara Nobuyuki
Priority: JP920347440   28.12.92

JP-A-6209096   26.07.94   197
Applicant: Fujitsu Ltd, Japan
Inventors: Nakamura Hiroko, Sudo Hajime, Arinaga Kenji, Fujiwara Koji, Kajiwara Nobuyuki.
Priority: JP930002300   11.01.93

JP-A-6236981   23.08.94   308
Applicant: Fujitsu Ltd, Japan
Inventors: Awamoto Kenji, Sakachi Yoichiro, Doi Shoji
Priority: JP930021592   10.02.93

JP-A-6237005   23.08.94   267
Applicant: Fujitsu Ltd, Japan
Inventors: Yamamoto Kosaku, Miyamoto Yoshihiro, Hikita Soichiro, Watanabe Yoshio
Priority: JP930020450   09.02.93

JP-A-6326342   25.11.94   373
Applicant: NEC Corp., Japan
Inventor: Ajisawa Akira
Priority: JP930112570   14.05.93

JP-A-7050330   21.02.95   289
Applicant: Fujitsu Ltd, Japan
Inventors: Fujiwara Koji, Kajiwara Nobuyuki, Arinaga Kenji, Sudo Hajime, Nakamura Hiroko
Priority: JP930194509   05.08.93

JP-A-7079008   20.03.95   201
Applicant: Fujitsu Ltd, Japan
Inventor:
Priority: JP930160947   30.06.93

JP-A-7094693   07.04.95   201
Applicant: Toshiba KK, Japan
Inventor:
Priority: JP930237060   24.09.93

JP-A-7111323   25.04.95   305
Applicant: NEC Corp., Japan
Inventor:
Priority: JP930253859   12.10.93

US-A-3806729   23.04.74   6
Applicant: Texas Instruments Incorporated, USA
Inventor: Caywood John Millard
Priority: US730355612   30.04.73

US-A-3808435   30.04.74   138; 274
Applicant: Texas Instruments Incorporated, USA
Inventors: Bate Robert T, Kinch Michael A, Buss Dennis D
Priority: US730365294   29.05.73

US-A-3842274   15.10.74   338
Applicant: The United States of America as represented by the Secretary of the Navy, USA
Inventors: Scharnhorst K Peter, Greene Richard F, Schoolar Richard B
Priority: US730416300   15.11.73

US-A-3852714   03.12.74   312
Applicant: Eocom Corporation, USA
Inventor: Carson John C
Priority: US720265144   22.06.72
Family: US-A-3970990   20.07.76

US-A-3902924   75.09.02   140
Applicant: Honeywell Inc., USA
Inventors: Maciolek Ralph B, Speerschneider Charles J
Priority: US730393264   30.08.73

US-A-3930161   30.12.75   138
Applicant: Societe Anonyme de Telecommunications, France
Inventors: Ameurlaine Jacques Francois, Sirieix Michel Benoit
Priority: FR730012067   04.04.73
Family: DE-A-2413256   10.10.74
        FR-A-2224748   31.10.74
        GB-A-1463611   02.02.77
        NL-A-7404036   08.10.74

US-A-3949223   06.04.76   141
Applicant: Honeywell Inc., USA
Inventors: Schmit Joseph L, Stelzer Ernest L
Priority: US730411970   01.11.73

US-A-3965568   29.06.76   89
Applicant: Texas Instruments Incorporated, USA
Inventor: Gooch Roland W
Priority: US730392148   27.08.73
        US750583927   05.06.75

US-A-3970990   20.07.76   315
Applicant: Grumman Aerospace Corporation, USA
Inventor: Carson John C
Priority: US720265144   22.06.72
            US740522476   11.11.74
Family: US-A-3852714   03.12.74

US-A-3980915   14.09.76   143
Applicant: Texas Instruments Incorporated, USA
Inventors: Chapman Richard A, Johnson Milo R, Morris Henry B
Priority: US740446185   27.02.74

US-A-3988774   26.10.76   253
Applicant: Societe Anonyme de Telecommunication, France
Inventors: Cohen-Solal Gerard David, Lussereau Alain Gilles
Priority: FR740027282   06.08.74
Family: DE-A-2517939   26.02.76
            FR-A-2281650   05.03.76
            GB-A-1508027   19.04.78
            JP-A-51036088 26.03.76
            NL-A-7506863   10.02.76

US-A-4025793   24.05.77   94
Applicant: Santa Barbara Research Center, USA
Inventors: Shaw John B, Blatt Peter R, Gesswein Francis I
Priority: US750623682   20.10.75

US-A-4081819   28.03.78   145
Applicant: Honeywell Inc., USA
Inventor: Wong Theodore T S
Priority: US770759922   17.01.77

US-A-4104674   01.08.78   339
Applicant: Honeywell Inc., USA
Inventors: Lorenze Jr Robert V, White William Joseph
Priority: US770766327   07.02.77
Family: US-A-4188709   19.02.80

US-A-4137625   06.02.79   341
Applicant: Honeywell Inc., USA
Inventor: White William J
Priority: US770829744   01.09.77
Family: US-A-4206470   03.06.80

US-A-4162507   24.07.79   95
Applicant: Licentia Patent-Verwaltungs G.m.b.H, FRG
Inventor: Fisher Konrad
Priority: DE772702571   22.01.77
Family: DE-A-2702571   27.07.78
            JP-A-53092683 14.08.78

US-A-4197633   15.04.80   340
Applicant: Honeywell Inc., USA
Inventors: Lorenze Robert V Jr, White William J
Priority: US770829745   01.09.77
Family: US-A-4286278   25.08.81

US-A-4228365   14.10.80   8
Applicant: The United States of America as represented by the Secretary of the Army, USA
Inventors: Gutierrez William A, Pollard John H
Priority: US780948129   03.10.78
Family: US-A-4273596   16.06.81

US-A-4229752   21.10.80   7
Applicant: Texas Instruments Incorporated, USA
Inventor: Hynecek Jaroslav
Priority: US780906385   16.05.78
Family: DE-A-2919522   22.11.79
            GB-A-2021313   28.11.79
            JP-A-55011394 26.01.80
            JP-A-57164567 09.10.82
            US-A-4994875   19.02.91

US-A-4231149   04.11.80   12
Applicant: Texas Instruments Incorporated, USA
Inventors: Chapman Richard A, Buss Dennis D, Kinch Michael A
Priority: US780950191   10.10.78
Family: US-A-4377904   29.03.83

US-A-4242149   30.12.80   146
Applicant: The United States of America as represented by the Secretary of the Army, USA
Inventors: Dunn Aubrey J, King Gerard J
Priority: US790062593   31.07.79

US-A-4259576   31.03.81   53
Applicant: The United States of America as represented by the Secretary of the Navy, USA
Inventors: Gridley C John, Weinberg Donald L
Priority: US790067646   17.08.79

US-A-4273596   16.06.81   10
Applicant: The United States of America as represented by the Secretary of the Army, USA
Inventors: Gutierrez William A, Pollard John H
Priority: US780948129   03.10.78
            US800109025   02.01.80
Family: US-A-4228365   14.10.80

US-A-4290844 22.09.81 317	**US-A-4429330** 31.01.84 73
Applicant: Carson Alexiou Corporation, USA	Applicant: Texas Instruments Incorporated, USA
Inventors: Rotolante Ralph A, Koehler Toivo	Inventor: Chapman Richard A
Priority: US790015070 26.02.79	Priority: US800141498 18.04.80
Family: US-A-4449044 15.05.84	

US-A-4290844 22.09.81   317
Applicant: Carson Alexiou Corporation, USA
Inventors: Rotolante Ralph A, Koehler Toivo
Priority: US790015070   26.02.79
Family: US-A-4449044   15.05.84

US-A-4304624 08.12.81   315
Applicant: Irvine Sensors Corporation, USA
Inventors: Carson John C, Dahlgren Paul F
Priority: US770855242   28.11.77
Family: US-A-4352715   05.10.82
          US-A-4354107   12.10.82

US-A-4311906 19.01.82   344
Applicant: Thomson-CSF, France
Inventors: Félix Pierre, Le Parquier Guy
Priority: FR790016560   27.06.79
Family: EP-A-0021910   07.01.81
          FR-A-2460079   16.01.81
          JP-A-56013781   10.02.81

US-A-4354107 12.10.82   315
Applicant: Irvine Sensors Corporation, USA
Inventors: Carson John C, Dahlgren Paul F
Priority: US770855242   28.11.77
          US800206994   14.11.80
Family: US-A-4304624   08.12.81
          US-A-4352715   05.10.82

US-A-4360732 23.11.82   54
Applicant: Texas Instruments Incorporated, USA
Inventors: Chapman Richard A, Lewis Jr Adam J, Hynecek Jaroslav, Kinch Michael A
Priority: US800159990   16.06.80

US-A-4369458 18.01.83   275
Applicant: Westinghouse Electric Corp., USA
Inventors: Thomas Richard N, Sopira Michael M
Priority: US800165158   01.07.80
Family: US-A-4416054   22.11.83

US-A-4403238 06.09.83   321
Applicant: Irvine Sensors Corporation, USA
Inventor: Clark Stewart A
Priority: US800213933   08.12.80

US-A-4427990 24.01.84   40
Applicant: Zaidan Hojin Handotai Kenkyu Shinkokai, Japan
Inventor: Nishizawa Jun-ichi
Priority: JP780086572   14.07.78
          JP780087988   18.07.78
Family: JP-A-55013924   31.01.80
          JP-A-55015229   02.02.80
          US-A-5019876   28.05.91

US-A-4429330 31.01.84   73
Applicant: Texas Instruments Incorporated, USA
Inventor: Chapman Richard A
Priority: US800141498   18.04.80

US-A-4439912 03.04.84   100
Applicant: The United States of America as represented by the Secretary of the Army, USA
Inventors: Pollard John H, Ramsey Jr John B
Priority: US820369774   19.04.82

US-A-4445269 01.05.84   99
Applicant: The United States of America as represented by the Secretary of the Army, USA
Inventor: Pollard John H
Priority: US810296751   27.08.81

US-A-4517464 14.05.85   157
Applicant: The United States of America as represented by the Secretary of the Air Force, USA
Inventors: Heath James E, Swenson Hilmer W
Priority: US820431866   30.09.82

US-A-4525921 02.07.85   320
Applicant: Irvine Sensors Corporation, USA
Inventors: Carson John C, Clark Stewart A
Priority: US810282459   13.07.81
          US830517221   25.07.83
Family: US-A-4646128   24.02.87

US-A-4532699 06.08.85   353
Applicant: Societe Anonyme de Telecommunications, France
Inventors: Bourdillot Michel, Gauthier André, Maillé Jacques, Pitault Bernard
Priority: FR820020031   30.11.82
Family: DE-A-3375679   17.03.88
          EP-A-0116791   29.08.84
          FR-A-2536908   01.06.84

US-A-4549195 22.10.85   101
Applicant: Westinghouse Electric Corp., USA
Inventor: Bluzer Nathan
Priority: US830485091   14.04.83

US-A-4551629 05.11.85   319
Applicant: Irvine Sensors Corporation, USA
Inventors: Carson John C, Clark Stewart A
Priority: US800187787   16.09.80
          US840572802   23.01.84
Family: US-A-4646128   24.02.87
          US-A-4672737   16.06.87

US-A-4553152 12.11.85   355
Applicant: Mitsubishi Denki Kabushiki Kaisha, Japan

Inventor: Nishitani Kazuo
Priority: JP820195441 06.11.82
Family: FR-A-2535899 11.05.84
JP-A-59084467 16.05.84

US-A-4566024 21.01.86 152; 278
Applicant: Societe Anonyme de Telecommunications, France
Inventors: Fleury Joel J, Maillé Jacques H P
Priority: FR820004259 12.03.82
Family: EP-A-0089278 21.09.83
FR-A-2523369 16.09.83
JP-A-58166763 01.10.83

US-A-4570329 18.02.86 105
Applicant: Honeywell Inc., USA
Inventors: Paine Christopher G, White William J, Resnick Susan J
Priority: US840641075 15.08.84

US-A-4583108 15.04.86 60
Applicant: Societe Anonyme de Telecommunications, France
Inventor: Sirieix Michel B
Priority: FR810022837 07.12.81
Family: CA-A-1186412 30.04.85
EP-A-0082035 22.06.83
FR-A-2517864 10.06.83
JP-A-58105673 23.06.83

US-A-4596930 24.06.86 158
Applicant: Licentia Patent-Verwaltungs-GmbH, FRG
Inventors: Steil Hans-Jörgen, Fibich Wolfgang
Priority: US830487025 21.04.83
Family: US-A-4737642 12.04.88

US-A-4618763 21.10.86 321
Applicant: Grumman Aerospace Corporation, USA
Inventor: Schmitz Charles E
Priority: US850722776 12.04.85
Family: US-A-4703170 27.10.87
US-A-4792672 20.12.88

US-A-4620209 28.10.86 162
Applicant: Texas Instruments Incorporated, USA
Inventors: Parker Sidney G, York Rudy L
Priority: US840656059 28.09.84
Family: JP-A-61228678 11.10.86

US-A-4646120 24.02.87 206
Applicant: The United States of America as represented by the Secretary of the Army, USA
Inventor: Hacskaylo Michael
Priority: US850714423 21.03.85
Family: US-A-4686761 18.08.87

US-A-4654686 31.03.87 106
Applicant: Texas Instruments Incorporated, USA
Inventor: Borrello Sebastian R
Priority: US850707319 01.03.85
Family: JP-A-61280684 11.12.86

US-A-4658277 14.04.87 76
Applicant: Texas Instruments Incorporated, USA
Inventor: Schiebel Richard A
Priority: US840611456 17.05.84

US-A-4660066 02.08.88 280
Applicant: Texas Instruments Incorporated, USA
Inventor: Reid Lee R
Priority: US820415787 08.09.82
US850767063 19.08.85

US-A-4665609 19.05.87 202; 255
Applicant: Thomson-CSF, France
Inventors: Henry Yves, Nicollet Andre, Villard Michel
Priority: FR830020842 27.12.83
Family: DE-A-3469247 10.03.88
EP-A-0148687 17.07.85
FR-A-2557371 28.06.85
JP-A-60169166 02.09.85

US-A-4695861 22.09.87 105
Applicant: Honeywell Inc., USA
Inventors: Paine Christopher G, White William J, Resnick Susan J
Priority: US850770893 21.10.85
US870013186 09.02.87

US-A-4720738 19.01.88 383
Applicant: Texas Instruments Incorporated, USA
Inventor: Simmons Arturo
Priority: US820416396 08.09.82
US860898890 21.08.86
Family: JP-A-59065474 13.04.84

US-A-4726885 23.02.88 388
Applicant: Texas Instruments Incorporated, USA
Inventors: Teherani Towfik H, Little D Dawn
Priority: US860827315 07.02.86
Family: US-A-4736104 05.04.88

US-A-4727406 23.02.88 150
Applicant: Rockwell International Corporation, USA
Inventor: Rode Jonathan P
Priority: US820348398 12.02.82

US-A-4731640 15.03.88 109
Applicant: Westinghouse Electric Corp., USA

Inventor:	Bluzer Nathan		
Priority:	US860864940	20.05.86	

US-A-4751560 14.06.88 166; 257
Applicant: Santa Barbara Research Center, USA
Inventor: Rosbeck Joseph P
Priority: US860832111 24.02.86

US-A-4761681 21.04.87 280
Applicant: Texas Instruments Incorporated, USA
Inventor: Reid Lee R
Priority: US820415783 08.09.82
Family: JP-A-59134989 02.08.84

US-A-4783594 08.11.88 217; 296
Applicant: Santa Barbara Research Center, USA
Inventors: Schulte Eric F, Kasai Ichiro
Priority: US870123426 20.11.87

US-A-4791467 13.12.88 17
Applicant: Commissariat A L'Energie Atomique, France
Inventors: Amingual Daniel, Félix Pierre
Priority: FR860000172 08.01.86
Family: EP-A-0229574 22.07.87
  FR-A-2592740 10.07.87
  JP-A-62160776 16.07.87

US-A-4807007 21.02.89 103
Applicant: Texas Instruments Corporation, USA
Inventors: Borrello Sebastian R, Roberts Charles G
Priority: US830538292 03.10.83
  US860820872 16.01.86
Family: JP-A-60157272 17.08.85

US-A-4868622 19.09.89 167
Applicant: Kabushiki Kaisha Toshiba, Japan
Inventor: Shigenaka Keitaro
Priority: JP860274313 18.11.86
Family: JP-A-63128677 01.06.88

US-A-4877752 31.10.89 320
Applicant: The United States of America as represented by the Secretary of the Army, USA
Inventor: Robinson William L
Priority: US880265107 31.10.88

US-A-4885619 05.12.89 14
Applicant: Santa Barbara Research Center, USA
Inventor: Kosai Kenneth
Priorities: US870088330 24.08.87
  US890355064 16.05.89

US-A-4910154 20.03.90 364
Applicant: Ford Aerospace Corporation, USA
Inventors: Zanio Ken, Bean Ross C
Priority: US880289959 23.12.88
Family: US-A-4965649 23.10.90

US-A-4912536 27.03.90 20; 70
Applicant: Northrop Corporation, USA
Inventor: Lou Liang-fu
Priority: US880181854 15.04.88

US-A-4935627 19.06.90 287
Applicant: Honeywell Inc., USA
Inventors: Zimmermann Peter H, Schmit Joseph L
Priority: US890322352 13.03.89

US-A-4943491 24.07.90 299
Applicant: Honeywell Inc., USA
Inventors: Norton Peter W, Stobie James A, Zimmermann Peter H
Priority: US890438241 20.11.89

US-A-4956555 11.09.90 176
Applicant: Rockwell International Corporation, USA
Inventor: Woodberry Frank J
Priority: US890374412 30.06.89

US-A-4956686 11.09.90 76
Applicant: Texas Instruments Incorporated, USA
Inventors: Borrello Sebastian R, Roberts Charles G
Priority: US860827207 04.02.86
  US870021373 03.03.87
Family: CN-A-1030140 04.01.89

US-A-4956695 11.09.90 320
Applicant: Rockwell International Corporation, USA
Inventors: Robinson William L, Roth Jr John C
Priority: US890351122 12.05.89

US-A-4972244 20.11.90 171; 224
Applicant: Commissariat a l'Energie Atomique, France
Inventors: Buffet Jean-Louis O, Laurent Jean-Yves, Rochas Jean-Luc
Priority: FR880008074 16.06.88
Family: DE-D-68911772D 10.02.94
  DE-T-68911772T 30.06.94
  EP-A-0350351 10.01.90
  FR-A-2633101 22.12.89

US-A-4989067 29.01.91 119
Applicant: General Electric Company, USA
Inventors: Noble Milton L, Milton Albert F, Endres Darrel W, Dietz Douglas W

Priority: US890374932 03.07.89

US-A-4992908 12.02.91 326
Applicant: Grumman Aerospace Corporation, USA
Inventor: Solomon Allen L
Priority: US890383429 24.07.89
Family: DE-A-4036093 14.05.92
US-A-5067233 26.11.91

US-A-5003364 26.03.91 360
Applicant: Licentia Patent-Verwaltungs-GmbH, FRG
Inventors: Kasper Erich, Kohlbacher Gerhard, Nothaft Peter
Priority: DE863622879 08.07.86
Family: DE-A-3622879 21.01.88

US-A-5006711 09.04.91 116
Applicant: Fujitsu Limited, Japan
Inventors: Hamashima Shigeki, Koseto Masaru, Nomura Shoji
Priority: JP880252818 05.10.88
Family: JP-A-2098965 11.04.90

US-A-5030828 09.07.91 238
Applicant: Grumman Aerospace Corporation, USA
Inventor: Solomon Allen L
Priority: US900542895 25.06.90

US-A-5075201 24.12.91 288
Applicant: Grumman Aerospace Corporation, USA
Inventor: Koh Wei H
Priority: US900607154 31.10.90

US-A-5081063 14.01.92 324
Applicant: Harris Corporation, USA
Inventors: Vonno Nicolaas V, Begley Patrick A
Priority: US890382388 20.07.89
Family: US-A-5185292 09.02.93

US-A-5113076 12.05.92 177
Applicant: Santa Barbara Research Center, USA
Inventor: Schulte Eric F
Priority: US890452891 19.12.89

US-A-5128749 07.07.92 326
Applicant: Grumman Aerospace Corporation, USA
Inventors: Hornback William B, Koh Wei H
Priority: US910681775 08.04.91
Family: US-A-5135556 04.08.92

US-A-5130259 14.07.92 67
Applicant: Northrop Corporation, USA
Inventor: Bahraman Ali

Priority: US880226501 01.08.88

US-A-5144138 01.09.92 80; 389
Applicant: Texas Instruments Incorporated, USA
Inventors: Kinch Michael A, Roberts C Grady
Priority: US890417931 06.10.89

US-A-5146303 08.09.92 120
Applicant: General Electric Company, USA
Inventors: Kornrumpf William P, Marcinkiewicz Walter M, Davern William E, Zieger Herbert C, Miles Jonathan R
Priority: US900504751 05.04.90
Family: US-A-5157255 20.10.92

US-A-5149956 22.09.92 179
Applicant: Santa Barbara Research Center, USA
Inventor: Norton Paul R
Priority: US910715086 12.06.91
Family: EP-A-0518243 16.12.92
IL-A-101822 07.10.94

US-A-5171994 15.12.92 67
Applicant: Northrop Corporation, USA
Inventor: Bahraman Ali
Priority: US860896358 13.08.86
US880226498 01.08.88

US-A-5177580 05.01.93 240
Applicant: Santa Barbara Research Center, USA
Inventors: Norton Paul R, Radford William A
Priority: US910643401 22.01.91

US-A-5189297 23.02.93 261
Applicant: Santa Barbara Research Center, USA
Inventor: Ahlgren William L
Priority: US880237806 29.08.88

US-A-5198370 30.03.93 187
Applicant: Mitsubishi Denki Kabushiki Kaisha, Japan
Inventors: Ohkura Yuji, Takiguchi Tohru
Priority: JP910115631 17.04.91
Family: JP-A-4318981 10.11.92

US-A-5254850 19.10.93 196
Applicant: Texas Instruments Incorporated, USA
Inventor: Dreiske Peter D
Priority: US920930881 14.08.92

US-A-5264699 23.11.93 302
Applicant: Amber Engineering Inc., USA
Inventors: Barton Jeffrey, Lockwood Arthur H
Priority: US910658985 20.02.91

Family: US-A-5308980   03.05.94

US-A-5279974   18.01.94   195
Applicant: Santa Barbara Research Center, USA
Inventor: Walsh Devin T
Priority: US920918957   24.07.92
Family: FR-A-2694134   28.01.94

US-A-5293036   08.03.94   176
Applicant: Santa Barbara Research Center, USA
Inventor: Norton Paul R
Priority: US890393343   11.08.89

US-A-5296384   22.03.94   266
Applicant: Santa Barbara Research Center, USA
Inventors: Cockrum Charles A, Gesswein Francis I, Schulte Eric F
Priority: US920917562   21.07.92
Family: US-A-5401986   28.03.95

US-A-5298733   29.03.94   197
Applicant: Texas Instruments Incorporated, USA
Inventors: Ehmke John C, Baker James C
Priority: US920988513   10.12.92
Family: JP-A-6283730   07.10.94

US-A-5300777   05.04.94   193
Applicant: Texas Instruments Incorporated, USA
Inventor: Goodwin Michael W
Priority: US920860733   26.03.92

US-A-5315147   24.05.94   367
Applicant: Grumman Aerospace Corporation, USA
Inventor: Solomon Allen L
Priority: US890412161   25.09.89

US-A-5318666   07.06.94   396
Applicant: Texas Instruments Incorporated, USA
Inventor: Elkind Jerome L, Orloff Glennis J, Smith Patricia B
Priority: US930049755   19.04.93

US-A-5365088   15.11.94   298
Applicant: Santa Barbara Research Center, USA
Inventor: Myrosznyk James M
Priority: US880227581   02.08.88

US-A-5373182   13.12.94   179
Applicant: Santa Barbara Research Center, USA
Inventor: Norton Paul R
Priority: US930003715   12.01.93

US-A-5376558   27.12.94   215; 294
Applicant: Fujitsu Limited, Japan
Inventors: Sudo Gen, Hikida Soichiro
Priorities: JP900403723   19.12.90
            US910806247   13.12.93
            US930041462   02.04.93
Family: JP-A-4218964   10.08.92

US-A-5384267   24.01.95   398
Applicant: Texas Instruments Incorporated, USA
Inventors: Hutchins Larry D, York Rudy L
Priority: US930140390   19.10.93

US-A-5401986   28.03.95   266
Applicant: Santa Barbara Research Center, USA
Inventors: Cockrum Charles A, Gesswein Francis I, Schulte Eric F
Priorities: US920917562   21.07.92
            US940188172   26.01.94
            US940270527   05.07.94
Family: US-A-5296384   22.03.94

US-A-5414294   09.05.95   198
Applicant: Santa Barbara Research Center, USA
Inventors: Granneman Russell D, McKeag William O
Priority: US930040710   31.03.93

WO-A-8603338   05.06.86   320
Applicant: Irvine Sensors Corporation, USA
Inventors: Belanger Robert J, Bisignano Alan G
Priority: US840674096   23.11.84
Family: EP-A-0204767   17.12.86
            JP-T-62500901T   09.04.87
            US-A-4617160   14.10.86
            US-A-4704319   03.11.87

WO-A-8707082   19.11.87   10
Applicant: Santa Barbara Research Center, USA
Inventors: Phillips James D, Casselman Thomas N, Koch Thomas L
Priority: US860860967   08.05.86
Family: EP-A-0270567   15.06.88
            JP-T-63503183T   17.11.88
            US-A-4952995   28.08.90

WO-A-8707083   19.11.87   260
Applicant: Santa Barbara Research Center, USA
Inventors: Rosbeck Joseph P, Kasai Ichiro
Priority: US860859674   05.05.86
Family: EP-A-0264437   27.04.88
            JP-B-7028049   29.03.95
            JP-T-63503266T   24.11.88
            US-A-4639756   27.01.87

WO-A-8807764   06.10.88   323
Applicant: Grumman Aerospace Corporation, USA
Inventor: Schmitz Charles E
Priority: US870034143   23.03.87
Family: AT-T-108577T   15.07.94
            CA-A-1264836   23.01.90
            DE-D-3750224D   18.08.94
            DE-T-3750224T   24.11.94
            EP-A-0308465   29.03.89
            JP-T-1502950T   05.10.89
            US-A-4792672   20.12.88

WO-A-8910007   19.10.89   170; 223
Applicant: Santa Barbara Research Center, USA
Inventors: Cockrum Charles A, Barton Jeffrey B, Schulte Eric F
Priority: US880178680   07.04.88
Family: DE-D-68917696D   29.09.94
            DE-T-68917696T   27.04.95
            EP-A-0374232   27.06.90
            JP-T-2503973T   15.11.90
            US-A-4956304   11.09.90

WO-A-9001802   22.02.90   286
Applicant: Honeywell Inc., USA
Inventor: Gurnee Mark N
Priority: US880230164   09.08.88
Family: US-A-4871921   03.10.89

WO-A-9006597   14.06.90   159
Applicant: The Secretary of State for Defence in Her Britannic Majesty's Government of the United Kingdom of Great Britain and Northern Ireland, GB
Inventors: Elliott Charles Thomas, White Anthony Michael
Priority: GB880028348   05.12.88
Family: DE-D-68909929D   18.11.93
            DE-T-68909929T   10.02.94
            EP-A-0401352   12.12.90
            ES-T-2044550T   01.01.94
            GB-A-2232816   19.12.90
            JP-T-3503226T   18.07.91
            US-A-5068524   26.11.91

WO-A-9202959   20.02.92   369
Applicant: Minnesota Mining and Manufacturing Company, USA
Inventors: Tran Nang Tri, Loeding Neil W, Nins David V
Priority: US900564632   08.08.90
Family: EP-A-0543951   02.06.93
            JP-T-5509204T   16.12.93
            US-A-5182624   26.01.93
            US-A-5235195   10.08.93
            US-A-5273910   28.12.93

WO-A-9417557   04.08.94   306
Applicant: Hughes Aircraft Company, USA
Inventors: Finnila Ronald M, Malloy Gerard T, Bendik Joseph J
Priority: US930006212   19.01.93
Family: GB-A-2279808   11.01.95

Inventor Index

Ahlgren William L
  US-A-5189297   23.02.93

Ajisawa Akira
  JP-A-4267562   24.09.92
  JP-A-4269869   25.09.92
  JP-A-4318979   10.11.92
  JP-A-5343729   24.12.93
  JP-A-6326342   25.11.94

Ameurlaine Jacques Francois
  US-A-3930161   30.12.75

Amingual Daniel
  US-A-4791467   13.12.88

Ard Christopher Kyle
  EP-A-0343738   29.11.89

Arinaga Kenji
  EP-A-0497326   05.08.92
  JP-A-3085762   10.04.91
  JP-A-6013642   21.01.94
  JP-A-6204449   22.07.94
  JP-A-6209096   26.07.94
  JP-A-7050330   21.02.95

Arques Marc
  FR-A-2526227   04.11.83

Ashley Timothy
  EP-A-0167305   08.01.86

Awamoto Kenji
  EP-A-0543537   26.05.93
  JP-A-6163865   10.06.94
  JP-A-6236981   23.08.94

Bahraman Ali
  US-A-5130259   14.07.92
  US-A-5171994   15.12.92

Baker Ian Martin
  EP-A-0007668   06.02.80
  EP-A-0061803   06.10.82
  EP-A-0475525   18.03.92
  EP-A-0481552   22.04.92
  GB-A-2239555   03.07.91

Baker James C
  US-A-5298733   29.03.94

Barton Jeffrey
  US-A-5264699   23.11.93
  WO-A-8910007   19.10.89

Bate Robert T
  US-A-3808435   30.04.74

Bean Ross C
  US-A-4910154   20.03.90

Begley Patrick A
  US-A-5081063   14.01.92

Belanger Robert J
  WO-A-8603338   05.06.86

Bendik Joseph J
  WO-A-9417557   04.08.94

Berger Jean-Luc
  FR-A-2526227   04.11.83

Bisignano Alan G
  WO-A-8603338   05.06.86

Blackburn Anthony
  EP-A-0061801   06.10.82
  EP-A-0061802   06.10.82

Blackman Maurice Victor
  EP-A-0007669   06.02.80
  GB-A-1559473   16.01.80
  GB-A-1559474   16.01.80

GB-A-1568958 11.06.80
GB-A-1600599 21.10.81
GB-A-2231199 07.11.90
GB-A-2231199 07.11.90

Blatt Peter R
US-A-4025793 24.05.77

Bluzer Nathan
US-A-4549195 22.10.85
US-A-4731640 15.03.88

Borrello Sebastian R
US-A-4654686 31.03.87
US-A-4807007 21.02.89
US-A-4956686 11.09.90

Bourdillot Michel
US-A-4532699 06.08.85

Brewitt Taylor Colin Raymond
GB-A-2260218 07.04.93

Buffet Jean-Louis O
US-A-4972244 20.11.90

Buss Dennis D
US-A-3808435 30.04.74
US-A-4231149 04.11.80

Carson John C
US-A-3852714 03.12.74
US-A-3970990 20.07.76
US-A-4304624 08.12.81
US-A-4354107 12.10.82
US-A-4525921 02.07.85
US-A-4551629 05.11.85

Casselman Thomas N
WO-A-8707082 19.11.87

Caywood John Millard
US-A-3806729 23.04.74

Chapman Richard A
EP-A-0044610 04.08.82
EP-A-0066020 08.12.82
US-A-3980915 14.09.76
US-A-4231149 04.11.80
US-A-4360732 23.11.82
US-A-4429330 31.01.84

Charlton David Eric
GB-A-1605321 19.07.89
GB-A-2241377 28.08.91

Clark Stewart A
US-A-4403238 06.09.83
US-A-4525921 02.07.85
US-A-4551629 05.11.85

Coale Cecil R Jr
EP-A-0137704 17.04.85

Cochran Bruce C
EP-A-0171801 19.02.86

Cockrum Charles A
EP-A-0406696 09.01.91
US-A-5296384 22.03.94
US-A-5401986 28.03.95
WO-A-8910007 19.10.89

Cohen-Solal Gerard David
US-A-3988774 26.10.76

Crimes Graham Joseph
GB-A-2231199 07.11.90
GB-A-2231199 07.11.90

D'Agostino Saverio A
EP-A-0405865 02.01.91

Dahlgren Paul F
US-A-4304624 08.12.81
US-A-4354107 12.10.82

Daiku Hiroshi
JP-A-3270269 02.12.91

Davern William E
US-A-5146303 08.09.92

Day Derek Joseph
GB-A-2007909 23.05.79

Dean Anthony Brian
EP-A-0068652 05.01.83
GB-A-2132017 27.06.84
GB-A-2207801 08.02.89

Dennis Peter Neil John
GB-A-2207801 08.02.89

Dietz Douglas W
US-A-4989067 29.01.91

Doi Shiyouji
JP-A-55102280 05.08.80
JP-A-55150279 22.11.80

Doi Shoji
JP-A-6163865 10.06.94
JP-A-6236981 23.08.94

Dreiske Peter D
US-A-5254850 19.10.93

Dunn Aubrey J
US-A-4242149 30.12.80

Dunn William Andrew Ernest
 EP-A-0475525   18.03.92

Dyson Clive Malcolm
 GB-A-2201834   07.09.88

Easton Brian Colin
 GB-A-2203757   26.10.88

Ehmke John C
 US-A-5298733   29.03.94

Elkind Jerome L
 US-A-5318666   07.06.94

Elliott Charles Thomas
 EP-A-0167305   08.01.86
 GB-A-1488258   12.10.77
 GB-A-1605321   19.07.89
 GB-A-2007909   23.05.79
 GB-A-2019649   31.10.79
 GB-A-2094548   15.09.82
 GB-A-2095906   06.10.82
 GB-A-2128019   18.04.84
 GB-A-2207801   08.02.89
 GB-A-2260218   07.04.93
 WO-A-9006597   14.06.90

Endo Yutaro
 JP-A-4318970   10.11.92
 JP-A-4337676   25.11.92
 JP-A-4337677   25.11.92

Endres Darrel W
 US-A-4989067   29.01.91

Farrow Robin Frederick Charles
 EP-A-0068652   05.01.83

Félix Pierre
 US-A-4311906   19.01.82
 US-A-4791467   13.12.88

Fibich Wolfgang
 US-A-4596930   24.06.86

Finnila Ronald M
 WO-A-9417557   04.08.94

Fisher Konrad
 US-A-4162507   24.07.79

Fleury Joel J
 US-A-4566024   21.01.86

Fujiwara Koji
 EP-A-0497326   05.08.92
 JP-A-6013642   21.01.94
 JP-A-6204449   22.07.94
 JP-A-6209096   26.07.94

 JP-A-7050330   21.02.95

Fukita Muneyoshi
 JP-A-3175682   30.07.91
 JP-A-4313267   05.11.92

Gauthier André
 FR-A-2593642   31.07.87
 US-A-4532699   06.08.85

Gesswein Francis I
 US-A-4025793   24.05.77
 US-A-5296384   22.03.94
 US-A-5401986   28.03.95

Gooch Roland W
 US-A-3965568   29.06.76

Goodwin Michael W
 US-A-5300777   05.04.94

Gordon Neil Thomson
 GB-A-2132017   27.06.84

Goto Junjiro
 JP-A-60028266   13.02.85
 JP-A-61147118   04.07.86
 JP-A-61222161   02.10.86
 JP-A-4253344   09.09.92

Granneman Russell D
 US-A-5414294   09.05.95

Greene Richard F
 US-A-3842274   15.10.74

Gridley C John
 US-A-4259576   31.03.81

Gurnee Mark N
 EP-A-0281026   07.09.88
 WO-A-9001802   22.02.90

Gutierrez William A
 US-A-4228365   14.10.80
 US-A-4273596   16.06.81

Hacskaylo Michael
 US-A-4646120   24.02.87

Hamashima Shigeki
 JP-A-57073984   08.05.82
 JP-A-61220459   30.09.86
 JP-A-61251167   08.11.86
 JP-A-63148677   21.06.88
 JP-A-63170960   14.07.88
 JP-A-1050557   27.02.89
 US-A-5006711   09.04.91

Hartley Martha A
   EP-A-0171801 19.02.86

Hashimoto Arihiro
   JP-A-2155269 16.06.90

Hayafuji Norio
   GB-A-2284930 21.06.95

Heath James
   US-A-4517464 14.05.85

Henry Yves
   US-A-4665609 19.05.87

Hikita Soichiro
   EP-A-0497326 05.08.92
   JP-A-60140869 25.07.85
   JP-A-62224982 02.10.87
   JP-A-62224983 02.10.87
   JP-A-62272564 26.11.87
   JP-A-63043366 24.02.88
   JP-A-63133580 06.06.88
   JP-A-63281460 17.11.88
   JP-A-63296272 02.12.88
   JP-A-1050560 27.02.89
   JP-A-1201971 14.08.89
   JP-A-1233777 19.09.89
   JP-A-2086177 27.03.90
   JP-A-3268463 29.11.91
   JP-A-3270269 02.12.91
   JP-A-6163969 10.06.94
   JP-A-6237005 23.08.94
   US-A-5376558 27.12.94

Hirota Koji
   JP-A-63170960 14.07.88
   JP-A-63229751 26.09.88

Hisa Yoshihiro
   GB-A-2229036 12.09.90
   GB-A-2238165 22.05.91
   GB-A-2261323 12.05.93
   JP-A-63116459 20.05.88
   JP-A-2214159 27.08.90

Hornback William B
   US-A-5128749 07.07.92

Horne Russell John
   EP-A-0510267 28.10.92

Hu Willliam C
   EP-A-0405865 02.01.91

Huang Chao
   EP-A-0445545 11.09.91

Hutchins Larry
   EP-A-0407062 09.01.91

   US-A-5384267 24.01.95

Hynecek Jaroslav
   EP-A-0044610 04.08.82
   US-A-4229752 21.10.80
   US-A-4360732 23.11.82

Imai Souichi
   JP-A-55102280 05.08.80
   JP-A-55150279 22.11.80
   JP-A-57024580 09.02.82
   JP-A-57091557 07.06.82

Irvine Stuart James Curzon
   GB-A-2146663 24.04.85

Ishii Jiyun
   JP-A-57169278 18.10.82

Ito Makoto
   JP-A-61220459 30.09.86
   JP-A-62104163 14.05.87
   JP-A-63070454 30.03.88

Ito Michiharu
   JP-A-61067958 08.04.86
   JP-A-63170960 14.07.88
   JP-A-63229751 26.09.88
   JP-A-1233777 19.09.89

Ito Yuichiro
   EP-A-0497326 05.08.92
   EP-A-0543537 26.05.93
   JP-A-61059771 27.03.86
   JP-A-1218062 31.08.89
   JP-A-1227472 11.09.89

Itou Makoto
   JP-A-57024580 09.02.82
   JP-A-57091557 07.06.82
   JP-A-60028266 13.02.85

Itou Michiharu
   JP-A-57073984 08.05.82
   JP-A-60180162 13.09.85
   JP-A-60193375 01.10.85

Jackson John
   GB-A-2208256 15.03.89

Jenkin Graham Thomas
   GB-A-2197984 02.06.88

Jenner Michael David
   EP-A-0007669 06.02.80
   EP-A-0561615 22.09.93
   GB-A-1559473 16.01.80
   GB-A-1559474 16.01.80
   GB-A-1568958 11.06.80
   GB-A-1600599 21.10.81
   GB-A-2231199 07.11.90

GB-A-2231199 07.11.90
GB-A-2241377 28.08.91
GB-A-2241605 04.09.91

Johnson Milo R
US-A-3980915 14.09.76

Jun-ichi Nishizawa
EP-A-0042218 23.12.81
EP-A-0042218 23.12.81

Kajiwara Nobuyuki
EP-A-0497326 05.08.92
JP-A-60140869 25.07.85
JP-A-6013642 21.01.94
JP-A-6204449 22.07.94
JP-A-6209096 26.07.94
JP-A-7050330 21.02.95

Kasai Ichiro
US-A-4783594 08.11.88
WO-A-8707083 19.11.87

Kasper Erich
US-A-5003364 26.03.91

Kawachi Tetsuya
JP-A-61220459 30.09.86
JP-A-62104163 14.05.87
JP-A-1050557 27.02.89
JP-A-3077373 02.04.91
JP-A-3219670 27.09.91

Kawatsu Yoshihei
JP-A-5055620 05.03.93

Kawazu Zempei
GB-A-2284930 21.06.95

Kinch Michael A
EP-A-0044610 04.08.82
EP-A-0416320 13.03.91
US-A-3808435 30.04.74
US-A-4231149 04.11.80
US-A-4360732 23.11.82
US-A-5144138 01.09.92
US-A-5144138 01.09.92

King Gerard J
US-A-4242149 30.12.80

King Raymond Edward John
GB-A-2241377 28.08.91

Kizuki Hirotaka
GB-A-2284930 21.06.95

Knowles Peter
GB-A-2197984 02.06.88

Koch Thomas L
WO-A-8707082 19.11.87

Koehler Toivo
US-A-4290844 22.09.81

Koh Wei H
US-A-5075201 24.12.91
US-A-5128749 07.07.92

Kohlbacher Gerhard
US-A-5003364 26.03.91

Komine Yoshiharu
JP-A-3104278 01.05.91
JP-A-4133363 07.05.92
JP-A-4199877 21.07.92

Kornrumpf William P
US-A-5146303 08.09.92

Kosai Kenneth
US-A-4885619 05.12.89

Koseto Masaru
JP-A-60028266 13.02.85
JP-A-61147118 04.07.86
JP-A-61222161 02.10.86
US-A-5006711 09.04.91

Kusner Ronald R
EP-A-0188241 23.07.86

Laurent Jean-Yves
US-A-4972244 20.11.90

Le Parquier Guy
US-A-4311906 19.01.82

Lenoble Claude
FR-A-2484705 18.12.81

Lewis Adam J
EP-A-0137988 24.04.85
EP-A-0173074 05.03.86
US-A-4360732 23.11.82

Little D Dawn
US-A-4726885 23.02.88

Lockwood Arthur H
US-A-5264699 23.11.93

Loeding Neil W
WO-A-9202959 20.02.92

Longerich Ernest P
EP-A-0405865 02.01.91

Lorenze Jr Robert V

US-A-4104674 01.08.78
US-A-4197633 15.04.80

Lou Liang-fu
US-A-4912536 27.03.90
US-A-4912536 27.03.90

Lussereau Alain Gilles
US-A-3988774 26.10.76

Luttmer Joseph D
EP-A-0407062 09.01.91
EP-A-0416320 13.03.91

Maciolek Ralph B
US-A-3902924 75.09.02

Maejima Yukihiko
JP-A-61214462 24.09.86
JP-A-62090986 25.04.87
JP-A-63005560 11.01.88
JP-A-63046765 27.02.88
JP-A-63147366 20.06.88
JP-A-63300559 07.12.88
JP-A-1061055 08.03.89
JP-A-1061056 08.03.89
JP-A-2248077 03.10.90
JP-A-2272765 07.11.90

Maekawa Toru
JP-A-57062563 15.04.82
JP-A-59047759 17.03.84
JP-A-61198787 03.09.86
JP-A-61251167 08.11.86
JP-A-61268075 27.11.86
JP-A-62036858 17.02.87
JP-A-62104163 14.05.87
JP-A-62165973 22.07.87
JP-A-63043366 24.02.88
JP-A-1050557 27.02.89
JP-A-1233777 19.09.89
JP-A-2155269 16.06.90
JP-A-3077373 02.04.91
JP-A-3133181 06.06.91
JP-A-3219670 27.09.91
JP-A-3241774 28.10.91

Maier Dr Horst
GB-A-2113467 03.08.83

Maillé Jacques
US-A-4532699 06.08.85
US-A-4566024 21.01.86

Malloy Gerard T
WO-A-9417557 04.08.94

Marciniec John W
EP-A-0171801 19.02.86

Marcinkiewicz Walter M
US-A-5146303 08.09.92

Martin Charles J
EP-A-0485115 13.05.92

Maruyama Kenji
JP-A-60193375 01.10.85
JP-A-61067958 08.04.86
JP-A-61128564 16.06.86
JP-A-63260171 27.10.88

Matthews Brian Edward
EP-A-0561615 22.09.93

McCullough John B
EP-A-0151311 14.08.85

McKeag William O
US-A-5414294 09.05.95

Migliorato Piero
EP-A-0068652 05.01.83

Miles Jonathan R
US-A-5146303 08.09.92

Milton Albert F
US-A-4989067 29.01.91

Mitsui Kotaro
GB-A-2284930 21.06.95

Miyamoto Keiji
JP-A-6204448 22.07.94

Miyamoto Yoshihiro
JP-A-57010986 20.01.82
JP-A-60074568 26.04.85
JP-A-60103666 07.06.85
JP-A-62224982 02.10.87
JP-A-62224983 02.10.87
JP-A-63133580 06.06.88
JP-A-63260171 27.10.88
JP-A-63281460 17.11.88
JP-A-63296272 02.12.88
JP-A-1201971 14.08.89
JP-A-1233777 19.09.89
JP-A-2086177 27.03.90
JP-A-2272766 07.11.90
JP-A-5226626 03.09.93
JP-A-6237005 23.08.94

Mizuguchi Kazuo
GB-A-2284930 21.06.95

Morris Henry B
US-A-3980915 14.09.76

Mortensen Craig A
    EP-A-0151311   14.08.85

Mullin John Brian
    GB-A-2146663   24.04.85

Munier Bernard
    FR-A-2494910   28.05.82

Myrosznyk James M
    US-A-5365088   15.11.94

Nagahama Kouki
    JP-A-57169278   18.10.82

Nagasaka Hiroo
    JP-A-58164261   29.09.83

Nakamura Hiroko
    JP-A-6013642   21.01.94
    JP-A-6204449   22.07.94
    JP-A-6209096   26.07.94
    JP-A-7050330   21.02.95

Naruse Yujiro
    JP-A-62011265   20.01.87
    JP-A-62022474   30.01.87

Nicollet Andre
    US-A-4665609   19.05.87

Nins David V
    WO-A-9202959   20.02.92

Nishino Hiroshi
    JP-A-6021419   28.01.94

Nishitani Kazuo
    JP-A-57169278   18.10.82
    US-A-4553152   12.11.85

Nishizawa Jun-ichi
    EP-A-0038697   28.10.81
    EP-A-0094973   30.11.83
    EP-A-0094974   30.11.83
    US-A-4427990   24.01.84

Noble Milton L
    US-A-4989067   29.01.91

Nomura Shoji
    JP-A-59112652   29.06.84
    JP-A-61147118   04.07.86
    JP-A-61222161   02.10.86
    US-A-5006711   09.04.91

Norton Paul R
    EP-A-0445545   11.09.91
    US-A-5149956   22.09.92
    US-A-5177580   05.01.93

US-A-5293036   08.03.94
US-A-5373182   13.12.94

Norton Peter W
    US-A-4943491   24.07.90

Nothaft Peter
    US-A-5003364   26.03.91

Ochi Seiji
    GB-A-2284930   21.06.95

Ohkura Yuji
    GB-A-2284930   21.06.95
    US-A-5198370   30.03.93

Ohmi Tadahiro
    EP-A-0038697   28.10.81

Oikawa Ryuichi
    JP-A-5343727   24.12.93

Okata Ryoji
    JP-A-63273365   10.11.88

Orent Thomas N
    EP-A-0407062   09.01.91

Orloff Glennis J
    US-A-5318666   07.06.94

Ozaki Kazuo
    JP-A-3241774   28.10.91
    JP-A-5175476   13.07.93
    JP-A-6163969   10.06.94

Paine Christopher G
    US-A-4570329   18.02.86
    US-A-4695861   22.09.87

Parker Sidney G
    US-A-4620209   28.10.86

Phillips James D
    WO-A-8707082   19.11.87

Pitault Bernard
    US-A-4532699   06.08.85

Pollard John H
    US-A-4228365   14.10.80
    US-A-4273596   16.06.81
    US-A-4439912   03.04.84
    US-A-4445269   01.05.84

Radford William A
    US-A-5177580   05.01.93

Ramsey Jr John B
    US-A-4439912   03.04.84

Readhead John Barry
 EP-A-0061801 06.10.82
 EP-A-0061802 06.10.82
 EP-A-0087842 07.09.83

Rebondy Jacques
 EP-A-0064918 17.11.82

Reboul Jean Philippe
 EP-A-0024970 11.03.81
 FR-A-2484705 18.12.81
 FR-A-2494910 28.05.82

Rees Huw David
 GB-A-2260218 07.04.93

Reid Lee R
 US-A-4660066 02.08.88
 US-A-4761681 21.04.87

Reine Marion B
 EP-A-0188241 23.07.86

Resnick Susan J
 US-A-4570329 18.02.86
 US-A-4695861 22.09.87

Roberts C G
 US-A-4807007 21.02.89
 US-A-4956686 11.09.90
 US-A-5144138 01.09.92

Robinson William L
 US-A-4877752 31.10.89
 US-A-4956695 11.09.90

Rochas Jean-Luc
 US-A-4972244 20.11.90

Rode Jonathan P
 US-A-4727406 23.02.88

Rokushiya Kiyoshi
 JP-A-60028266 13.02.85
 JP-A-61147118 04.07.86
 JP-A-61222161 02.10.86

Rosbeck Joseph P
 EP-A-0406696 09.01.91
 US-A-4751560 14.06.88
 WO-A-8707083 19.11.87

Roth Jr John C
 US-A-4956695 11.09.90

Rotolante Ralph A
 US-A-4290844 22.09.81

Saito Tetsuo
 JP-A-60180162 13.09.85
 JP-A-3077373 02.04.91
 JP-A-3108371 08.05.91
 JP-A-3219670 27.09.91
 JP-A-3241774 28.10.91
 JP-A-3268463 29.11.91
 JP-A-3270269 02.12.91

Sakachi Yoichiro
 JP-A-6163865 10.06.94
 JP-A-6236981 23.08.94

Sakai Hiroshi
 JP-A-59061080 07.04.84
 JP-A-62013085 21.01.87

Scharnhorst K Peter
 US-A-3842274 15.10.74

Schiebel Richard A
 US-A-4658277 14.04.87

Schmit Joseph L
 US-A-3949223 06.04.76
 US-A-4935627 19.06.90

Schmitz Charles E
 US-A-4618763 21.10.86
 WO-A-8807764 06.10.88

Schoolar Richard B
 US-A-3842274 15.10.74

Schulte Eric F
 EP-A-0137988 24.04.85
 EP-A-0173074 05.03.86
 EP-A-0288792 02.11.88
 US-A-4783594 08.11.88
 US-A-5113076 12.05.92
 US-A-5296384 22.03.94
 US-A-5401986 28.03.95
 WO-A-8910007 19.10.89

Schulz Dr Max
 GB-A-2113467 03.08.83

Shaw John B
 US-A-4025793 24.05.77

Shibata Shusaku
 JP-A-61222161 02.10.86

Shigenaka Keitaro
 JP-A-6089991 29.03.94
 US-A-4868622 19.09.89

Simmons A
 EP-A-0407062 09.01.91
 US-A-4720738 19.01.88

Singh Surendra Pratap
    EP-A-0293216   30.11.88

Sirieix Michel B
    US-A-4583108   15.04.86
    US-A-3930161   30.12.75

Smith Patricia B
    US-A-5318666   07.06.94

Solomon Allen L
    EP-A-0317083   24.05.89
    EP-A-0317084   24.05.89
    US-A-4992908   12.02.91
    US-A-5030828   09.07.91
    US-A-5315147   24.05.94

Sopira Michael M
    US-A-4369458   18.01.83

Speerschneider Charles J
    US-A-3902924   75.09.02

Steil Hans-Jörgen
    US-A-4596930   24.06.86

Stelzer Ernest L
    US-A-3949223   06.04.76

Stobie James A
    US-A-4943491   24.07.90

Strong Roger Lynn
    EP-A-0366886   09.05.90

Sudo Gen
    EP-A-0497326   05.08.92
    US-A-5376558   27.12.94

Sudo Hajime
    JP-A-63150976   23.06.88
    JP-A-63170961   14.07.88
    JP-A-63268271   04.11.88
    JP-A-1050560   27.02.89
    JP-A-2303160   17.12.90
    JP-A-3268463   29.11.91
    JP-A-3270269   02.12.91
    JP-A-6013642   21.01.94
    JP-A-6125108   06.05.94
    JP-A-6204449   22.07.94
    JP-A-6209096   26.07.94
    JP-A-7050330   21.02.95

Sugiura Risa
    JP-A-6089991   29.03.94

Sugiyama Iwao
    JP-A-3219670   27.09.91
    JP-A-5129580   25.05.93
    JP-A-5315580   26.11.93
    JP-A-6021419   28.01.94

Sulzbach Frank C
    EP-A-0416299   13.03.91

Swenson Hilmer W
    US-A-4517464   14.05.85

Tadahiro Ohmi
    EP-A-0042218   23.12.81
    EP-A-0042218   23.12.81

Takada Yutaka
    JP-A-3175682   30.07.91

Takami Akihiro
    GB-A-2261767   26.05.93
    GB-A-2284930   21.06.95

Takashige Tamamushi
    EP-A-0042218   23.12.81
    EP-A-0042218   23.12.81

Takeuchi Youji
    JP-A-60102530   06.06.85

Takigawa Hiroshi
    JP-A-55102280   05.08.80
    JP-A-55150279   22.11.80
    JP-A-57031170   19.02.82
    JP-A-57062563   15.04.82
    JP-A-57073984   08.05.82
    JP-A-58171848   08.10.83
    JP-A-60180162   13.09.85
    JP-A-60182766   18.09.85
    JP-A-60193375   01.10.85
    JP-A-61198787   03.09.86
    JP-A-2272766   07.11.90

Takiguchi Toru
    GB-A-2246907   12.02.92
    JP-A-2213174   24.08.90
    JP-A-3196568   28.08.91
    JP-A-4288882   13.10.92
    US-A-5198370   30.03.93

Tamamushi Takashige
    EP-A-0038697   28.10.81

Tanaka Masahiro
    JP-A-1050557   27.02.89
    JP-A-2155269   16.06.90
    JP-A-3108371   08.05.91
    JP-A-3133181   06.06.91
    JP-A-6163969   10.06.94

Tanigawa Kunihiro
  JP-A-57031170 19.02.82
  JP-A-59047759 17.03.84
  JP-A-59112652 29.06.84
  JP-A-62013085 21.01.87
  JP-A-62272564 26.11.87
  JP-A-1050560 27.02.89

Teherani Towfik H
  US-A-4726885 23.02.88

Tew Claude E
  EP-A-0137704 17.04.85
  EP-A-0137988 24.04.85

Thomas Richard N
  US-A-4369458 18.01.83

Timlin Harold A
  EP-A-0485115 13.05.92

Tran Nang Tri
  WO-A-9202959 20.02.92

Tsugami Mari
  GB-A-2284930 21.06.95

Tunnicliffe Jean
  GB-A-2146663 24.04.85

Udagawa Kenji
  JP-A-63310165 19.12.88

Ueda Tomoshi
  JP-A-57062563 15.04.82
  JP-A-57073984 08.05.82
  JP-A-58171848 08.10.83
  JP-A-60180162 13.09.85
  JP-A-60182766 18.09.85
  JP-A-60193375 01.10.85
  JP-A-62104163 14.05.87
  JP-A-63170960 14.07.88
  JP-A-63281460 17.11.88
  JP-A-63296272 02.12.88
  JP-A-1050557 27.02.89

Ueda Toshiyuki
  JP-A-1233777 19.09.89
  JP-A-2155269 16.06.90
  JP-A-3108371 08.05.91
  JP-A-3133181 06.06.91
  JP-A-3241774 28.10.91
  JP-A-5226626 03.09.93
  JP-A-6163969 10.06.94

Villard Michel
  EP-A-0024970 11.03.81
  EP-A-0064918 17.11.82
  US-A-4665609 19.05.87

Vonno Nicolaas V
  US-A-5081063 14.01.92

Wakayama Hiroyuki
  JP-A-1218062 31.08.89
  JP-A-1227472 11.09.89
  JP-A-2303160 17.12.90

Walsh Devin T
  US-A-5279974 18.01.94

Wan Chang F
  EP-A-0407062 09.01.91
  EP-A-0416299 13.03.91

Watanabe Shuji
  JP-A-61059771 27.03.86

Watanabe Yoshio
  JP-A-6163969 10.06.94
  JP-A-6237005 23.08.94

Weinberg Donald L
  US-A-4259576 31.03.81

Whiffin Peter Arthur Charles
  GB-A-2203757 26.10.88

White Anthony Michael
  EP-A-0068652 05.01.83
  GB-A-2095906 06.10.82
  GB-A-2128019 18.04.84
  GB-A-2260218 07.04.93
  WO-A-9006597 14.06.90

White William J
  EP-A-0281026 07.09.88
  US-A-4104674 01.08.78
  US-A-4137625 06.02.79
  US-A-4197633 15.04.80
  US-A-4570329 18.02.86
  US-A-4695861 22.09.87

Williams Gerald Martin
  EP-A-0068652 05.01.83
  GB-A-2132017 27.06.84

Withers R B
  EP-A-0007667 06.02.80
  EP-A-0007668 06.02.80

Wong Theodore T S
  US-A-4081819 28.03.78

Woodberry Frank J
  US-A-4956555 11.09.90

Wotherspoon John Thomas Mclean
EP-A-0062367 13.10.82
EP-A-0062367 13.10.82
GB-A-2207802 08.02.89
GB-A-2207802 08.02.89

Yamagata Toshio
JP-A-62071270 01.04.87

Yamamoto Kosaku
JP-A-61220459 30.09.86
JP-A-63170960 14.07.88
JP-A-63229751 26.09.88
JP-A-3077373 02.04.91
JP-A-3219670 27.09.91
JP-A-3241774 28.10.91
JP-A-5175476 13.07.93
JP-A-6237005 23.08.94

Yamamoto Tamotsu
JP-A-3077373 02.04.91
JP-A-3108371 08.05.91
JP-A-5175476 13.07.93
JP-A-5226626 03.09.93

Yamamoto Toshiro
JP-A-61059771 27.03.86

Yasumura Kenji
JP-A-3175682 30.07.91
JP-A-4313267 05.11.92

Yasunaga Masatoshi
JP-A-4293240 16.10.92

York Rudy L
EP-A-0407062 09.01.91
US-A-4620209 28.10.86
US-A-5384267 24.01.95

Yoshida Yasuaki
GB-A-2246662 05.02.92
GB-A-2274739 03.08.94
JP-A-2009180 12.01.90

Yoshikawa Mitsuo
JP-A-57073984 08.05.82
JP-A-58171848 08.10.83
JP-A-60180162 13.09.85
JP-A-60182766 18.09.85
JP-A-60193375 01.10.85
JP-A-61067958 08.04.86
JP-A-1205476 17.08.89
JP-A-1228180 12.09.89

Young Ian Malcolm
GB-A-2132017 27.06.84

Young Miriam F
EP-A-0050512 28.04.82

Zanio Ken
US-A-4910154 20.03.90

Zieger Herbert C
US-A-5146303 08.09.92

Zimmermann Peter H
US-A-4935627 19.06.90
US-A-4943491 24.07.90

Company Index

Amber Engineering Inc., USA
US-A-5264699   23.11.93

Carson Alexiou Corporation, USA
US-A-4290844   22.09.81

Cincinnati Electronics Corporation, USA
EP-A-0485115   13.05.92

Commissariat A L'Energie Atomique, France
US-A-4791467   13.12.88
US-A-4972244   20.11.90

Eocom Corporation, USA
US-A-3852714   03.12.74

Ford Aerospace Corporation, USA
US-A-4910154   20.03.90

Fujitsu Limited, Japan
EP-A-0497326   05.08.92
EP-A-0543537   26.05.93
JP-A-55102280   05.08.80
JP-A-55150279   22.11.80
JP-A-57010986   20.01.82
JP-A-57024580   09.02.82
JP-A-57031170   19.02.82
JP-A-57062563   15.04.82
JP-A-57073984   08.05.82
JP-A-57091557   07.06.82
JP-A-58171848   08.10.83
JP-A-59047759   17.03.84
JP-A-59061080   07.04.84
JP-A-59112652   29.06.84
JP-A-60028266   13.02.85
JP-A-60074568   26.04.85
JP-A-60103666   07.06.85
JP-A-60140869   25.07.85
JP-A-60180162   13.09.85
JP-A-60182766   18.09.85
JP-A-60193375   01.10.85
JP-A-61059771   27.03.86
JP-A-61067958   08.04.86

JP-A-61128564   16.06.86
JP-A-61147118   04.07.86
JP-A-61198787   03.09.86
JP-A-61220459   30.09.86
JP-A-61222161   02.10.86
JP-A-61251167   08.11.86
JP-A-61268075   27.11.86
JP-A-62013085   21.01.87
JP-A-62036858   17.02.87
JP-A-62104163   14.05.87
JP-A-62165973   22.07.87
JP-A-62224982   02.10.87
JP-A-62224983   02.10.87
JP-A-62272564   26.11.87
JP-A-63043366   24.02.88
JP-A-63070454   30.03.88
JP-A-63133580   06.06.88
JP-A-63148677   21.06.88
JP-A-63150976   23.06.88
JP-A-63170960   14.07.88
JP-A-63170961   14.07.88
JP-A-63229751   26.09.88
JP-A-63260171   27.10.88
JP-A-63268271   04.11.88
JP-A-63281460   17.11.88
JP-A-63296272   02.12.88
JP-A-1050557   27.02.89
JP-A-1050560   27.02.89
JP-A-1201971   14.08.89
JP-A-1205476   17.08.89
JP-A-1218062   31.08.89
JP-A-1227472   11.09.89
JP-A-1228180   12.09.89
JP-A-1233777   19.09.89
JP-A-2086177   27.03.90
JP-A-2155269   16.06.90
JP-A-2272766   07.11.90
JP-A-2303160   17.12.90
JP-A-3077373   02.04.91
JP-A-3085762   10.04.91
JP-A-3108371   08.05.91
JP-A-3133181   06.06.91
JP-A-3219670   27.09.91

JP-A-3241774   28.10.91
JP-A-3268463   29.11.91
JP-A-3270269   02.12.91
JP-A-4253344   09.09.92
JP-A-5129580   25.05.93
JP-A-5175476   13.07.93
JP-A-5226626   03.09.93
JP-A-5315580   26.11.93
JP-A-6013642   21.01.94
JP-A-6125108   06.05.94
JP-A-6163865   10.06.94
JP-A-6163969   10.06.94
JP-A-6021419   28.01.94
JP-A-6204449   22.07.94
JP-A-6209096   26.07.94
JP-A-6236981   23.08.94
JP-A-6237005   23.08.94
JP-A-7050330   21.02.95
JP-A-7079008   20.03.95
US-A-5006711   09.04.91
US-A-5376558   27.12.94

General Electric Company, USA
US-A-4989067   29.01.91
US-A-5146303   08.09.92

Grumman Aerospace Corporation, USA
EP-A-0317083   24.05.89
EP-A-0317084   24.05.89
US-A-3970990   20.07.76
US-A-4618763   21.10.86
US-A-4992908   12.02.91
US-A-5030828   09.07.91
US-A-5075201   24.12.91
US-A-5128749   07.07.92
US-A-5315147   24.05.94
WO-A-8807764   06.10.88

Harris Corporation, USA
US-A-5081063   14.01.92

Honeywell Inc., USA
EP-A-0050512   28.04.82
EP-A-0151311   14.08.85
EP-A-0171801   19.02.86
EP-A-0188241   23.07.86
EP-A-0281026   07.09.88
US-A-3902924   75.09.02
US-A-3949223   06.04.76
US-A-4081819   28.03.78
US-A-4104674   01.08.78
US-A-4137625   06.02.79
US-A-4197633   15.04.80
US-A-4570329   18.02.86
US-A-4695861   22.09.87
US-A-4935627   19.06.90
US-A-4943491   24.07.90
WO-A-9001802   22.02.90

Hughes Aircraft Company, USA
EP-A-0405865   02.01.91
WO-A-9417557   04.08.94

Irvine Sensors Corporation, USA
US-A-4304624   08.12.81
US-A-4354107   12.10.82
US-A-4403238   06.09.83
US-A-4525921   02.07.85
US-A-4551629   05.11.85
WO-A-8603338   05.06.86

Licentia Patent-Verwaltungs-GmbH, FRG
GB-A-2113467   03.08.83
US-A-4162507   24.07.79
US-A-4596930   24.06.86
US-A-5003364   26.03.91

Minnesota Mining and Manufacturing Company, USA
WO-A-9202959   20.02.92

Mitsubishi Denki Kabushiki Kaisha, Japan
GB-A-2229036   12.09.90
GB-A-2238165   22.05.91
GB-A-2246662   05.02.92
GB-A-2246907   12.02.92
GB-A-2261323   12.05.93
GB-A-2261767   26.05.93
GB-A-2274739   03.08.94
GB-A-2284930   21.06.95
JP-A-57169278   18.10.82
JP-A-63116459   20.05.88
JP-A-63273365   10.11.88
JP-A-2009180   12.01.90
JP-A-2213174   24.08.90
JP-A-2214159   27.08.90
JP-A-3104278   01.05.91
JP-A-3175682   30.07.91
JP-A-3196568   28.08.91
JP-A-4133363   07.05.92
JP-A-4199877   21.07.92
JP-A-4288882   13.10.92
JP-A-4293240   16.10.92
JP-A-4313267   05.11.92
JP-A-4318970   10.11.92
JP-A-4337676   25.11.92
JP-A-4337677   25.11.92
JP-A-5055620   05.03.93
US-A-4553152   12.11.85
US-A-5198370   30.03.93

Mullard Limited, GB
GB-A-1559473   16.01.80
GB-A-1559474   16.01.80
GB-A-1568958   11.06.80

NEC Corp., Japan
JP-A-61214462   24.09.86
JP-A-62071270   01.04.87

JP-A-62090986 25.04.87
JP-A-63005560 11.01.88
JP-A-63046765 27.02.88
JP-A-63147366 20.06.88
JP-A-63300559 07.12.88
JP-A-1061055 08.03.89
JP-A-1061056 08.03.89
JP-A-2248077 03.10.90
JP-A-2272765 07.11.90
JP-A-4267562 24.09.92
JP-A-4269869 25.09.92
JP-A-4318979 10.11.92
JP-A-5343727 24.12.93
JP-A-5343729 24.12.93
JP-A-6204448 22.07.94
JP-A-6326342 25.11.94
JP-A-7111323 25.04.95

Nikon Corp., Japan
JP-A-63310165 19.12.88

Northrop Corporation, USA
US-A-4912536 27.03.90
US-A-4912536 27.03.90
US-A-5130259 14.07.92
US-A-5171994 15.12.92

Philips Electronic and Associated Industries Limited, GB ;
N.V. Philips' Gloeilampenfabrieken, The Netherlands
EP-A-0007667 06.02.80
EP-A-0007668 06.02.80
EP-A-0007669 06.02.80
EP-A-0061801 06.10.82
EP-A-0061802 06.10.82
EP-A-0061803 06.10.82
EP-A-0062367 13.10.82
EP-A-0087842 07.09.83
EP-A-0343738 29.11.89
EP-A-0475525 18.03.92
EP-A-0481552 22.04.92
EP-A-0510267 28.10.92
EP-A-0561615 22.09.93
GB-A-1600599 21.10.81
GB-A-1605321 19.07.89
GB-A-2201834 07.09.88
GB-A-2203757 26.10.88
GB-A-2207802 08.02.89
GB-A-2208256 15.03.89
GB-A-2231199 07.11.90
GB-A-2239555 03.07.91
GB-A-2241377 28.08.91
GB-A-2241605 04.09.91

Rockwell International Corporation, USA
US-A-4727406 23.02.88
US-A-4956555 11.09.90
US-A-4956695 11.09.90

Santa Barbara Research Center, USA
EP-A-0406696 09.01.91
EP-A-0445545 11.09.91
US-A-4025793 24.05.77
US-A-4751560 14.06.88
US-A-4783594 08.11.88
US-A-4885619 05.12.89
US-A-5113076 12.05.92
US-A-5149956 22.09.92
US-A-5177580 05.01.93
US-A-5189297 23.02.93
US-A-5279974 18.01.94
US-A-5293036 08.03.94
US-A-5296384 22.03.94
US-A-5365088 15.11.94
US-A-5373182 13.12.94
US-A-5401986 28.03.95
US-A-5414294 09.05.95
WO-A-8707082 19.11.87
WO-A-8707083 19.11.87
WO-A-8910007 19.10.89

Selenia Industrie Elettroniche Associate S.p.A., Italy
GB-A-1597581 09.09.81
GB-A-1337968 21.11.73

Semiconductor Research Foundation, Japan
EP-A-0038697 28.10.81
EP-A-0042218 23.12.81
EP-A-0094973 30.11.83
EP-A-0094974 30.11.83

Societe Anonyme de Telecommunications, France
FR-A-2593642 31.07.87
US-A-3930161 30.12.75
US-A-3988774 26.10.76
US-A-4532699 06.08.85
US-A-4566024 21.01.86
US-A-4583108 15.04.86

Texas Instruments Incorporated, USA
EP-A-0044610 04.08.82
EP-A-0066020 08.12.82
EP-A-0137704 17.04.85
EP-A-0137988 24.04.85
EP-A-0173074 05.03.86
EP-A-0288792 02.11.88
EP-A-0366886 09.05.90
EP-A-0407062 09.01.91
EP-A-0416299 13.03.91
EP-A-0416320 13.03.91
US-A-3806729 23.04.74
US-A-3808435 30.04.74
US-A-3965568 29.06.76
US-A-3980915 14.09.76
US-A-4229752 21.10.80
US-A-4231149 04.11.80
US-A-4360732 23.11.82

US-A-4429330 31.01.84
US-A-4620209 28.10.86
US-A-4654686 31.03.87
US-A-4658277 14.04.87
US-A-4660066 02.08.88
US-A-4720738 19.01.88
US-A-4726885 23.02.88
US-A-4761681 21.04.87
US-A-4807007 21.02.89
US-A-4956686 11.09.90
US-A-5144138 01.09.92
US-A-5254850 19.10.93
US-A-5298733 29.03.94
US-A-5300777 05.04.94
US-A-5318666 07.06.94
US-A-5384267 24.01.95

The Marconi Company Limited, GB
EP-A-0293216 30.11.88
GB-A-2197984 02.06.88

The Secretary of State for Defence, GB
EP-A-0068652 05.01.83
EP-A-0167305 08.01.86
GB-A-1488258 12.10.77
GB-A-2007909 23.05.79
GB-A-2019649 31.10.79
GB-A-2094548 15.09.82
GB-A-2095906 06.10.82
GB-A-2128019 18.04.84
GB-A-2132017 27.06.84
GB-A-2146663 24.04.85
GB-A-2207801 08.02.89
GB-A-2260218 07.04.93
WO-A-9006597 14.06.90

The United States of America as represented by the Secretary of the Air Force, USA ;
The United States of America as represented by the Secretary of the Army, USA ;
The United States of America as represented by the Secretary of the Navy, USA
US-A-3842274 15.10.74
US-A-4228365 14.10.80
US-A-4242149 30.12.80
US-A-4259576 31.03.81
US-A-4273596 16.06.81
US-A-4439912 03.04.84
US-A-4445269 01.05.84
US-A-4517464 14.05.85
US-A-4646120 24.02.87
US-A-4877752 31.10.89

Thomson-CSF, France
EP-A-0024970 11.03.81
EP-A-0064918 17.11.82
FR-A-2484705 18.12.81
FR-A-2494910 28.05.82
FR-A-2526227 04.11.83
US-A-4311906 19.01.82

US-A-4665609 19.05.87

Toshiba Corp., Japan ;
JP-A-58164261 29.09.83
JP-A-62011265 20.01.87
JP-A-62022474 30.01.87
JP-A-6089991 29.03.94
JP-A-7094693 07.04.95
US-A-4868622 19.09.89

Westinghouse Electric Corp., USA
US-A-4369458 18.01.83
US-A-4549195 22.10.85
US-A-4731640 15.03.88

Yokokawa Hokushin Denki KK, Japan
JP-A-60102530 06.06.85

Zaidan Hojin Handotai Kenkyu Shinkokai, Japan
US-A-4427990 24.01.84

Subject Index

Generic terms are shown in *italics*.

—A—

absorption (*e.g.* absorber, absorbing)
    layer, 8; 176; 210; 212; 220; 230; 235; 241; 244; 245; 253
    region, 209; 227; 231
    structure, 215; 220; 221
    substrate, 211
ambipolar drift, **23**; 382
ambipolar drift field imager, 23
amorphous material, 373
anodic
    oxide (*also* oxidation), 63; 103; 165; 202; 250; 255; 386; 389
    selenization, 388
    sulphide, 80; 258; 285
    surface treatment, 92
antimony, Sb, 387
aperture plate, 116
auxiliary region, 189
avalanche breakdown, 6
avalanche multiplication, 20; 70

—B—

beam lead interconnect, 119
bipolar transistor, 43
bismuth, Bi, 387
bonding agent (conductive), 103
buffer board, 323
buffer layer, 240; 261; 298; 363; 371
buried channel, 7; 10

—C—

capacitive coupling, 338; 345
CdMnTe, 111
CdTeSe, 67; 266
channel stop, 15; 16; 17; 57; 74; 104; 280; 386
charge coupled device, CCD, **6**; 52
charge coupled device imager, 6
charge imaging matrices, 73
charge injection device, CID, 52
charge injection device imagers, 52
colour operation (*see* multi-spectral imager)
cooling (simulated), 159; 377
copper layer, 300
cross-talk, 202

—D—

detectivity, 100
diffusion of indium, 149; 165; 187; 228; 280; 360

—E—

evaporation of Hg, 16; 17; 62; 187; 247; 248

—F—

field gate (*see* field plate)

field plate (*also* field gate) 57; 63; 64; 74; 76; 78; 104; 166; 257; 386
flexible membrane, 302
flip-chip technique, 139; **274**

—G—

gallium arsenide, GaAs, 180; 232; 267; 295; 356; 359; 364; 371
glycol phthalate, 161
guard
    electrode, 60
    plate, 166; 257
    ring, 12

—H—

hexagonal arrangement, 375
HgCdZnTe, 170; 223; 264

—I—

indium-filled tubes, 299
induced junction, 150; 166; 187; 242; 257
infrared background radiation, 6; 16; 59; 74; 139; 177; 274
infrared shield (*e.g.* opaque field plate), 6; 12; 53; 56; 57; 60; 61; 62; 74; 76; 78; 119; 162; 182; 196; 202; 208; 212; 216; 225; 254; 256; 281; 294; 367
injection pulse, 53; 57
insulator walls, 308
interconnection
    layer, 112
    pattern, 89
    structure, 120
interface states (*also* surface energy states), 14; 167
ionizing radiation, 108; 157

—L—

light signature, 198
loophole technique (*see* through hole technology)

—M—

meandering current path, 93

meandering drift path (*also* serpentine structure), 27; 30
Metal-Insulator-Semiconductor, MIS
    device, 193
    electrode, 57; 74; 82
    photocapacitor, 63; 75; 77; 80; 103; 106; 139; 155; 203; 274; 356; 384; 388; 389
    switch, 81
molybdenum, 100; 266; 320; 360; 393; 398
multi-spectral imager (*e.g.* two colour operation), 10; 76; 80; 158; 162; 173; 175; 176; 177; 179; 193; 341; 350; 379; 380; 389

—N—

neural network, 176
nickel, 56; 58; 74; 76; 77; 106; 265; 387

—O—

overflow, 10; 15

—P—

parallel-in to serial-out conversion, 23
polycrystalline (HgCdTe), 205; 349

—Q—

quantum differential detector, QDD, 139; 274

—R—

ramped gate voltage, 74; 104
random access imager, 73
RC-network, 201
recombination layer, 66
reflection prevention, 70; 100; 142; 144; 195; 196; 198; 219; 226; 251; 358
refractory metal, 393
refresh transistor, 44; 45
resonant cavity structure, 111

—S—

Schottky barrier (*e.g.* Schottky diode), 10; 100; 143; 356; 359

serpentine structure (*see* meandering drift path)
shape-memory alloy, 282
side-wall electrode, 92
silicon-on-sapphire, SOS, 362
silicone rubber, 145
surface energy states (*see* interface states)
skimmed off, 59
spacer, 290; 291
stacked silicon chips, 319; 321; 324; 326
staggered configuration, 60
static induction transistor, **40**; 345
static induction transistor imagers, 40
sweep-out, 93
switching transistor, 43

—T—

temporary substrate, 160; 168; 194; 307
test element group, 181

thermal background radiation (*see* infrared background radiation)
thin film transistor, 369
through hole technology (*also* loophole technique), 33; **374**
time-delay and integration, TDI, 15
titanium oxynitride, TiNO, 387

—V—

virtual electrode, 7
visible radiation, 180

—W—

wax, 91; 307

—Z—

Z-technology, 312